Histology, Ultrastructure and Molecular Cytology of Plant-Microorganism Interactions

Edited by

MICHEL NICOLE
ORSTOM,
Montpellier, France

and

VIVIENNE GIANINAZZI-PEARSON
INRA/CNRS,
Dijon, France

KLUWER ACADEMIC PUBLISHERS
DORDRECHT / BOSTON / LONDON

A C.I.P. Catalogue record for this book is available from the Library of Congress.

ISBN 0-7923-3886-3

Published by Kluwer Academic Publishers,
P.O. Box 17, 3300 AA Dordrecht, The Netherlands.

Kluwer Academic Publishers incorporates
the publishing programmes of
D. Reidel, Martinus Nijhoff, Dr W. Junk and MTP Press.

Sold and distributed in the U.S.A. and Canada
by Kluwer Academic Publishers,
101 Philip Drive, Norwell, MA 02061, U.S.A.

In all other countries, sold and distributed
by Kluwer Academic Publishers Group,
P.O. Box 322, 3300 AH Dordrecht, The Netherlands.

Printed on acid-free paper

Printed in the Netherlands

CONTENTS

ACKNOWLEDGEMENTS

The following institutions, research organizations, local authorities and private companies are gratefully acknowledged for their generous support

Agropolis

CIRAD

Conseil Général de l'Hérault

Conseil Régional Languedoc Roussillon

Drukker International

INRA, Département de Pathologie Végétale

Jeol Europe (S.A.)

Leica

Montpellier District

ORSTOM, Commission Scientifique

ORSTOM, Département MAA

ORSTOM DIST

PREFACE

Plants interact with a large number of microoganisms which have a major impact on their growth either by establishing mutually beneficial symbiotic relationships or by developing as pathogens at the expense of the plant with deleterious effects. These microorganisms differ greatly not only in their nature (viruses, phytoplasmas, bacteria, fungi, nematodes,...) but also in the way they contact, penetrate and invade their host.

Histology and cytology have brought an essential contribution to our knowledge of these phenomena. They have told us for instance, how specialized structures of the pathogen are often involved in the adhesion and penetration into the plant, how the interface between both organisms is finely arranged at the cellular level, or what structural alterations affect the infected tissues. They have thus set the stage for the investigations of the underlying molecular mechanisms could be undertaken. Such investigations have been remarkably successful in the recent years, expanding considerably our understanding of plant-microorganism interactions in terms of biochemical changes, rapid modifications of enzymatic activities, coordinated gene activation, signal reception and transduction. Biochemistry, molecular biology and cellular physiology have taken precedence in the phytopathologist's set of methods. Although very efficient tools, they have inevitably led to some underrating of the topographic dimension of the interactions. Yet, considering the mere example of plant defense responses, it is well known that there are cell-to-cell heterogeneities, that different tissues do not react in the same way and that the spatial gradient of responses is just as important as its timing in determining the outcome of the interaction.

In recent years, cytologists have made significant progress in developing methods that bridge the gap with biochemistry and molecular biology. For instance, it is possible to locate, and to some extent quantitate, molecules such as polymers, proteins and nucleic acids at the cytological level with a high specificity, based on the specific recognition between enzyme and substrate, antibodies and antigens or between complementary strands of nucleic acids. Molecular cytology is now ready to play its due part in the unravelling of the complexities of plant-microorganism interactions. It is the merit of Michel Nicole and Vivienne Gianinazzi-Pearson to have convinced a number of cytologists in Plant Pathology from several countries to meet in Montpellier on November 1994 in order to draw up the state of the art. Hosted by the Société Française de Phytopathologie, it was probably one of the very few meetings in the last years specifically devoted to this topic, and a very successful one. Most of the major lectures presented at the meeting have been collected in this book together with a few additional contributions, providing a braod overview of this evolving field of research.

Pierre RICCI
President for the Société Française de Phytopathologie

IMAGE ANALYSIS IN BIOLOGY

C. SOUCHIER
Centre Commun de Quantimétrie
et laboratoire de Cytologie Analytique
8 avenue Rockefeller, 69373 Lyon CEDEX 08, France

Introduction

Image analysis [1,2] is used to extract quantitative information from images of any source. Progress has resulted from advances in computing science, and nowadays image analysis is widely used for increasingly more sophisticated tasks in the laboratory and in industry, in material science as well as in biology.

This article explains how image analysis works and describes a few applications in biology [3,4], mainly in the quantitative microscopy field. The image is first digitized, segmented in order to select objects of interest, and then measured. Image analysis provides precise and objective data on morphology, cytochemistry and in situ hybridization reactions, cell patterns. Moreover, image analysis provides useful tools for numerical image acquisition and enhancement, and makes possible automatic recognition of events.

1. Material and methods

1.1. IMAGE ANALYSIS SYSTEM

An image analysis system (Figure 1) includes: 1) an input system constituted by a camera fitted to a microscope or a macroviewer, 2) a control and interactive system with screen, mouse and keyboard, 3) a computer associated to specialized processor cards, 4) output devices: printer, magnetic media such as disk, floppy disk, WORM. All images are displayed on the screen and make possible visual control throughout analysis. The inputs may be transmission, epi-fluorescence, inverted or electronic

M. Nicole and V. Gianinazzi-Pearson (eds.), Histology, Ultrastructure and Molecular Cytology of Plant-Microorganism Interactions, 1-20.

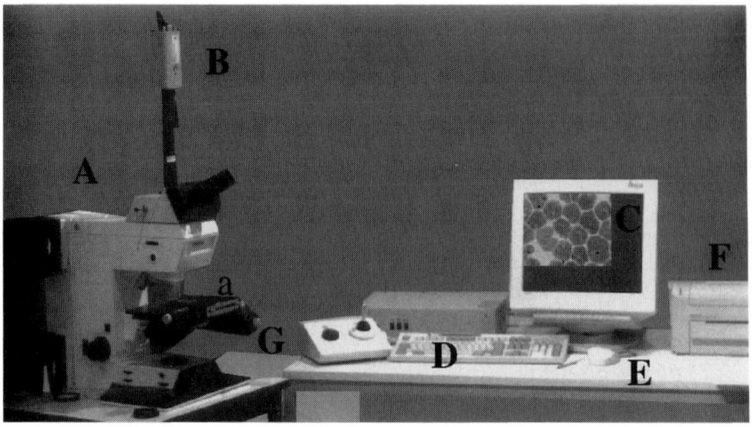

Figure 1. Leica QTM 600 image analysis system
A – microscope with automatic stage (a), B – color CCD camera
C – screen, D – keyboard, E – mouse, F – printer
G – control and specialized processors

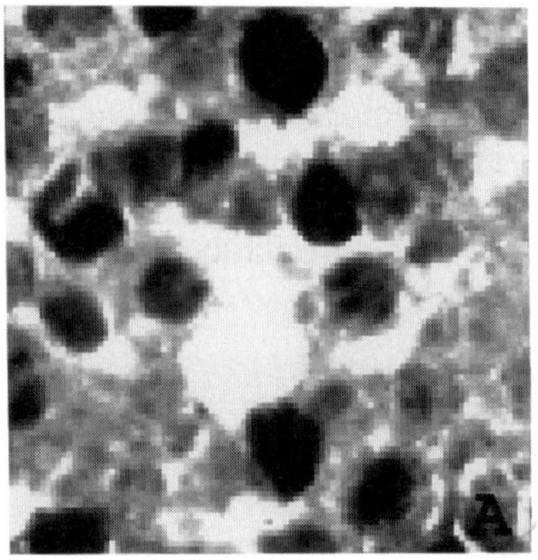

9	9	5	4	2	10	9	10
9	14	10	5	3	15	7	9
6	5	6	5	4	6	7	13
6	4	6	11	5	5	7	10
7	6	6	14	12	6	13	10
7	10	8	12	7	10	9	10
8	8	9	6	2	6	10	10
5	4	10	12	7	5	7	9

B

Figure 2. Image acquisition
A – optical image, B – digitized image, The image is divided into 8x8 pixels and each point is coded into 16 grey levels from 0 (black) to 15 (white)

microscopes or macroviewer. Moreover, microscope lamp, stage, and focus may be motorized and software controlled. The camera may be monochrome or colour and may be as SIT, intensified or cooled CCD cameras, adapted to low light illumination. Furthermore, images already digitized by another system may be read providing image format is compatible. A large variety of images may thus be analyzed and image analysis may be considered as an open method.

1.2. IMAGE ANALYSIS PRINCIPLE

First, to an optical image, a digitized image is associated. Image is acquired with a CCD camera and is converted to a grey or colour image (Figure 2). The image is divided into a large number of points (512X512 for example) and each point is coded into 256 grey levels (8 bits, standard value) or 256 X 256 X 256 colours (3X8 grey images). Indeed, a colour image may be divided into three grey images: the red, green and blue images. The elementary number is called a pixel.

Second, softwares are applied to the data. Image analysis system may provide two types of softwares.

The first one corresponds to a ready-to-use interactive software that includes procedures and allows the users to build up their own applications. Such procedures make it possible to acquire images, to transform them and to perform measurements. The biologist must only know what each instruction performs and has only to organize them according to his own application. Previous computer science knowledge is not necessary.

The second one is software application. This corresponds to an already existing application such as DNA quantification or immunohistochemistry evaluation. These applications might, for example, have been implemented by the image analysis company or by the laboratory expert. Running an application is very easy. Knowing the application conditions is all one needs to obtain correct data.

1.3. IMAGE ANALYSIS STEPS

Image analysis consists of a sequence of three main steps (Figure 3). First, the image is acquired. Second, the image is processed, segmented in order to successively identify the different objects of interest. Image processing includes both grey and binary image transformations and operations. It is a difficult step, and preparative methods are essential in order to make it easier. Third, measurements are performed and data are stored and analyzed. Such measurements may be volume density of tissue components, number, size, shape and colour of cells, DNA content and texture of nuclei, intensity and spatial distribution of immunoenzymatic, immunofluorescent

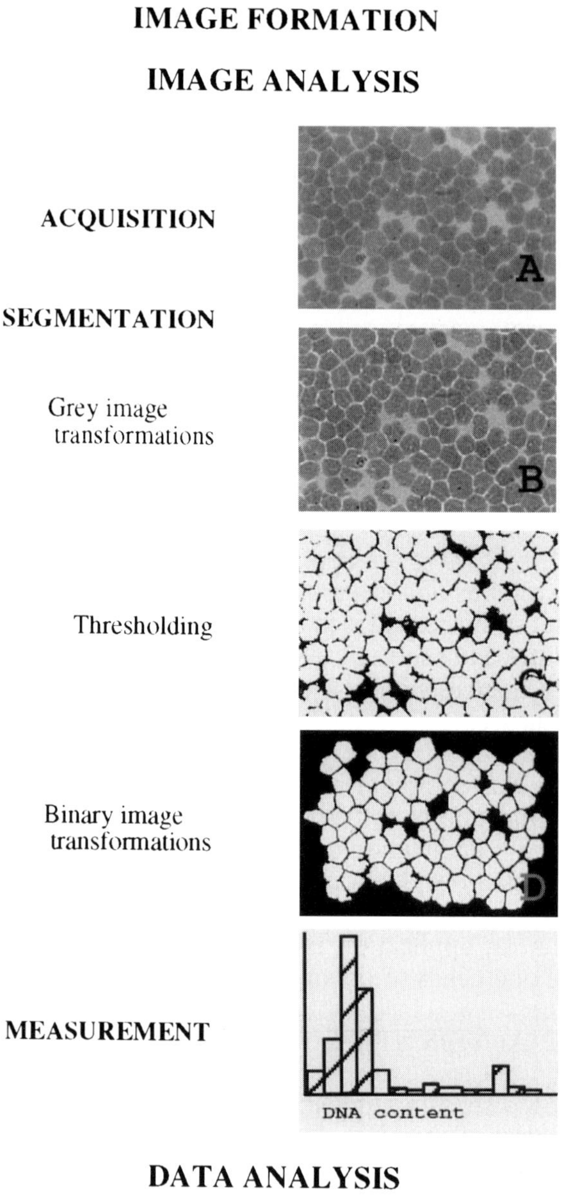

Figure 3. Image analysis steps
A – original image, B – grey enhanced image, C – binary image,
D – binary transformed image used for morphological measurements and as mask for densitometric measurements performed on the original image

or autoradiographic labellings.

1.4. IMAGE ANALYSIS TRANSFORMATIONS

Image segmentation includes image operations and transformations both on grey and binary images. Image transformations may be divided into point to point, local and global transformations [1,2,5,6].

In a point transformation, each pixel is transformed independently of other pixels. Thresholding that transforms a grey image into a binary image (Figure 3B-C) or stretching that extends acquired grey values at maximum (Figure 4), are point transformations.

Look Up Tables (LUTs), which associate an output value to an input value and remap the image grey levels, perform point transformations. LUTs may also be used only before display without modifying the image data. For example, LUTs are used to display pseudo-colour image. In this case, three output values (red, green, blue) are associated to one input value.

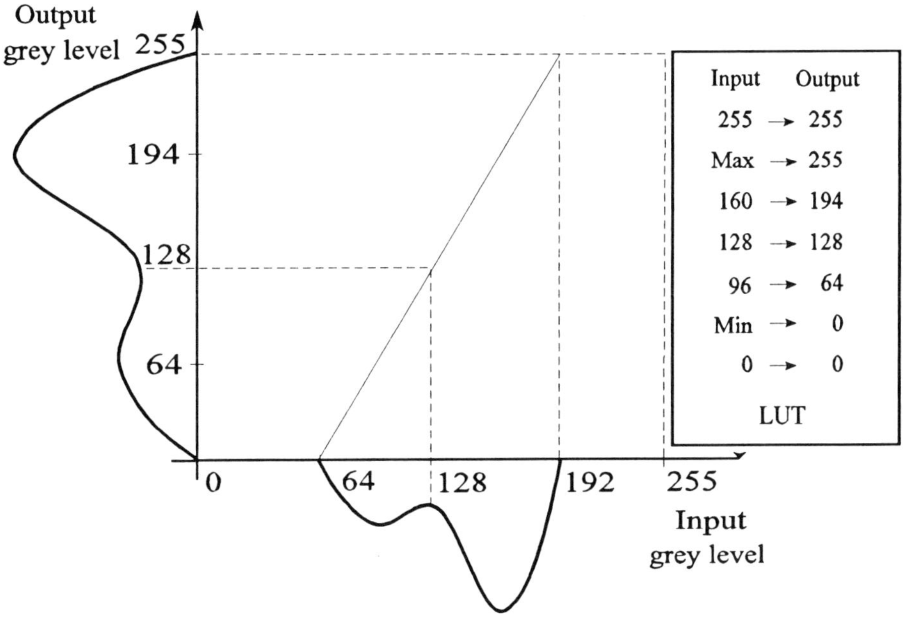

Figure 4. Stretching transformation
Grey output = (grey input - min) x 255 / (max - min)

In a local transformation, each pixel is transformed according to a rule that involves not only the transformed pixel but also close pixels (Figure 5). Many different local transformations exist and most of them are mathematical morphology or convolution filters.

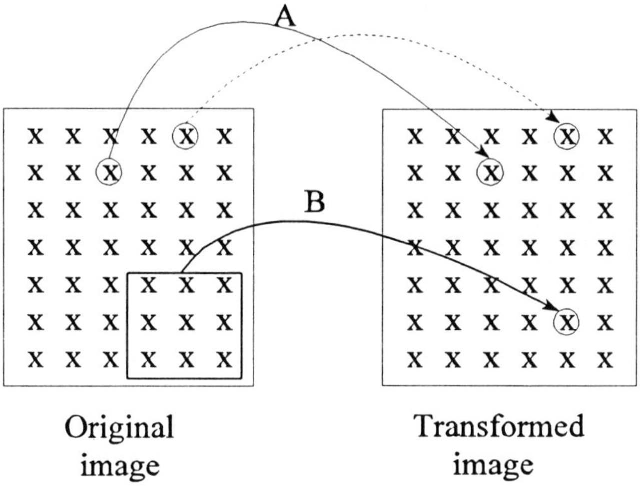

Figure 5. Point or local transformation
A - point transformation, B - local transformation

The basic transformations of mathematical morphology [2,7] are erosion and its complementary operation: dilation (Figure 6). A structuring element defines the point to transform and its neighbours. With plane structuring elements, erosion replaces each pixel by the minimum neighbour value and dilation replaces each pixel by the maximum value. Powerful transformations such as opening, closing, geodesic dilations (Figure 6F) or top-hat transformations (Figure 6I) are merely successions of erosions or/and dilations, associated to operations between images (binary operations: and, or; grey operations: minimum or maximum). Mathematical transformation may be extended to three dimensions [8].

Convolutions [1,5,6] (Figures 7, 8) replace each pixel by the result of additions defined by a kernel operator that states the weight of the central transformed pixel and of its neighbours. Gaussian average filter, a low pass filter, or Laplace filter, a high pass filter, are basic transformations. As previously, convolutions may be the result of basic convolutions and operations (arithmetic, minimum, maximum operations).

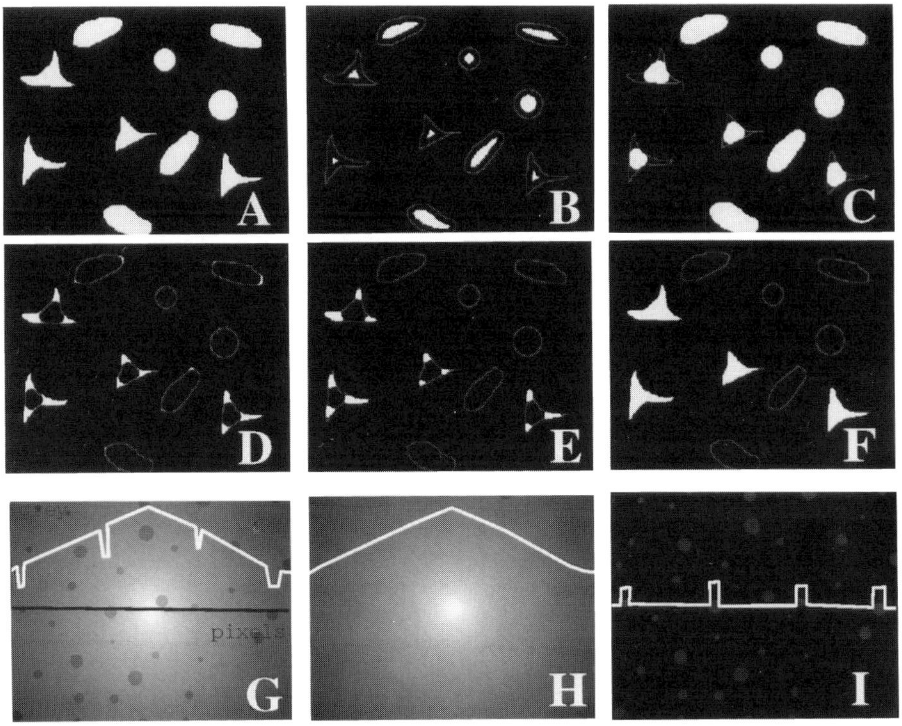

Figure 6. Mathematical morphology
A .. F – Binary transformations
A – original image, B – eroded image, C – image after B dilation or A opening
D – intersection between A and non C, E – image after small D opening
F – geodesic dilation of E in A = successive (dilation size 1 and intersection)

G.. H – Grey transformations
G – original image and grey profile across the line
H – closed image of G = dilation and then erosion of G
I – top hat transformed image = H–G = closed–original
Thresholding may be performed on I but not on G

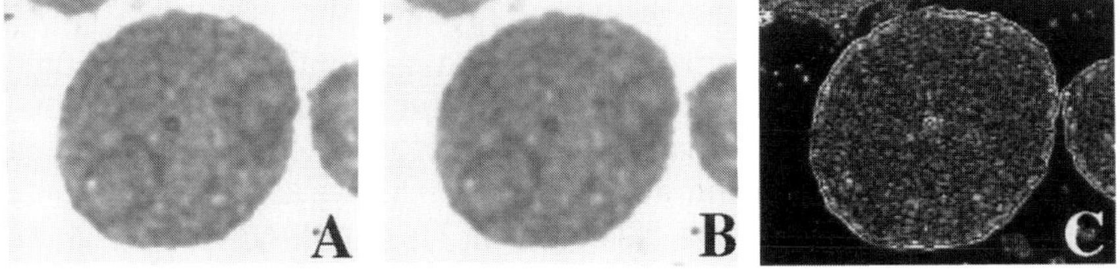

Figure 7. Convolution
A – original image, B– image transformed by gaussian filter
C – image transformed by laplace filter and then contrast enhanced

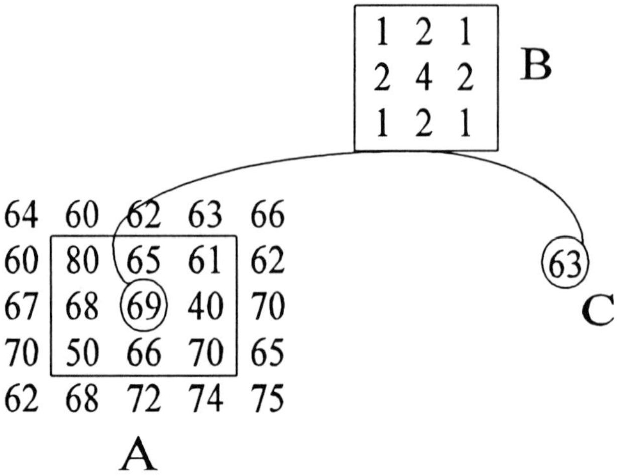

Figure 8. Convolution principle
Gaussian average filter
A - image part, B - kernel, C - transformed point

In a global transformation such as Fourier transform [6], the image is first transformed in another space such as frequency space.

1.5. STEREOLOGY

Image analysis performs most measurements on 2D images and stereology [9,10] is used for 3D interpretation of data. Unbiased 3D measurements are difficult to obtain, and the interpretation of 2D measurements may be carried out carefully. For instance, the nuclear size measured on histological section is only profile size, and the result even depends on section thickness. Stereology method based on test systems may be efficiently used [11] and may even be the best method when image segmentation is difficult to perform. Moreover, test systems may be drawn with the image analysis tools and applied directly on the video image.

1.6. LASER SCANNING CONFOCAL MICROSCOPY

New image analysis perspectives are offered by confocal laser scanning microscopy (CLSM). In CLSM, the image is illuminated and acquired pixel per pixel (Figure 9) [12-15]. Imaging may be performed in fluorescence and in reflective mode. The main advantages of CLSM are 1) reduction of contribution of out-of-focus structures, 2) improvement of X/Y/Z resolution, 3) possibility to acquire optical serial sections. It improves visualization of fine details such as cytoskeleton elements [16-20] or in situ

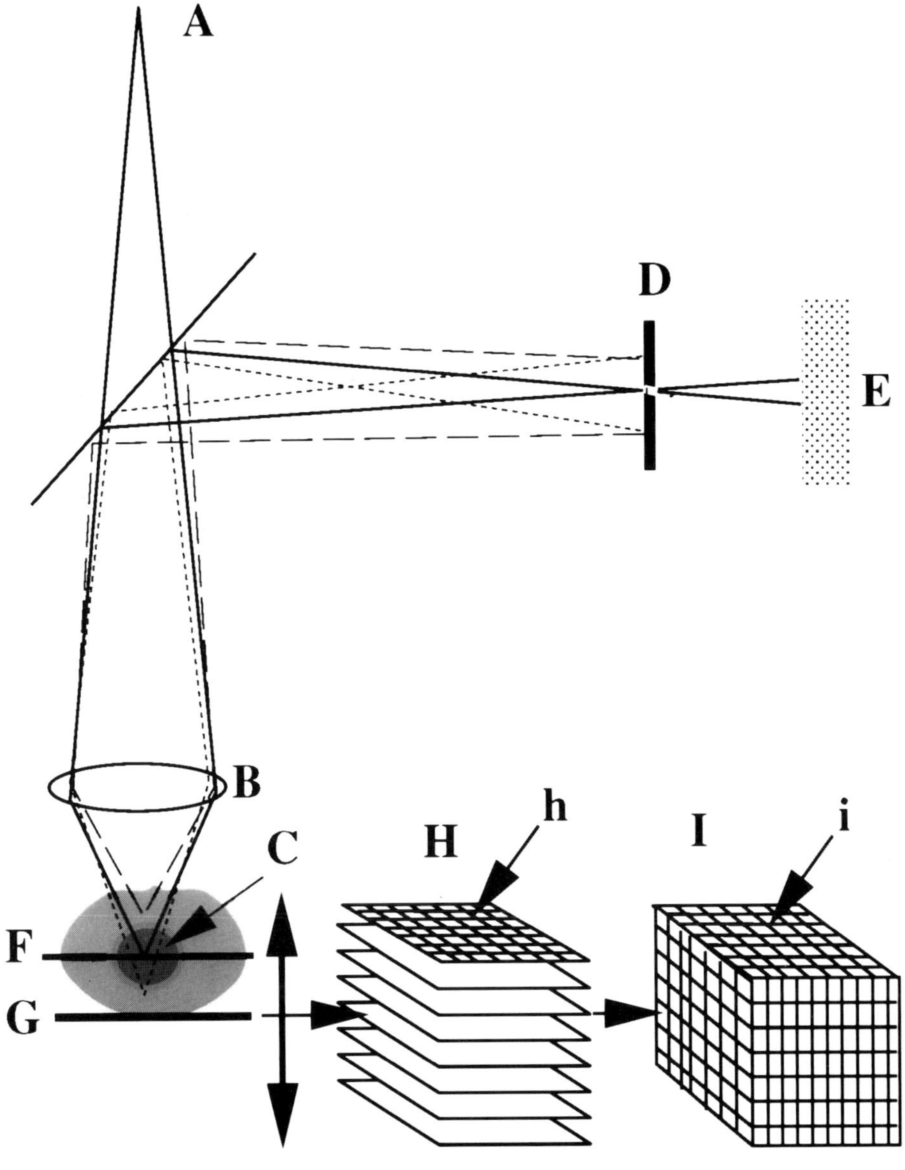

Figure 9 – Confocal Laser Scanning Microscope
A – laser beam, B – lens, C – confocal spot, D – pinhole, E – detector, F – focal plane, G – stage plane, H – optical sections and h pixel, I – 3D image and i voxel
Only signal emanating from the confocal spot reaches the detector. Signals from planes above or below the focal plane are stopped by the pinhole. The focal plane is scanned point by point by the laser beam, the stage is then moved, another plane is focussed and acquired.

hybridization [21,22] and makes possible three dimensional visualization [23,24] and quantification [25-27].

2. Applications

A large variety of images may be acquired, and thus different objects may be analyzed: histological sections, smears, imprints, electron micrographs, but also grains, leaves, photographs, electrophoresis gels.

2.1. IMAGE ENHANCEMENT

Image processing makes it possible to restore images as this should have been obtained with an acquisition system free of defects. Electronic noise, observed on digitized images acquired in epi-fluorescence, CLSM or electron microscopy, may be removed by averaging successive images (Figure 10). Shading due to uneven illumination may be corrected.

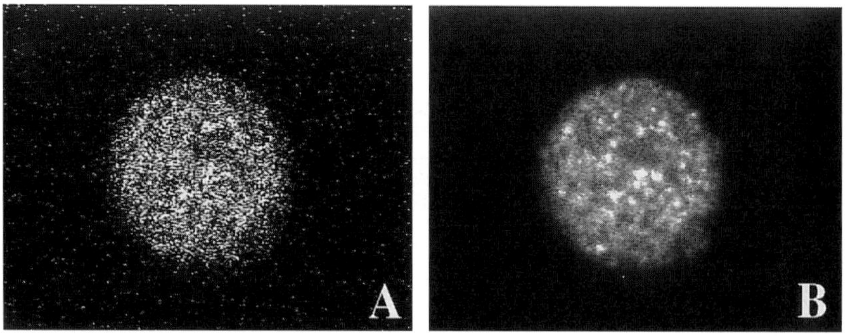

Figure 10. Image averaging
Cyclin A labelling image acquired using CLSM
A - 1 image, B - 32 images were averaged, noise was removed

Image processing makes it possible to enhance images in order to point out certain details in them. Contrast may be improved by simply extending the grey values between black and white. Pseudo-colour images may be used to better display variations of intensity within monochrome images such as autoradiograms, calcium images [28,29], and CLSM fluorescent intensity images [18-21].

Furthermore, images may be combined and reconstructed from several images. Images of double or triple fluorescent labelling may be easily acquired, aligned when necessary and superimposed. It is often difficult to take such images straight under the microscope, even when double or triple band fluorescent filter sets may be applied. Image processing is in development for cytogenetics by Fluorescent In Situ Hybridization (FISH) [30,31]. Moreover, composite images representative of entire sections may be built from contiguous microscopic fields and they may reflect the spatial distribution of immunohistochemical or in situ hybridization labellings [32,33]. Furthermore, CLSM makes possible three-dimensional visualization of surfaces, thick histological sections or cells [21,22,26].

All images may be stored on computer disks and this may be useful for image data banks, image transfer, for an educational or communication purposes.

2.2. IMAGE MEASUREMENTS

Image quantification is the main purpose of image analysis. With image analysis, tissue components, cells and organelles may be measured, and several microscopic modes may be used (Figure 11). At macroscopic level, measurements may be performed without object manipulation. The image is the intermediate between the object to be measured and the system. With image analysis, multiparametric data may be acquired on a cell population or even a sub-cellular population. All steps are under visual control.

Field and feature measurements can be obtained. Features can be numbered and individually described. Besides morphological data such as size and shape, densitometric and intensity data, data concerning texture or cell patterns can be obtained. With care taken in sample preparation and in image formation, objective, reproducible and precise data is acquired and allows us to describe samples.

Such data might be used to compare different cell types, vegetal species, experimental and environmental conditions, growth or kinetic times and to evaluate relationships between function and structure. Most studies are related to a problem of counting, size, shape, colour description, DNA content and texture, immunohistochemistry or in situ hybridization labelling, Calcium, pH or chlorophyll evaluation and dynamic process.

2.2.1. *Counting*

Counting is most often associated to morphometric or densitometric measurements. However, the enumeration of cells in culture or of microorganisms on cells or aquatic, soil or food samples may be the main purpose [4,34].

2.2.2. *Morphometry and colour description*

Image analysis is a powerful technique to analyze size and shape. Parameters and features are evaluated without interactions with each other unlike visual observation. For instance, small objects are always viewed

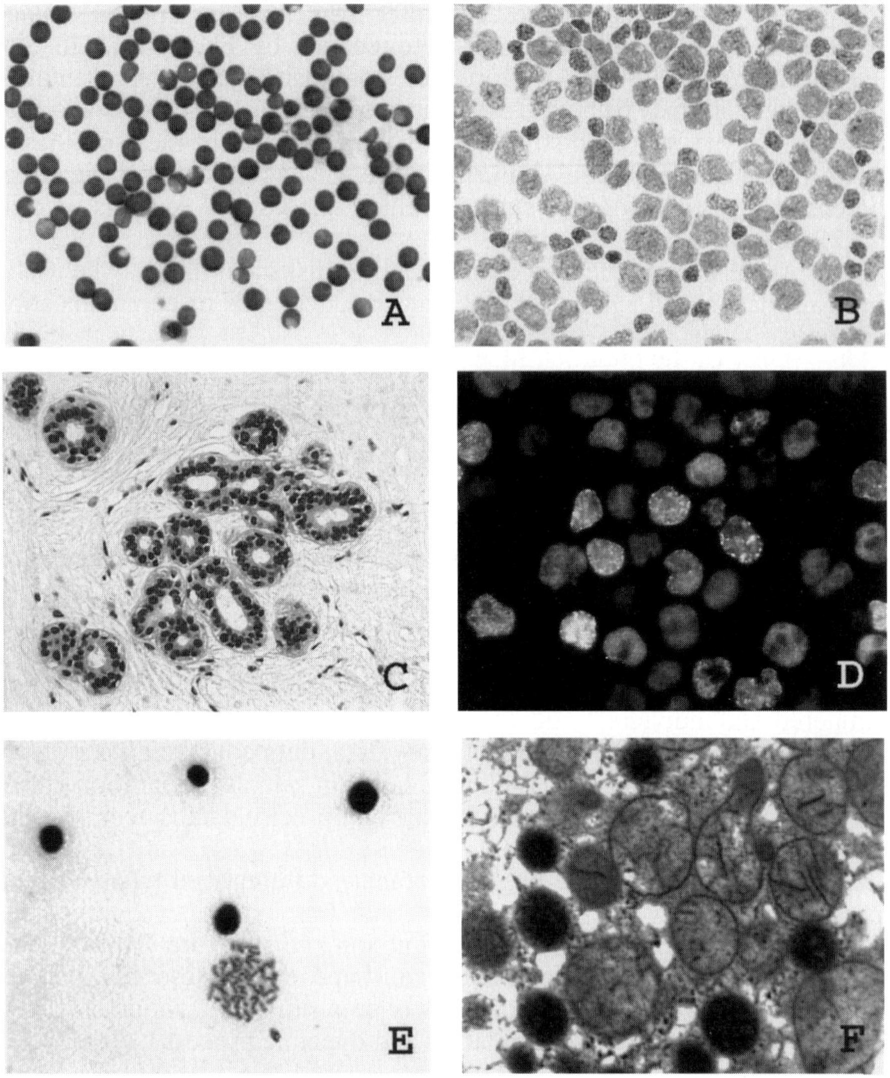

Figure 11. Applications
A – lentil morphometry, B – DNA content on Feulgen stained nuclei
C – 5–methylcytosine labelling histochemistry, D – PCNA labelling cytofluorometry
E – metaphase recognition, F – peroxisome granulometry

smaller when their neighbours are large than when they also are small.

Tissue component may be evaluated by area density ($A_A=V_V$): ratio of the tissue component area under the reference tissue space. Granulometry may be easily performed. Moreover, shape may be described by variables such as roundness ($perimeter^2*4\pi/area$), elongation (maximum Feret diameter/minimum Feret diameter), convexity (area/convex area) shape factors [1,2]. For instance, morphometry variables were obtained on nuclei, root cortex cells [35], phytolitts [36], fungal spores [37], mycelia [38], bacteria [39], microorganism pellets [40]. In addition, brightness or even colour may be quantified. Root nodules of the Common bean were classified by size and surface brightness [41], leaf colour [42] and heterogeneity of the radiation field within leaves were evaluated [43].

2.2.3. *DNA content and texture*

One of the main applications in biology is DNA content evaluation on feulgen stained specimens [44-46] (Figure 11A). Integrated Optical Density (IOD) is measured, DNA histograms are obtained and ploidy, proliferation, and perturbations of cell cycle are evaluated (Figure 12).

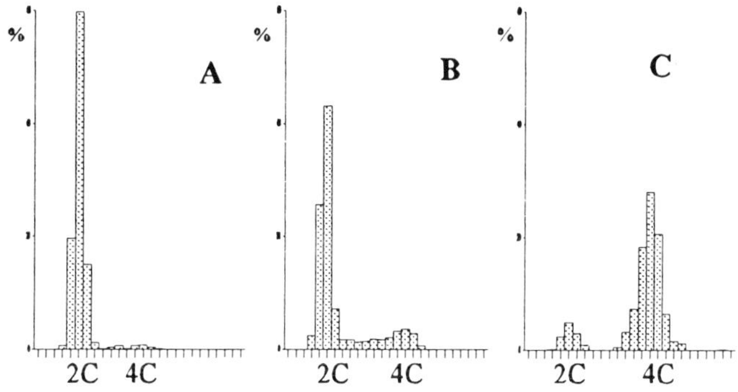

Figure 12. DNA histograms
A - diploid, B - diploid and proliferative, C - tetraploid

Feulgen method is the main staining, but fluorescent stains such as Hoechst 33258 and 33342, DAPI, Propidium Iodide may be used too, and total fluorescent intensity may be measured. Nuclear [47] and even mitochondrial [48] DNA were quantified. Image and flow DNA cytometry [49,50] are close methods. Image cytometry can select cells on visual, morphologic or immunologic criteria, measure small samples and acquire multiple data on the same cells. However, the method is significantly slower and not as largely diffused as flow cytometry.

Besides DNA content, size, shape, texture that is more or less homogeneous, more or less in clumps, more or less eccentric may be described [26, 51-53].

2.2.4. *Immunology or in situ hybridization labelling*

The second main application in biology is immunochemistry and in situ hybridization quantification (Figure 11C, D). This covers a large variety of questions and even specimens. Indeed, several approaches can be used for the localization of probes bound to DNA or RNA sequences and of antibodies bound to proteins.

First, the input system may be adapted to non-isotopic or isotopic labelling of the probes. At the optical level, with enzymatic method, such as peroxidase or phosphatase alkaline labelling, monochrome or colour cameras are used. With fluorescent method, high sensitivity or integrated cooled CCD cameras are needed. With colloidal gold labelling or autoradiography labelling, the choice of the camera depends on the microscope illumination mode (transmission, dark field or epipolarization). For the autoradiogram reading, the input system is a monochrome camera fitted to a macroviewer.

As usual, the main difficulty is image segmentation. Labelling may be restricted to a certain number of foci, as it is observed on PCNA or BrdUrd [54] images. The labelling measurement may be the labelled foci or be extended to the entire labelled cell. The first segmentation allows us to evaluate the labelled area percent distribution, cell by cell. Otherwise, the second segmentation may be better adapted for estimating labelled tissue component in histology. However, it may be difficult to extend a nuclear labelling such as Ki67 or BrdUrd to the entire cell including the cytoplasm. With autoradiography, even when grain segmentation may be significantly improved with epipolarization or dark field mode, clusters are still difficult to separate. Grains may be indirectly enumerated by total area or intensity measurement divided by mean value acquired on isolated grains. On autoradiograms, frames are most often used to define the measurement zones.

Measurements may be counting, area or intensity estimation [39,55-58]. At electronic level, specific methods may be used to compute grain source densities from autoradiography [10,59]. Moreover, spatial distributions of cells in a tissue section [60-62], of probes intra-nuclei [22] or along chromosomes, of intra-cellular components [26] and colocalization of labels [63] may be efficiently studied by 2D or even 3D image analysis [25].

2.2.5. *Fluorescence imaging techniques : Ca^{++}, chlorophyll measurements*

Imaging and quantification of ions or pH imply mono or dual wavelength fluorescent dyes [29,64]. With dual fluorescent dyes such as Fura 2, Indo 1, the Ca^{++} measurement is performed at two excitation or emission wavelengths, the ratio is proportional to the ion concentration and the standardization is better than with mono fluorescent dye such as Fluo 3. Ratio imaging may also be applied to chlorophyll imaging [65]. Many different cell components may be specifically stained [66]. Moreover, other fluorescence techniques than ratio imaging have been developed. Microspectrofluorometry involves multiple measurements throughout the excitation or emission spectra [67]. Fluorescence resonance energy transfer imaging (FRET) implies acceptor fluorochrome excitation by donor emission in close proximity with the acceptor [68]. Time resolved fluorescence microscopy makes possible using fluorescence staining [69] or delay luminescing immunolabels [70] to reduce autofluorescence or to differentiate tissue components provided that they differ in fluorescence or luminescence decay time. Furthermore, besides spectrofluorometry, image spectrometry may be used, even in the infrared bandwidth. Chemical constituents of wheat were thus recorded [71].

2.2.6. *Dynamic study*

Dynamic applications are possible. Successive images of the same microscopic field may be recorded and analyzed. Fading curves [72], induction curves of chlorophyll fluorescence [73] were thus obtained. Bacteria growth and related morphological changes and patterns under controlled environmental conditions [74], changes in Ca^{++} and pH during exposure to unilateral illumination and gravity in maize coleoptiles [28], cell movement [75] and chloroplast motility [76] were thus described. Image analysis may be efficiently associated to techniques such as microinjection and photobleaching. Dynamics of actin microfilaments and tubulins were thus recorded in living cells after microinjection with either FITC and rhodamine phalloidin [16,17] or carboxifluorescein tubulin [20]. In Fluorescence Redistribution After Photobleaching (FRAP), a brief intense pulse of laser is used to bleach a fluorophore in a localized region of a cell or tissue, and the image analysis system makes it possible to follow and quantify the recovery of fluorescence from the other unbleached parts into the bleached region. Microtubule turnover in living plants was thus quantified [18].

2.3. AUTOMATIC RECOGNITION

With image analysis, it may be possible to carry out an exhaustive or a systematic screening of a slide and to recognize automatically a structure. In such work, an exhaustive analysis is performed. No-one is needed, the stage and the focus work alone, and the image analysis system can work during the night. Stage with multiple slides may be used. The event to be sought out can even be rare. This possibility is illustrated by the example of automatic recognition of metaphases (Figure 11E). Metaphases are recognized automatically and precise locations of metaphases are recorded. It may be possible to relocate even further the metaphase at higher magnification.

Conclusion

Image analysis provides data on organelles, cells and tissues, and may be used to objectively compare specimens. Efficiency for research was demonstrated. Future developments should improve image segmentation in order to process automatically, quickly and in a user-friendly way a larger variety of images than nowadays. Progress should be made especially for 3D CLSM images, in order to obtain automatically 3D unbiased data without model assumptions.

Acknowledgments

This work was supported by grants from ARC and "Ligue contre le Cancer" (Ain, Drôme, Saône & Loire). The author would like to thank A.M. Manel, M.P. Pages (Debrousse Hospital), A. Niveleau (Pasteur Institute), L. Frappart (Edouart Herriot Hospital), C. Dumontel (A. Carrel Faculty) for providing specimens of Figures 3 and 11, M. Benchaib, R. Delorme, M. Pluvinage, C. Masson (Analytical Cytology) for their help in preparing the manuscript, and D. Clausen for reviewing its English version.

References

1. Russ, J.C.: *Computer-assisted microscopy, the measurement and analysis of images*, Plenum Press, London, 1990.
2. Souchier, C.: L'analyse d'images, *Techniques de l'Ingénieur, Traité Analyse Chimique et Caractérisation* **7** (1991), 855-1-855-18.
3. Baak, J.P.A.: *Manual of quantitative pathology in cancer diagnosis and prognosis*, Springer Verlag, London, 1991.
4. Caldwell, D.E., Korber, D.R., and Lawrence J.R.: Confocal laser microscopy and digital image analysis in microbial ecology, *Advances in Microbial Ecology* **12** (1992), 1-67.
5. Pratt, W.K.: *Digital image processing*, 2nd edition, J. Wiley (ed.), 1990.
6. Sonka, M., Hlavac, V., and Boyle, R.: *Image processing, analysis and machine vision*, Chapman et Hall (eds.), 1993.
7. Coster, M., and Chermant, J.L.: *Précis d'analyse d'images*, CNRS, 1989.
8. Meyer, F.: Mathematical Morphology: from two dimensions to three dimensions, *J. Microsc.* **165** (1992), 5-28.
9. Cruz Orive, L.M., and Weibel, E.R.: Recent stereological methods for cell biology: a brief review, *Am. J. Physiol.* **258** (1990), L148-L156.
10. Cau, P.: *Microscopie quantitative, stéréologie, autoradiographie et immunocytochimie quantitatives*, INSERM, 1990.
11. Kubínová L.: Recent stereological methods for the measurement of leaf anatomical characteristics - estimation of volume density, volume and surface area, *J. Exp. Bot.* **44** (1993), 165-173.

12. Pawley, J.B.: *Handbook of biological confocal microscopy*, Plenum Press, New York, 1995.
13. Kwon, Y.H., Wells, K.S., and Hoch, H.C.: Fluorescence confocal microscopy: applications in fungal cytology, *Mycologia* **85** (1993), 721-733.
14. Paddock, S.W.: To boldly glow .... Applications of laser scanning confocal microscopy in developmental biology, *BioEssays* **16** (1994), 357-365.
15. Bryon, P.A., Delorme, R., and Souchier, C.: La microscopie confocale à balayage laser et ses applications hématologiques, *Rev. fr. laboratoires* **275** (1995), 37-43.
16. Zhang, D., Wadsworth, P., and Hepler, P.K.: Dynamics of microfilaments are similar, but distinct from microtubules during cytokinesis in living, dividing plant cells, *Cell Motil. Cytoskel.* **24** (1993), 151-155.
17. Meindl, U., Zhang, D., and Hepler, P.K.: Actin microfilaments are associated with the migrating nucleus and the cell cortex in the green alga Micrasterias - studies on living cells, *J. Cell Sci.* **107** (1994), 1929-1934.
18. Hush, J.M., Wadsworth, P., Callaham, D.A., and Hepler, P.K.: Quantification of microtubule dynamics in living plant cells using fluorescence redistribution after photobleaching, *J. Cell Sci.* **107** (1994), 775-784.
19. Llyod, C.V., Venverloo, C., Goodbody, K.C., and Shaw, P.J.: Confocal laser microscopy and three-dimensional reconstruction of nucleus-associated microtubules in the division plane of vacuolated plant cells, *J. Microsc.* **166** (1992), 99-109.
20. Foissner, I., and Wasteneys, G.O.: Injury to Nitella internodal cells alters microtubule organization but microtubules are not involved in the wound response, *Protoplasma* **182** (1994), 102-114.
21. Harders, J., Lukács, N., Robert-Nicoud, M., Jovin, T.M., and Riesner, D.: Imaging of viroids in nuclei from tomato leaf tissue by *in situ* hybridization and confocal laser scanning microscopy, *EMBO J.* **8** (1989), 3941-3949.
22. Montijn, M.B., Houtsmuller, A.B., Oud, J.L., and Nanninga, N.: The spatial localization of 18 S rRNA genes, in relation to the descent of the cells, in the root cortex of Petunia hybrida, *J. Cell Sci.* ***107*** (1994), 457-467.
23. Shaw P.J.: Computer Reconstruction in 3-Dimensional Fluorescence Microscopy, Electronic Light Microscopy, in : *Techniques in Modern Biomedical Microscopy* (1993), 211-230.
24. Van der Voort, H.T.M., Messerli, J.M., Noordmans, H.J., and Smeulders, A.W.M.: Volume visualization for interactive microscopic image analysis, *Bioimaging* **1** (1993), 20-29.
25. Rigaut, J.P., Carvajal-Gonzalez, S., and Vassy, J.: 3-D image cytometry, in A. Kriete (ed.), *Visualization in biomedical microscopies*, VCH (1992), pp. 205-248.
26. Delorme, R., Souchier, C., Ffrench, M., and Bryon, P.A.: Confocal image analysis of three-dimensional intracellular protein distribution: application to cyclin A distribution in lymphoid cells, *Bioimaging* **2** (1994), 69-77.
27. Thoni, C., and Schnepf, E.: Nuclear and organelle DNA replication during spore germination in bryophytes and Equisetum, *Botanica Acta* **107** (1994), 210-217.
28. Gehring, C.A., Williams, D.A., Cody, S.H., and Parish, R.W.: Phototropism and geotropism in maize coleoptiles are spatially correlated with increases in cytosolic free

calcium, *Nature* **345** (1990), 528-530.
29. Read, N.D., Allan, W.T.G., Knight, H., Knight, M.R., Malhó R., Russell, A., Shacklock, P.S., and Trewavas, A.J.: Imaging and measurement of cytosolic free calcium in plant and fungal cells, *J. Microsc.* **166** (1992), 57-86.
30. Leitch, A. R., Schwarzacher, T., and Leitch, I. J.: The use of fluorochromes in the cytogenetics of the small-grained cereals (Triticeae), *Histochem. J.* **26** (1994), 471-479.
31. Fukui, K., Ohmido, N., and Khush, G.S.: Variability in rDNA Loci in the genus Oryza detected through fluorescence in situ hybridization, *Theor. Appl. Genet.* **87** (1994), 893-899.
32. Souchier, C., Ffrench, M., Berger, F., Scoazec, J.Y., and Bryon, P.A.: Image analysis applied to proliferating cells in malignant lymphoma, *Cytometry* **9** (1988), 201-205.
33. Westerkamp, D., and Gahm, T.: Non-distorted assemblage of the digital images of adjacent fields in histological sections, *Anal. Cell. Pathol.* **5** (1993), 235-247.
34. Evans-Hurrell, J.A., Adler, J., Denyer, S., Rogers, T.G., and Williams, P.: A method for the enumeration of bacterial adhesion to epithelial cells using image analysis, *FEMS Microbiol. Lett.* **107** (1993), 77-82.
35. Baluška, F., Brailsford, R.W., Hauskrecht, M., Jackson, M.B., and Barlow, P.W.: Cellular dimorphism in the maize root cortex: involvement of microtubules, ethylene and gibberellin in the differentiation of cellular behaviour in postmitotic growth zones, *Bot. Acta* **106** (1993), 394-403.
36. Ball, T.B., and Brotherson J.D.: The effect of varying environmental conditions on phytolith morphometries in two species of grass (Bouteloua curtipendula and Panicum virgatum), *Scanning Microsc.* **6** (1992), 1163-1181.
37. Paul, G.C., Kent, C.A., and Thomas, C.R.: Viability testing and characterization of germination of fungal spores by automatic image analysis, *Biotechnol. Bioeng.* **42** (1993), 13-23.
38. Tucker, K.G., Kelly, T., Delgrazia, P., and Thomas, C.R.: Fully automatic measurement of mycelial morphology by image analysis, *Biotechnol.Prog.* **4** (1992), 353-359.
39. Wilkinson, M.H.F., Jansen, G.J., Van der Waaij, D.: Computer processing of microscopic images of bacteria: morphometry and fluorometry, *Trends in Microbiology* **2** (1994), 485-489.
40. Nielsen, J., Johansen, C.L., Jacobsen, M., Krabben, P., and Villadsen, J.: Pellet formation and fragmentation in submerged cultures of Penicillium chrysogenum and its relation to penicillin production, *Biotechnol. Prog.* **11** (1995) 93-98.
41. Vikman, P.A., and Vessey, J.K.: Ontogenetic changes in root nodule subpopulations of Common bean (Phaseolus vulgaris L.). III Nodule formation, growth and degradation, *J Exp. Bot.* **44** (1993), 579-586.
42. Seracu, D., and Baiulescu, G.E.: Color measurement in leaf diagnosis, *Anal. Lett.* **26** (1993), 2349-2359.
43. Myers, D.A., Vogelmann, T.C., Bornman, J.F.: Epidermal focussing and effects on light utilization in Oxalis acetosella, *Physiologia Plantarum* **91** (1994), 651-656.
44. Böcking, A., Giroud, F., and Reith, A.: Consensus report of the ESACP task force on standardization of diagnostic DNA image cytometry, *Anal. Cell. Pathol.* **8** (1995), 67-74.

45. Driss-Ecole, D., Schoëvaërt, D., Noin, M., and Perbal G.: Densitometric analysis of nuclear DNA content in lentil roots grown in space, *Biol. Cell* **81** (1994), 59-64.
46. Cremonini, R., Colonna, N., Stephani, A., Galasso, I., and Pignone, D.: Nuclear DNA content, chromatin organization and chromosome banding in brown and yellow seeds of Dasypyrum villosum (L.) P. Candargy, *Heredity* **72** (1994), 365-373.
47. Vigo, J., Salmon, J.M., Lahmy, S., and Viallet, P.: Fluorescent image cytometry: from qualitative to quantitative measurements, *Anal. Cell. Pathol.*, **3** (1991), 145-165.
48. Satoh, M., Nemoto, Y., Kawano, S., Nagata, T., Hirokawa, H., and Kuroiwa, T.: Organization of heterogeneous mitochondrial DNA molecules in mitochondrial nuclei of cultured tobacco cells, *Protoplasma* **175** (1993), 112-120.
49. Marie, D., and Brown, S.C.: A cytometric exercise in plant DNA histograms, with 2C values for 70 species, *Biol. Cell* **78** (1993), 41-51.
50. Fouchet, P., Jayat, C., Héchard, Y., Ratinaud, M.H., and Frelat, G.: Recent advances of flow cytometry in fundamental and applied microbiology, *Biol. Cell* **78** (1993), 95-109.
51. Young, I.T., Verbeek, P.W., and Mayall B.H.: Characterization of chromatin distribution in cell nuclei, *Cytometry* **7** (1986), 467-474.
52. Haralick, R.M., Shanmugam, K., and Its'Hak Dinstein: Textural features for image classification, *IEE Transactions on systems, man and cybernetics* **3** (1973), 610- 621.
53. Galloway, M.M.: Texture analysis using gray level run lengths, *Computer Graphics and image processing* **4** (1975), 172-1' 9.
54. Sparvoli, E., Levi, M., and Rossi, E.: Replicon clusters may form structurally stable complexes of chromatin and chromosomes, *J. Cell Sci.* **107** (1994), 3097-3103.
55. Le Guellec, D., Mallein-Gerin, F., Treilleux, I., Bonaventure, J., Peysson, P., and Herbage, D.: Localization of the expression of type I, II and III collagen genes in human normal and hypochondrogenesis cartilage canals, *Histochem. J.* **26** (1994), 695-704.
56. Nico, A., and Schellart, M.: Automatic grain counting in autoradiographs by computerized pattern recognition, in Donat-P. Häder (ed.), *Image analysis in biology*, CRC Press, London (1992), pp. 271-286.
57. Bacus, S., Fowers, J.L., Press, M.J., Bacus, J.W., and McCarty, K.S.: The evaluation of estrogen receptor in primary breast carcinoma in computer assisted image analysis, *AJCP* (1988), 233-239.
58. Slarew, R.J., Bodmer, S.C., and Pertschuk, L.P.: Quantitative imaging of imunohistochemical (PAP) estrogen receptor staining patterns in breast cancer sections, *Cytometry* **11** (1990), 359-378.
59. Williams, M.A.: Autoradiography: its methodology at the present time, *J. Microsc.* **128** (1982), 79-94.
60. Dussert, C., Rasigni, G., Rasigni, M., and Palmari J.: Minimal spanning tree: a new approach for studying order and disorder, *Phys. Rev. B.* **34** (1986), 3528-3531.
61. Marcepoil, R., and Usson, Y.: Methods for the study of cellular sociology: voronoï diagrams and parametrization of the spatial relationships, *J. Theor. Biol.* **154** (1992), 359-369.
62. Raymond, E., Raphael, M., Grimaud, M., Vincent, L., Binet, J.L., and Meyer, F.: Germinal center analysis with the tools of mathematical morphology on graphs, *Cytometry* **14** (1993), 848-861.

63. Hassan, B., Errington, R.J., White, N.S., Jackson, D.A., and Cook, P.R.: Replication and transcription sites are colocalized in human cells, *J. Cell Sci.* **107** (1994), 425-434.
64. Dunn, K.W., Mayor, S., Myers, J.N., and Maxfield, F.R.: Applications of ratio fluorescence microscopy in the study of cell physiology, *FASEB J.* **8** (1994), 573-582.
65. Lang, M., Lichtenthaler, H.K., Sowinska, M., Summ, P., and Heisel, F.: Blue, green and red fluorescence signatures and images of Tobacco leaves, *Bot. Acta* **107** (1994), 230-236.
66. Haugland, R.P.: *Handbook of fluorescent probes and research chemicals*, Molecular probes, 1992.
67. Salmon, J.M., Vigo, J., and Viallet, P.: Resolution of complex fluorescence spectra recorded on single unpigmented living cells using a computerised method, *Cytometry* **9** (1988), 25-32.
68. Jovin, T.M., and Arndt-Jovin, D.J.: FRET microscopy: digital imaging of fluorescence resonance energy transfer. Application in cell biology, in E. Kohen and J.G. Hirschberg (eds.), *Cell Structure and Function by Microspectrofluorometry*, Academic Press (1989), pp. 99-117.
69. Tian, R., and Michael, A.J.R.: Time-resolved fluorescence microscopy, in R.J. Cherry (ed.), *New techniques of optical microscopy and microspectroscopy - Topics in Molecular and Structural Biology* ***15***, Macmillan Press (1991), pp. 177-198.
70. Verwoerd, N.P., Hennink, E.J., Bonnet, J., Van der Geest, C.R.G., and Tanke, H.J.: Use of ferro-electric liquid crystal shutters for time-resolved fluorescence microscopy, *Cytometry* **16** (1994), 113-117.
71. Robert, P., Bertrand, D., Devaux, M.F., and Sire, A.: Identification of chemical constituents by multivariate near-infrared spectral imaging, *Analyt. Chem.* **64** (1992), 664-667.
72. Longin, A., Souchier, C., Ffrench, M. and Bryon, P.A.: Comparison of anti-fading agents used in fluorescence microscopy: image analysis and laser confocal microscopy study, *J. Histochem. Cytochem.* **41** (1993), 1833-1840.
73. Plieth, C., Tabrizi, H., and Hansen, U.P.: Relationship between banding and photosynthetic activity in Chara corallina as studied by the spatially different induction curves of chlorophyll fluorescence observed by an image analysis system, *Physiologia Plantarum* **91** (1994), 205-211.
74. James, G.A., Korber, D.R., Caldwell, D.E., and Costerton J.W.: Digital image analysis of growth and starvation responses of a surface-colonizing Acinetobacter sp., *J. Bacteriol.* **177** (1995), 907-915.
75. Killich, T., Plath, P.J., Wei, X., Bultmann, H., Rensing, L., and Vicker, M.G.: The locomotion, shape and pseudopodial dynamics of unstimulated Dictyostelium cells are not random, *J. Cell Sci.* **106** (1993), 1005-1013.
76. Menzel, D.: An interconnected plastidom in Acetabularia: Implications for the mechanism of chloroplast motility, *Protoplasma* **179** (1994), 166-171.

In situ Hybridization to RNA in Plant Biology

Judy BRANGEON
Institut de Biotechnologie des Plantes
Université Paris-Sud, Bât. 630, 91405 Orsay,
France

1. Introduction

The technique of *in situ* hybridization (ISH) is an elegant synthesis of molecular biology and cytology. *In situ* hybridization makes possible the detection and localization of specific nucleic acid sequences within tissue sections or whole mount preparations of single cells, organelles or chromosomes. It is based on the phenomenon elucidated by Watson and Crick of base pairing between two complementary nucleic acids. The technique works as follows—a labelled single-stranded fragment of DNA or RNA (probe), exogenously applied, will hybridize to a complementary sequence on cellular DNA or RNA (target), forming stable hybrids. The hybrids are then located by using a system of detection (visible marker), which can be coupled to the label on the probe and visualized in the microscope. If the probe is labelled with a radioisotope, the hybrid is detected by autoradiograhic procedures, whereas nonradioactive labels such as biotin, bromo-deoxyuridine, or digoxigenin are invariably detected by histochemical techniques (enzyme reactions, affinity or immunocytochemical (ICC) methods). In this way the observer is able to determine where a particular DNA or RNA sequence is located with respect to cell morphology. ISH was practiced initially at the light microscopic (LM) level, but recent development in non-isotopic probe technology has extended the method to more routine use at the electron microscopic level [41,59]. Considerable literature of hybridizing to RNAs or DNAs in plant systems at the LM level are available and should serve as a convenient stepping stone towards ultrastructural applications [see references in 41].

The kinetics of *in situ* hybridization depends on probe to tissue penetration and diffusion, accessibility of targets to probe, and the hybridization reaction itself. These requirements, in turn, depend on fixation conditions which should give acceptable structural preservation and good nucleic acid retention while permitting probe diffusion throughout the preparation.

Plant material presents morphological features which constitute drawbacks for certain technical aspects of ISH in that 1) most tissue is composed of a highly

M. Nicole and V. Gianinazzi-Pearson (eds.), Histology, Ultrastructure and Molecular Cytology of Plant-Microorganism Interactions, 21-42.

heterogeneous cell population which necessitates compromises in fixation/permeabilization procedures 2) cell walls constitute barriers to probe penetration adding to those of cell and organelle boundary membranes and 3) vacuoles, fragile membrane-bound compartments, must remain intact to avoid leakage /diffusion of *in situ* targets.

This chapter focuses on applications of ISH technique to plant material at both the light and electron microscopic level based on our experience and will evoke some advantages and disadvantages of different procedures.

2. Historical Background

In situ hybridization studies began with the work by Gall and Pardue [17] and John et al.[29] in 1969, using radioactively labelled ribosomal RNA probes to visualize genes coding for 18S and 28S ribosomal RNA in *Xenopus* oocytes at LM cellular and chromosomal levels. Subsequent applications were developed to cell smears [51], chromosome spreads [27,36], cryostat and vibrotome sections [36] and to sections of both paraffin-embedded and to a lesser extent resin-embedded samples [37]. Up until 1984, only radiolabelled probes and autoradiographic detection techniques were used; these early ISH experiments mainly made use of the pre-embedding method [18,20,21,22, 27,45,46,61] and whole-mount chromosomes or cell spreads [42,51].

The technique adapted to the EM level for the detection of hybridized nucleic acid sequences in ultrathin sections is more recent [1,26,34,47,53,57,58]. Binder et al.[2] were the first to test biotinylated probes for visualizing hybridized targets on ultrathin sections of Lowicryl K4M-embedded material. Their two-step immunocytochemical system of detection employing anti-biotin antibodies combined with protein A-gold complexes greatly enhanced the labelling.

In plant biology, the method has been routinely used at light microscopic resolution, mainly with ^{35}S- or ^{3}H-probes [10,13,24,30,31,35,48,62]. The introduction of non-radioactive probes and the vast array of immuno- or affinity detection systems now available will undoubtedly help the development of EM ISH [41]. An elegant study has been carried out by McFadden and collaborators on ultrathin resin sections using a double labelling method with biotin-and digoxigenin-probes for the simutaneous detection of two specific rRNAs [39,40,41]. At present, the most frequently used labels for EM ISH are biotin and digoxigenin mediated systems, which are easy to use, commercially available and of high sensitivity.

3. Cytological Considerations in *in situ* Hybridization

ISH uses the same general molecular biological principles as those used in solution hybridization or Northern blot analysis. However, nucleic acids within cells and tissues may not behave in the same way as those in a homogeneous solution. Targets are distributed in patterns in tissues; for Northern blot or solution hybridizations, a target in low abundance can be concentrated, but *in situ* it must be detected at the endogenous levels of individuel cells. In cytological preparations, hybridization is carried out in a less favorable heterogeneous environment composed of protein, lipid and nucleic acid components of the cell. Cellular material consists of a dense gel-like cytosol through which runs an array of membranes, ie. endoplasmic reticulum. The cytosol is enclosed by delimiting membranes (tonoplasts and plasmalemma) as are the organelles, all of which constitute barriers to probe penetration and its access to targets. Protein associations may interfere with conditions of hybridization and also may mask the target sites. Added to biological barriers, fixatives impede diffusion by cross-linking nucleic acids to each other and to associated proteins.

In situ hybridization methods initially developed for animal systems were transposed as such to plant systems. However, plant cells are characterized by the presence of two unique and specific compartments: 1) *the cell wall*, an extra-cellular matrix consisting of cellulose and pectic substance often impregnated with lignins and specialized proteinaceous material. It is a relatively thin envelope in young dividing cells but can become massive and elaborate in mature specialized cells, constituting a rigid, semi-permeable barrier 2) *the vacuole*, a water-filled membrane-bound compartment, which in differentiated tissue can occupy the greater part of the plant cell. This centrally located compartment contains a high solute concentration which maintains cell turgidity and positions the cytoplasm against the cell wall. The osmolarity must be maintained during tissue preparatory steps in order to avoid deleterious affects to cell structure. Another feature which sets plant systems apart from animal systems is the existence of permanently embryonic regions termed *meristems*. These zones of continuous dividing cells are often distributed throughout tissue and consequently can be contiguous to "juvenile" and/or mature cells contributing to a heterogenous tissue pattern. Numerous *air spaces* are interpersed among cells in certain types of tissue and these interfere with the free movement and penetration of solutions among cells.

Given these constraints, the combination of molecular biology and cytology in ISH presents challenges and technical compromises must be found empirically. The investigator must adjust to maximum localization and abundance of mRNA as a function of the tissue, its fixation and preparatory steps, probe and labelling systems and hybridization requirements.

4. Methodology

4.1 FIXATION

Fixation procedures for ISH have been reported in detail for animal systems and should be consulted when choosing a fixative [28 and references in 38]. There are two types of fixatives, precipitating fixatives such as acetic acid-alcohol mixtures and aldehydes. For plant material, aldehydes (glutaraldehyde and paraformaldehyde) are the fixatives of choice.

Optimal fixation for *in situ* hybridization has two goals: 1) to preserve and immobilize the target and prevent its extaction or leakage into other subcellular compartments during subsequent processing 2) to give good preservation of tissue morphology. This is almost impossible to achieve and compromises have to be accepted. RNA targets, which are held in a 3-dimensional array attached to proteins and other molecules, are particularly difficult to retain. This must be accomplished without overfixation which can form a highly cross-linked tissue matrix hindering accessibility of target RNA to the probe. Under-fixed samples will give little or no signal even for highly abundant targets due to loss of RNA.

Pretreatment of samples with a detergent or proteinase digestion is a standard procedure in almost all protocols to incease probe penetration and accessibility. This renders the matrix formed between fixative and cell components porous, permeabilizes cell membranes and digests proteins which may mask targets.

These two steps are interrelated; the extent of fixation as well as the choice of a fixative will have an effect on the optimum extent of protein digestion during pre-hybridization. Keep in mind that conditions necessary for stabilization of morphology and retention of targets are at odds with permeabilizing conditions which may induce leaching of RNA and loss of tissue morphology.

4.2 EMBEDDING METHOD

There are two ISH approaches that have been successfully employed in plant systems to detect cell nucleic acid targets: the pre-embedding method and the post-embedding method, both of which have their strengths and limitations. The choice of the method is governed by the starting material (fixation and permeabilization requirements), the type and quantity (copy numbers) of targets, the ease with which probes and/or tags (i.e. probe length, marker sizes) can enter and diffuse within preparations.

The *pre-embedding* approach starts out with small samples (blocks) of pre-fixed tissue, which must be permeabilized by a "pore-forming" treatment in order to faciliate the probe penetration and its accessibility to targets throughout the volume of the tissue block. Once inside the cells the probe drives the reaction and virtually all specific targets would be bound, greatly increasing the signals. The labeled probes are detected either by autoradiography on sections or by the subsequent diffusion of a visible marker into the tissue block. The tissue is then embedded in a resin for sectioning and observation. The pre-embedding method has been extensively used in animal systems with both chemically fixed and frozen preparations [20,21,22, 54, 57,61]. It has been shown that probe and marker systems can be diffused into tissue with minimum disruption of cell morphology. In that cell barriers in most plant systems are relatively sturdy, the treatment for rendering cells porous without cell damage can be delicate [4,5].

The *post-embedding* method also starts out with a light fixation procedure; samples are then processed, embedded in a wax or resin and semi-thin or ultra-thin sections are prepared. Probe-target formation (hybridization) and its detection takes place on the sectioned material. Sensitivity can be rather low in that essentially only targets exposed at the surface of the section are accessible for hybridizing. This is less of a problem for wax-embedding as wax can be removed from sections. For resin-embedding, the problem can be circumvented by use of hydrophilic acrylic resins such as LR (London

Resin) White or LR Gold; these resins are semi-permeable and allow some probe penetration and thus their accessibility to targets deeper within the section. Hybridization on ultra-thin sections has been used in both animal and plant systems using low-temperature embedding in Lowicryl K4M [1, 2,53,57] and room temperature embedding in LR Gold [38,40] or LR White [33,58].

5. Results

Our laboratories have used ISH at both the optical and electron microscopic level using either radioactive or non-radioactive probes. Examples at the resolution of light microscope deal with 1) the localization of two protein kinase transcripts on wax sections of petunia flower buds using ^{35}S-labelled RNA probes [13] and 2) the localization of two subunit mRNAs of ADPglucose pyrophosphorylase in resin-embedded maize kernels with digoxigenin-tagged DNA probes. For electron microscopic ISH, we illustrate examples of pre-embedding and post-embedding procedures obtained with resin-embedded tobacco leaves using biotinylated cDNA probes for the detection of 3) mRNAs coding for the small subunit and large subunit composites of a chloroplastic enzyme, ribulose bisphosphate carboxylase [5,6].

5.1. EXAMPLES OF ISH ON WAX-EMBEDDED MATERIAL USING RADIOACTIVE-LABELLED RNA PROBES

Several studies have shown that protein kinases are involved in developmental processes in both animal and plant systems [8,43].Two different subgroups, designated PSK4 and PSK6, are known to exhibit differential expression patterns in petunia flower buds throughout their development. To study the tissue-specific distribution of their mRNAs in the whole flower, and in particular in the ovules and stamens, an *in situ* method was applied to petunia flower buds at different stages of male and female gametogenesis.

5.1.1. Plant material

Petunia hybrida lines H1F4 and BBF1 and the hybrid dBBF1 were grown and maintained *in vitro* on 0.5 x MS medium and subsequently propagated in the greenhouse at 25°C under a 16 h photoperiod. Flower buds from stage 1 (less than 0.5 cm long) to mature stage (one day before anthesis) were excised whole; ovaries were dissected from opening flowers for ovule digestion and embryo sac isolation [16].

5.1.2. *In situ* hybridization

Plant material was fixed in a paraformaldehyde/glutaraldehyde mixture, embedded in wax and cut into 7 µm sections (see protocols in appendix). Hybridization was performed with ^{35}S-labelled single stranded antisense RNA. Single stranded RNA probes were used because of several advantages [41,44,50,55] 1) high stability of RNA:RNA hybrids 2) strand-specific probes of defined size 3) the absence of competing strands in the hybridization mixture 4) the possibility of employing a posthybridization RNase treatment to digest the unbound probes, thus reducing background and 5) the DNA-coding strand (sense RNA) can be used as a negative control probe. Riboprobes can be labelled with a range of isotopes or non-isotopic reporter molecules [see reference in 3,9,19]. In this study, ^{35}S labelling was chosen for its high specific activity and its high sensitivity for detecting low levels of RNA and for providing acceptable resolution in relatively thick sections (7μm.). Protocols for the probe preparation, labelling and hybridization conditions are presented in the appendix.

5.1.3. Results

Figure 1A is a bright-light photograph of a longitudinal section of a petunia flower bud at low magnification. Wax-embedding gives acceptable morphological preservation of large samples such as whole flower buds and allows reliable collection of serial sections. Wax also has the advantage that it can be removed from the tissue after sectioning, allowing maximum probe penetration. A serial section of Figure 1, incubated with ^{35}S labelled-PSK4 antisense probe is illustrated in Figure 1B. Sections were photographed either by dark-field or epipolarization microscopy; silver grains corresponding to regions containing RNA/RNA hybrids are visible as bright white dots. A high magnification view shows intense hybridization of PSK4 mRNAs in the periphery of the placenta (arrow Fig.1C). A section through an anther at a young developmental stage after incubation with PSK6 anti-mRNA probe and photographed by bright-light microscopy is shown in Figure1 D; the arrow indicates a region in the stamen, the tapetum, showing intense signal. A high magnification of this region is visualised in epipolarization optics (Fig.1E). *In situ* hybridization using PSK4 and PSK6 control (sense) probes was carried out on serial sections of all preparations; only low background levels were observed.

This method proved suitable for the detection of tissue-specfic and organ-specific transcripts of several genes implicated in developmental processes during flower organogenesis and gametogenesis at a resolution of overall tissue morphology. However, it is often desirable to obtain maximal resolution, on a cell to cell basis, particularly in early stages of differentiation where distinctive cells and/or tissues are in close proximity or in complex tissue composed of a heterogeneous cell population. We have recently used resin-embedded material and non-isotopic labelling with the view to obtain resolution at the cell and subcellular levels with LM.

5.2. EXAMPLES OF ISH ON RESIN-EMBEDDED MATERIAL USING DIGOXIGENIN-LABELLED cDNA PROBES

ADP-glucose pyrophosphorylase, an enzyme composed of two subunit types, is active in all starch-synthesizing organs and is considered to play a pivotal role in plant starch biosynthesis. During the grain-filling period in maize kernels, starch accumulation and

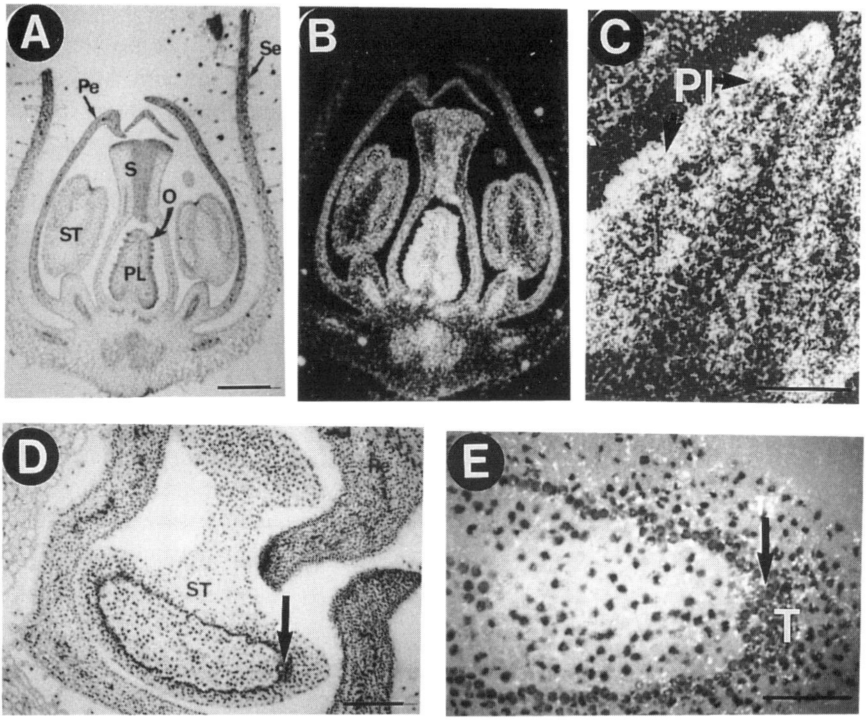

Figure 1. Localization of PSK mRNAs in *Petunia hybrida* floral bud using ^{35}S-labelled single stranded RNA probes. (A) Bright-field photograph of a longitudinal section through a young flower bud (B) Dark field view of same section; signals visible as white dots are localized in the ovule. Bar = 500 µm (C) High magnification dark-field photograph of a section through placenta and ovule primordia showing intense hybridization of PSK4 anti-mRNA in the periphery of the placenta. Bar = 100µm. (D) Bright field view of a section through anthers at a young developmental stage; arrowhead indicates the tapetum region showing hybridization. (E) High magnification of tapetum visualized by epipolarization optics; silver grains indicating hybrid formation are visible as white dots. Bar = 50 µm. O, ovule; Pe, petal; Placenta, Pl; Se, sepal; ST, stamen; T, tapetum. (from V.Decroocq-Ferrant)

ADP-glucose pyrophosphorylase activity have been correlated. In an attempt to understand kernel development in relation to the concurrent events of synthesis and storage of starch, localization of the two composite subunits of the enzyme, designated BT2 and SH2, and the corresponding mRNAs were detected by immunocytochemistry and *in situ* hybridization at different developmental stages during grain filling.

5.2.1. Plant material

Maize (Zea mays L.) plants were grown in an experimental field. Grains were excised from the ear at 2-3 day intervals starting at 5 days after pollination (DAP) up to 30 DAP.

5.2.2. *In situ* hybridization

Whole or half maize kernels were fixed in a phosphate-buffered paraformaldeyde/glutaraldehyde mixture and embedded in LR White resin (London Resin Co.); sections cut at 1μm were placed on Biobond-coated slides. The endogenous mRNAs were hybridized with digoxigenin-labelled cDNA probes and subsequently visualized using silver-enhanced anti-digoxigenin antibodies conjugated to gold (protocol is detailed in the appendix). LR white resin was chosen because high-resolution light microscopy requires thin sections (0.5μm-1μm) and these are not readily prepared from wax-embedded samples. LR White is easy to section and in general infiltrates well into plant material giving excellent morphological preservation. A factor that indirectly affects resolution is the quality of the tissue histology. Another advantage is that the same fixed and embedded material can be used for both immunocytochemistry and *in situ* hybridization at the light and electron microscopic level. The enzyme was first localized to the stroma region of amyloplasts in the endosperm using immunocytochemical techniques; ISH was then performed on serial sections. Immuno-detection of digoxigenin-labelled probes permitted a rapid and sensitive approach. Gold markers were visualized at the light microscope level using silver enhancement ; in this procedure, a shell of metallic silver builds up around each individual gold particle, increasing the signal and rendering it visible in LM. It is beneficial to use small (1 or 5nm) gold probes as this gives a more intense silver staining due to the higher number of gold tags on the probe.The degree of silver enhancement can be monitored under a light microscope [56].

5.2.3. Results

A global view of a portion of maize kernel shows the outer degenerating nucellus, the meristematic aleurone layers and the inner starch-containing endosperm.. Probing with digoxigenin-labelled BT2 cDNA revealed the presence of transcripts only in the inner endosperm cells; labelling is visible as black deposits (Fig.2A, arrows). Hybridized transcripts can be more readily seen at high magnification; an example of positively-labelled cytosol in endosperm cells is shown in Figures 2B and 2C. Labelling appears as black dots under bright-light microscopy and white dots under epipolarization optics. A control preparation probed with a digoxigenin-tagged pBR cDNA probe showed no labelling over that of background (not shown). We have recently applied this same approach to the localization of mitochondrial mRNAs in tobacco pollen grains. These preliminary results show a sparse labelling distribution (corresponding to mitochondria) (Fig.3A,B) compared to the denser labelling pattern of cytoplasmic transcripts. Verification at the EM level will be carried out on the same preparations.

This method combined with high resolution LM can readily distinguish signals associated with intra-cellular sub-structure. On low magnification views, signals are less visible, depending on their intensity; however these may be amplified by increasing gold tags and silver-enhancement time.

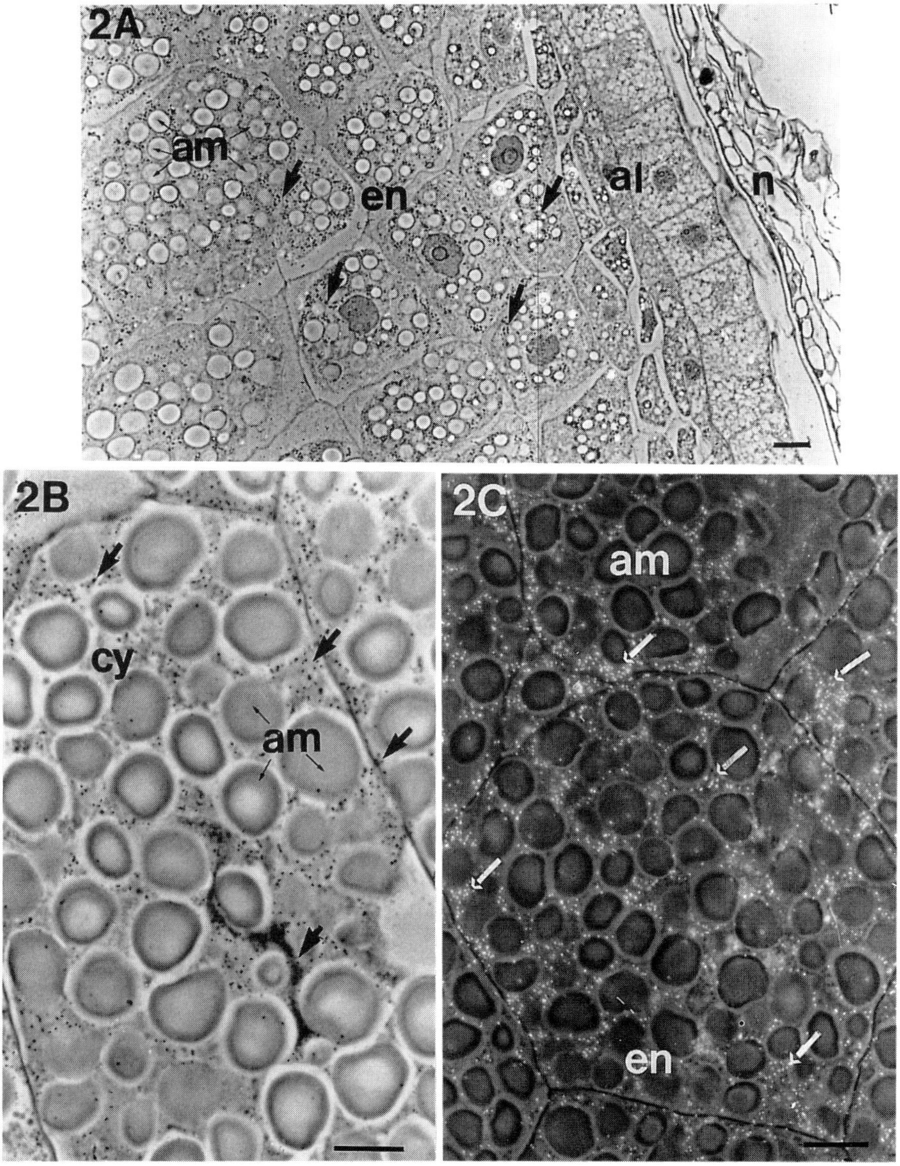

Figure 2. Localization of ADP-glucose pyrophosphorylase mRNAs in *Zea mays* kernel using digoxigenin-labelled cDNA probes revealed by silver-enhanced immuno-gold (A) Bright-field view of a portion of a kernel showing nucellus, aleurone layers and endosperm; labelling is visible as black deposits (arrows) in endosperm layer. (B) (C) High-magnifications of probed endosperm cells; labelling is seen in cytosol as black dots under bright-field and white dots under epipolarization. am,amyloplast; al, aleurone; en, endosperm; n, nucellus. Bar = 1μ.

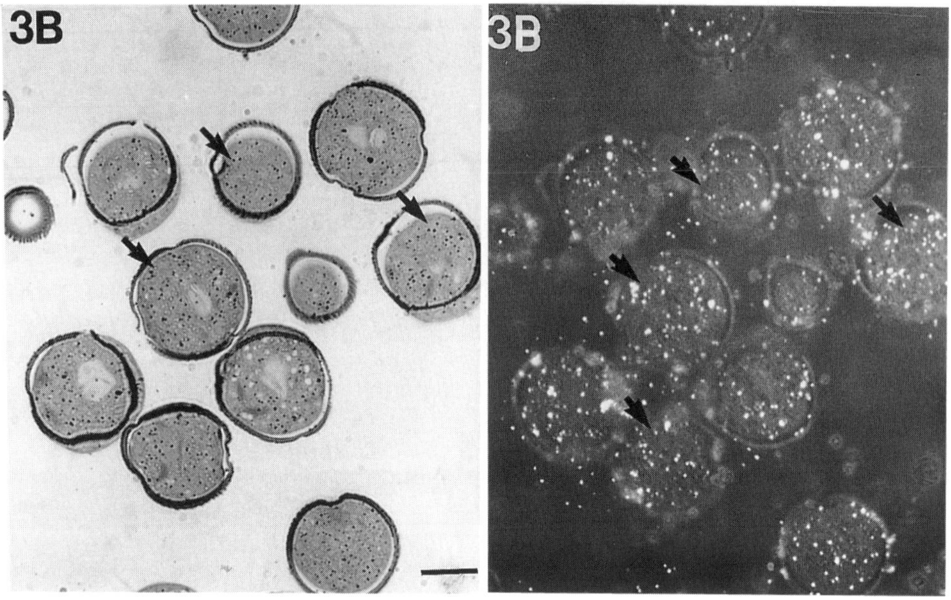

Figure 3. Localization of mitochondrial mRNA in tobacco pollen using the same technique as in Figure 2 (A) Sparse labelling pattern (mitochondria) visualized in bright-field as black dots; (B) Labelling under epipolarization optics of same section appears as white dots. Bar = 1μ.

5.3. EXAMPLES OF PRE-EMBEDDING AND POST-EMBEDDING ULTRASTRUCTURAL ISH WITH RESIN-EMBEDDED MATERIAL USING BIOTINYLATED cDNA PROBES

As our initial goal was to establish a reliable and reproducible technology to locate cell RNAs at the EM level, we chose as a model an enzyme/mRNA system in which targets are abundant and localization to specific subcellular compartment is well-documented. Ribulose 1,5-bisphosphate carboxylase, a chloroplastic enzyme, is a key enzyme of CO^2 fixation in C3 plants. It is composed of two types of subunits, the small subunit (SSU) is nuclear-encoded and synthesized as a precursor on free cytoplasmic polysomes whereas the large subunit (LSU) is encoded and synthesized inside the chloroplasts where final enzyme assembly also occurs [15]. Prior to undertaking the ISH study, immuno-gold cytochemistry [25] was used to visualize the target proteins confined exclusively to the stroma matrix of wild-type tobacco chloroplasts; an enzyme-deficient tobacco mutant was totally immuno-negative and served as a control [5].

5.3.1. Plant material

Tobacco plants (*Nicotiana tabacum*) were grown in a growth chamber under a 16h day/8h night regime. Samples from young developing leaves and from fully-expanded leaves were excised from lateral apices.

5.3.2. *In situ* hybridization

The *pre-embedding* method for EM ISH was applied to paraformaldehyde-fixed leaf material. Leaf pieces (100µm) were permeabilized in a pronase B solution, incubated with the labelled probe and in a second step with the marker. The whole block was then embedded in a resin for sectioning.The endogenous mRNAs were hybridized with biotin-labelled LSU and SSU probes and subsequently visualized using avidin-ferritin conjugates [4]. Protocols for the procedures are given in the appendix. For ultrastructural hybrid detection, biotin-labelling was chosen. Biotinylated probes have the advantage of being more stable than radiolabeled probes, usable up to several months [14]; the coupled affinity and/or immuno-detection systems are commercially available and less time-consuming than histoautoradiographic techniques (protocols take two days). Sensitivity can be increased by amplifying the detection system giving high specificity at organelle levels. For EM ISH, a major problem concerns the "thickness" of the preparations in which only a small portion of the cell nucleic acids are present, and the high specific activity of the probes as the number of hybrids formed at one site is necessarily low; hence the need for an efficient detection system. The second problem deals with properties of the fixative which must preserve the integrity of sub-cellular membrane structure to assure correct identification of organelles.

Results

Pre-embedding Method. Cell architecture in pronase-treated, probed leaves is relatively well preserved and cell components-nucleus, plastids, and endoplasmic reticulum-are readily recognizable, however membrane systems appear whitish and lack detail [4]. A plastid in a young cell probed with a biotinylated LSU probe and subsequently with avidin-ferritin is shown in Figure 4A; ferritin markers (arrows) indicating the target LSU mRNAs are exclusively located in the plastid stroma (st), while the surrounding cytosol (cy), is unlabelled. A portion of an SSU-probed cell is shown in Figure 4B; a homogenous distribution of ferritin throughout the cytoplasm indicates the SSU messengers, the plastid is ferritin-free. This specificity was observed in all neighboring cells although some internal cells of thicker leaf slices were unlabelled, undoubtedly due to low probe penetration. The sections were unstained as contrasting with either uranyl acetate or lead citrate led to electron-dense cytoplasmic background which interfered with viewing the ferritin particles. Although these markers have a low electron density and are not easily visible in the electron microscope, their small size (6nm) is advantageous for cell penetration. Background labelling was minimal and comparable to that in control samples hybridized with biotinylated pBR322 probe and incubated with

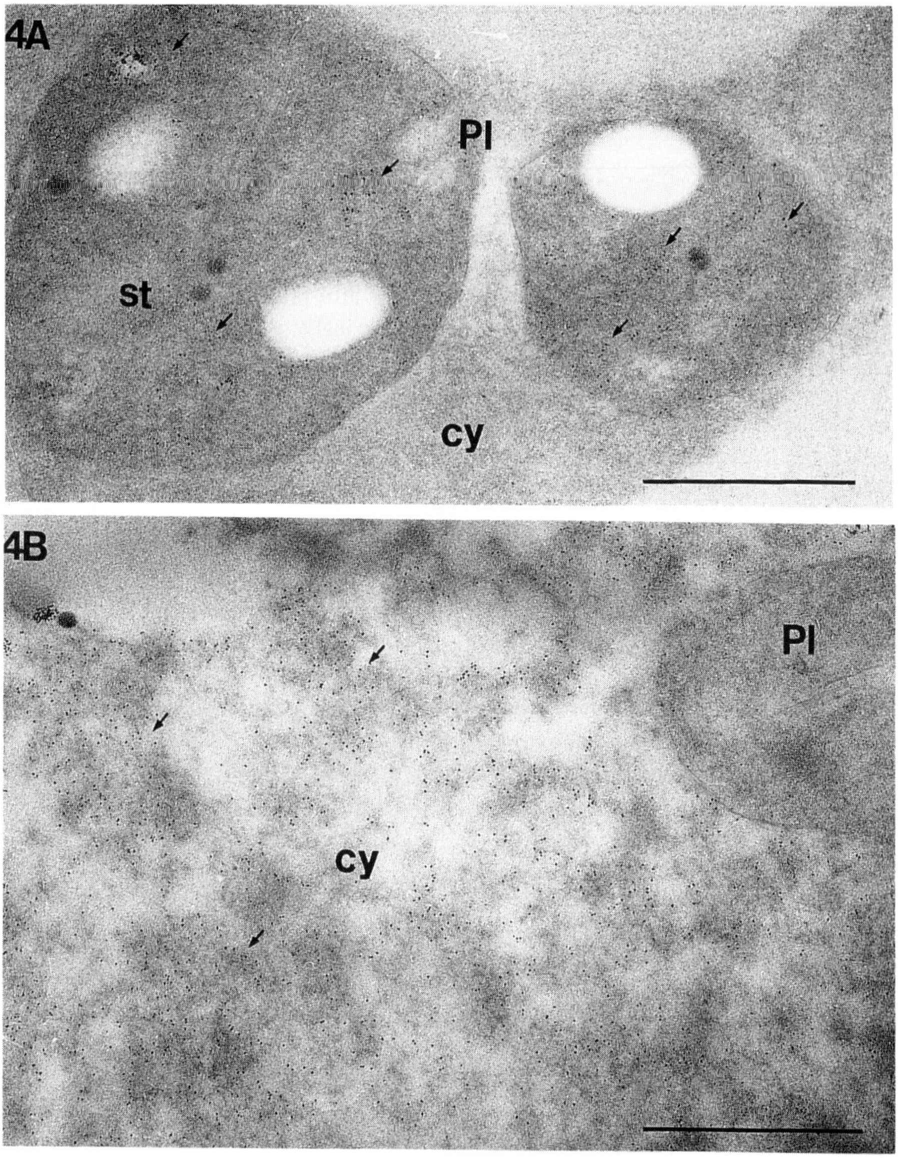

Figure 4. Ultrastructural localization of LSU and SSU mRNAs in young tobacco leaf using biotinylated cDNA probes and avidin-ferritin conjugates with a pre-embedding method (A) In LSU-probed plastid, ferritin markers indicating the target mRNA are located only in plastid stroma.; cytoplasm is untagged. (B) In an SSU-probed cell, distribution of ferritin throughout cytoplasm indicates the SSU targets; plastid is ferritin-free.cy,cytosol p,plastid; st,stroma Bar = 1μ.

avidin-ferritin or treated only with avidin-ferritin (not shown).

It was noted that in some SSU-probed cells, ferritin labelling was associated with the nuclei, suggesting that SSU pre-messenger RNAs were targetted.To confirm this observation, isolated nuclei were prepared (see protocol) and incubated with either the LSU or SSU probes. Ferritin tagging was dense within SSU-probed nuclei and absent after LSU probing, coherent with the fact that SSU is nuclear-coded and LSU chloroplast-encoded. It is suggested that ISH can be used to detect un-processed mRNA within nuclei.

For the pre-embedding method, permeabilization is essential in order to assure the accessibility of the probe/tag to the target nucleotides. The amount of enzymatic digestion or non-ionic detergents varies depending upon the type of tissue [7]. A major drawback which we have encountered with this approach was the variability of probe penetration or diffusion leading to a heterogeneity in labelling particularly in highly differentiated leaf tissue. If deproteinization was increased, this inevitably led to a deterioration of cell morphology. On the other hand, hybridization of pronase-treated young tissue, isolated organelles or protoplast preparations yielded homogeneous specific labelling, with no detrimental effect on structures. In the case of protoplasts, post-fixation with osmium (see protocol) was employed, considerably improving thylakoid membrane resolution (Fig 5 A,B).

Attempts to replace the avidin-ferritin tags with streptavidin-gold complex were unsuccessful. It may well be that the large molecular size of the streptavidin-gold complex (as is the case for antibody-gold conjugates) prevents its movement through the tissue. The use of small oligoprobes [32,52] in conjunction with small colloidal gold markers (1-5 nm), now commercially available, would certainly improve probe/tag penetration and reduce the need of rather severe permeabilization procedures, thus extending the pre-embedding method to more wide-spread use.

Post-embedding Method. All plant material will not be easily amenable to the pre-embedding approach. As an alternative method, we have tried a *post-embedding* procedure, using the same material and probes as above. This method is based on previously published procedures on resin-embedded sectioned material [40,57,58]. Ultra-thin sections of aldehyde-fixed, LR white embedded tobacco leaves were probed with biotinylated probes; the hybrids were subsequently detected by an anti-biotin antibody and a protein A-gold complex as described in the protocol in appendix. These samples showed improved ultrastructural preservation, probably due to the limited protease digestion needed on sections (Fig. 6). It is our experience that enzymatic digestion is not always necessary, probably depending on target abundance. Figure 6A shows gold tagging randomly distributed throughout the cytosol in an SSU-probed young leaf cell; a positively LSU-labelled plastid is illustrated in Figure 6B and other cell components are unlabelled. This method has several advantages: good fine structural morphology, immuno-detection is rapid and gold markers are readily visible in EM. It is possible to detect two different targets on a single section by using collodial gold markers of different sizes. We have successfully used the post-embedding method for the detection of glutamine synthetase transcripts expressed in transgenic tobacco and for sucrose synthase mRNA in developing maize; however for low copy nucleic acid

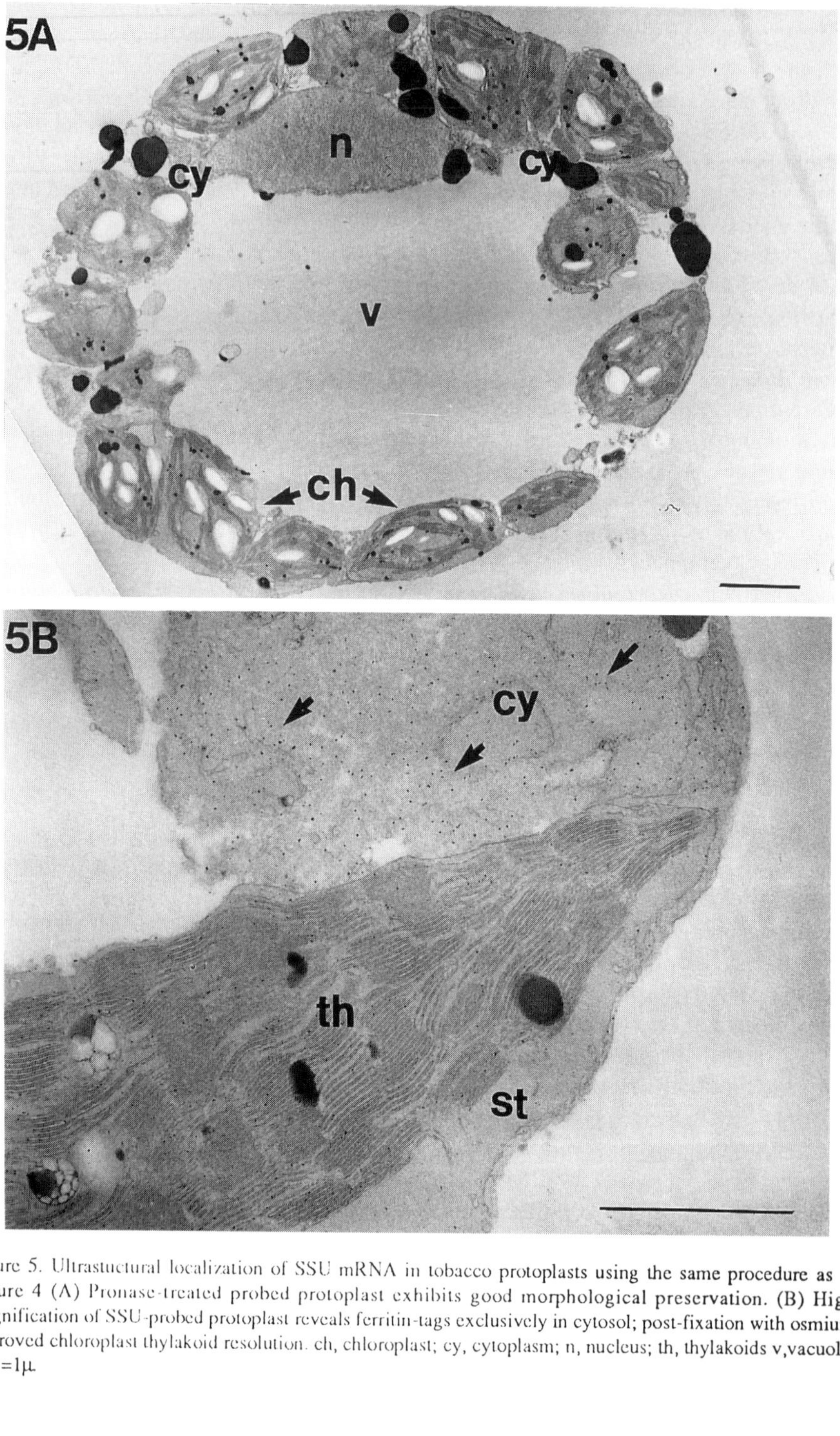

Figure 5. Ultrastuctural localization of SSU mRNA in tobacco protoplasts using the same procedure as in Figure 4 (A) Pronase-treated probed protoplast exhibits good morphological preservation. (B) High magnification of SSU-probed protoplast reveals ferritin-tags exclusively in cytosol; post-fixation with osmium improved chloroplast thylakoid resolution. ch, chloroplast; cy, cytoplasm; n, nucleus; th, thylakoids v,vacuole. Bar =1μ.

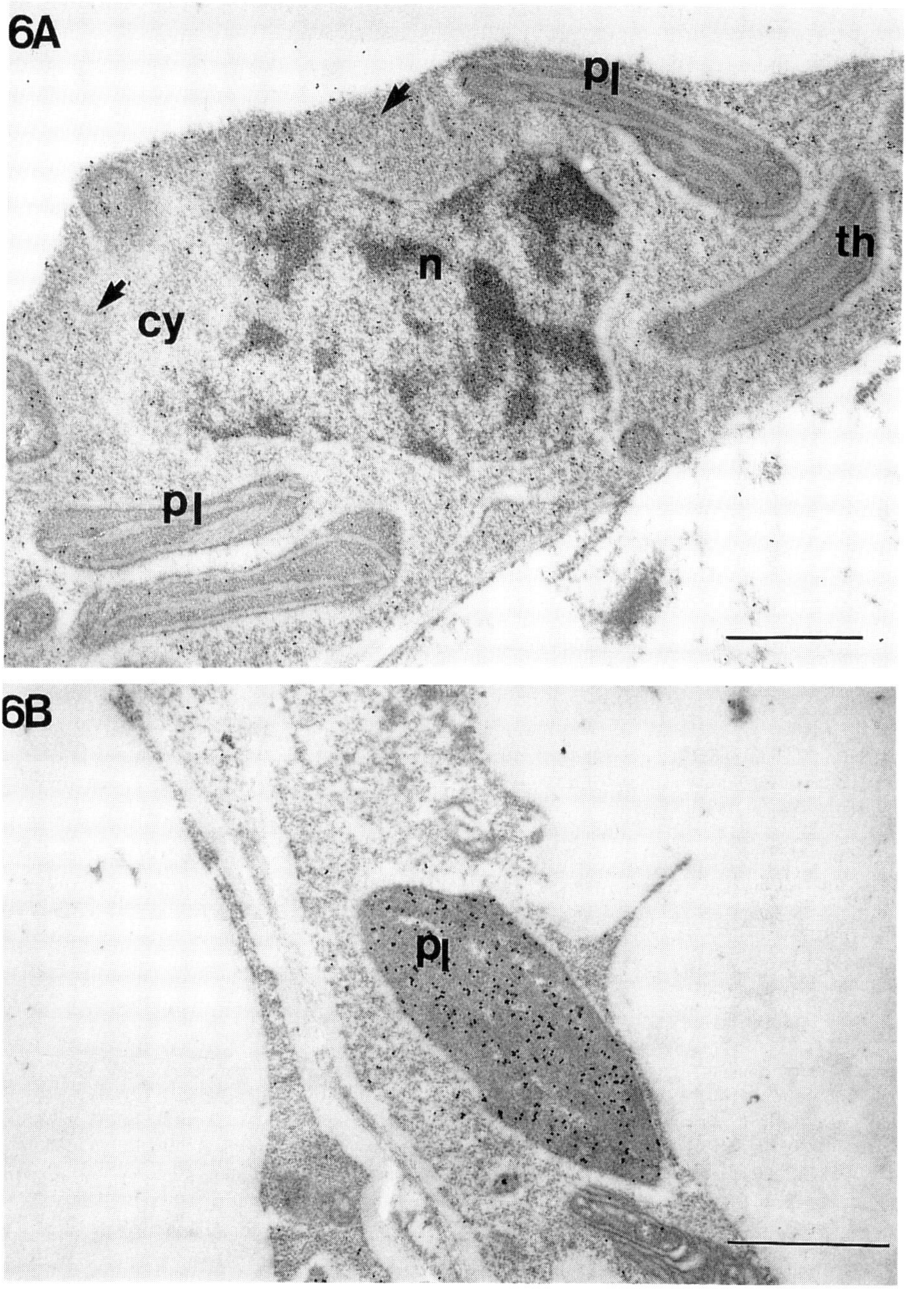

Figure 6. Ultrastructural localization of LSU and SSU mRNAs in young tobacco leaf using biotinylated cDNA probes, immuno-detected with anti-biotin and protein A-gold with a post-embedding method (A) Gold particles distributed in cytoplasm indicate target SSU mRNA; (B) Gold-tagged plastid reveals target LSU mRNA. cy, cytoplasm; n, nucleus; pl,plastid.

detection, technical modifications may be necessary to increase signal intensity. We are now experimenting with digoxigenylated-RNA probes which are purported to increase sensitivity and specificty by 5-fold [41]. Labelling by random-priming increases the number of digoxigenin reporter molecules on probes; hence these are available for subsequent immuno-detection. RNA probes are reported to be highly specific, form stable hybrids and thus more suitable for detecting low copy number RNA.

6. Conclusions

These results demonstrate the application of *in situ* hybridization to detect cellular mRNA molecules, in particular, their accumulation within sub-cellular compartments. The ultimate usefulness of ISH depends on the accuracy with which markers reflect local concentrations of target nucleotides, so that gene products can be topologically and chronologically evaluated. The possibility of relating cellular structures to molecular aspects of gene expression is a powerful tool, which can contribute to the understanding of plant development and plant response mechanisms.

7. Appendix: Material and Methods

Protocol for LM ISH on wax-embedded sections using ^{35}S-labelled RNA probes

a. Preparation and labelling of RNA probes

PSK4 and PSK6 sequences used as probes were amplified using the RT- PCR specific primers to produce fragments of 600 bp and 500bp, respectively. Probe-generated fragments were subcloned into the pBluescript-KS vector and used for the synthesis of RNA probes. ^{35}S labelling was carried out by *in vitro* transcription of the linearized plasmid as follows:

1. Add the following to a microcentrifuge tube on ice : 1-2 µg of linearized plasmid,1 X buffer T7 (5X buffer T7--200mM Tris-HCl pH8,100mM MgCl2, 25 mM DTT,500µg/ml BSA), 1mM rNTP (moins rUTP), 3µl rUTP-S^{35} (10^3Ci/mmol, Amersham),10 U enzyme T7 RNA polymerase. Incubate for 1 h at 37°C.
2. Stop the reaction by adding 10U DNAse (RNase-free); 10µg of tRNA (carrier) is added prior to purification by phenol/chloroform. Non-incorporated nucleotides are eliminated by column filtration (Sephadex).
3. Test labelling in a scintillation counter and dilute probe to 10^5cpm/µl.

b. Fixation and preparation of sections

1. Flower buds, excised ovaries and ovules are fixed in 4% paraformaldehyde and 0.25% glutaraldehyde in 1x PBS (130mM NaCl, 7mM Na_2HPO_4, 3mM NaH_2PO,0.27mM KCl, pH7.4) for 1 hour under vacuum. Renew the fixative and fix for 2-3 hours at room temperature. Anthers may be pre-embedded in an agarose block to avoid loss of pollen grains during subsequent steps. Rinse in 1X PBS, then in H_20.
2. Dehydrate progressively in a graded ethanol series, and then in an ethanol/xylene series: 3:1 (3 volumes of ethanol for 1 volume of xylene, 1:1,1:3, then into pure xylene.
3. Infiltrate progressively with melted, filtered paraffin (Paraclean,Klinipath) at 42° C; infiltration for 4 to 16 hours is necessary to insure complete penetration of tissue. Infiltrate in pure paraffin at 60°C for 48h, changing paraffin baths 4 X.
4. Cut routine sections (7-10 µm) and place them on coated (poly-L-lysine) slides
5. Bake the sections for 30 min (42°C°), dewax in xylene and rehydrate them in a graded ethanol series.
6. Treat sections with Proteinase K (1µg/ml) in a buffer of 100mM Tris-HCl (pH7.5), 50 mM EDTA for 30 min at 37°. Rinse in distilled water and then in 2X SSC for 5 min.

7. Dehydrate in a graded ethanol series; sections are air-dried and are ready for ISH or can be stored at -20°C.

c. Hybridization and post-hybridization washes

1. Hybridize using probes diluted in hybridization buffer (0.01M Tris-HCl, pH 7.5; 12.5% Denhardts solution, 2 X SSC, 50% formamide, 0.5 % SDS, 250μg/ml salmon sperm DNA .

2. Cover sections with coverslips, seal with Kleer tak rubber cement (Mecanorama) and hybridize overnight at 42°C in a sealed box.

3. Remove coverslips by immersion in 4 X SSC; renew washing solution 3 X 10 mn.

4. Incubate with RNAse A (Sigma) at 37°C for 30 min (50μg/ml in 500mM NaCl, 1mM EDTA, 10mM Tris-HCl, pH 7.5, followed by 4 washes in same buffer.

5. Dehydrate sections in a decreasing gradient of ethanol and 0.3M ammonium acetate to limit RNA loss.

6. Air dry sections prior to application of emulsion.

d. Autoradiography

1. Dip each slide in emulsion (LM-1 emulsion-Amersham), avoiding air bubbles; air-dry for 1h in dark at room temperature.

2. Place in slide box, cover and store at 4°C for 2 weeks.

3. Develop in D-19 Kodak 5-10min at 15°C, following instruction manual.

Protocol for LM ISH on LR-White-embedded sections using digoxigenylated-DNA probes

a. Labelling of DNA probes

Sh2 and Bt2 cDNA clones were kindly provided by L.C. Hannah. Hybridization was carried out with inserts excised by an EcoR1 restriction enzyme. Contol hybridization was performed with a PUC19 probe. Random priming was carried out with digoxigenin-dUTP using a Boehringer Mannheim kit following the instruction manual:

1. Use 0.5-1μg linearized DNA in 10μl sterilized distilled water.

2. Denature in a boiling water bath for 10 min and chill on ice.

3. Add 2μl hexanucleotide mixture (10 X conc.) 2μl DIG-DNA labelling mixture (10 X conc.) and bring volume up to 19μl with sterile distilled water.

4. Add 1μl Klenow enzyme, mix, centrifuge briefly and incubate either 2h at 37°C or overnight at room T°.

5. Stop labelling procedure by heating in boiling water for 10 min.

6. Add water and 20 X SSC to a final concentration of 100μl with a final concentration of 5 X SSC; add 1μl of 10% SDS to give 0.1% SDS.

7. Store at -20°C.

b. Fixation and embedding

Follow fixation and embedding protocol for post-embedding EM ISH (below) with the exception: for large samples ie. whole or half maize kernels, increase fixation time to overnight and extend LR White infiltration up to two days. Cut sections (1-2μm) and place on Biobond-treated (Biocell) slides.

c. Pre-hybridization and hybridization

1. Wet sections with distilled water.

2. Digest with Proteinase K (10μg/ml) in a buffer of 100mM Tris-HCl, pH 7.5, 50m M EDTA for 30 min ; fix in a 4% paraformaldehyde solution for 10 min to stop the protease reaction; wash in 2 X SSC for 5-10 min.

3. Prehybridize sections in hybridization buffer of 50% formamide/, 10% dextran sulfate, 6% Denhardts 2 X SSC and sonicated salmon sperm DNA at 250 μl/ml to block non-specific binding for 2 h.

4. Prepare probe-hybridization mixture-2 μg/ml of digoxigenylated probe in hybridization buffer; denature double-stranded probes prior to use by boiling for 2 min, immediately cool to 0°C in ice where kept until use.

5. Cover sections with probe-mixture; coverslip the sections, seal with rubber cement and hybridize overnight at selected T°(40°C-42°C) in a sealed humid chamber.

6. Remove coverslips by washing several times in 2 X SSC.

d. Immuno-gold detection

1. Pre-incubate grids in 0.1 M Tris-HCl, 0.1 M NaCl, 1 % BSA and 1 % Tween 20, pH 7.2 for 30 min. Drain off excess solution.

2. Incubate with sheep anti-dig antibody conjugated to 1nm gold diluted 1/100 in pre-incubation buffer containing 0.1 % BSA and 0.1 % Tween 20.

3. Rinse in Tris-HCl (as above) without BSA for 4 x 5 min. Drain off excess buffer. Wash with PBS 3 x 5 min.

4. Silver-enhance gold particles using silver enhancement kit (Biocell); coverslip sections and monitor under microscope until black deposits appear.

5. Rinse by dipping grids several times in distilled H_2O.

6. Counterstain lightly with blue toluidene.

3. Protocol for EM pre-embedding method of ISH using biotinylated DNA probes

a. Preparation and biotinylation of DNA probes

SSU (small subunit) probe from tobacco was kindly provided by Dr. J. Fleck (Clone ps TV 34). The hybridization was performed with the insert (543 BP) excised by CFol restriction enzyme. The LSU probe from tobacco chloroplast DNA was cloned (clone pt B1) by Shinozaki and Sugiura[49] ; the BAM H1 fragment of the 3' coding region was used for hybridization. Control hybridizations were carried out with a pBR322 probe.

The nick translation with biotinylated UTP (11 atom linkers) was performed with a Bethesda Research Laboratories (BRL) kit following the instruction manual..

1. Prepare reaction mixture comprised of 5 µg DNA, 10 units of DNA polymerase I, 100 pg of DNAse, 0.4 mM each of dATP, dGTP, dCTP and biotin-II-dUTP in a final volume of 50 µl of 100 mM Tris-HCl, pH 7.8, and 10 µg/ml bovine serum albumin (BSA), according to Harris and Croy [23].

2. Load reaction mixture on a 5 to 10 ml Sephadex G-50 superfine column equilibrated with 1 x SSC (0.15 M sodium chloride, 0.015 M sodium citrate pH 7.0) containing 0.1 % (v/v) SDS. A plastic pipette fitted with siliconized glass wool is a satisfactory column. Collect excluded fractions in 1.5 ml polypropylene microcentrifuge tubes. The presence of 0.1 % SDS in elution buffer will reduce non specific binding of the biotin-labeled DNA to the plastic tubes.

3. To verify biotin incorporation, spot 1-2 µl of each fraction onto nitrocellulose paper and visualize by streptavidin coupled to alkaline phosphate as described in DNA detection instructions of BRL kit.

4. Store aliquots of biotinylated probes in DNA dilution buffer (provided in BRL kit) at -20°C until use.

b. Pre-fixation and permeabilization.

1. Cut thin slices (about 200 µm) of leaf material in a droplet of freshly made 4 % paraformaldehyde in Na-phosphate-sucrose buffer (P-S: 0.05 M phosphate buffer, 0.05 M sucrose, pH 7) for 5 h at room temperature. The addition of sucrose to the buffer maintains osmolarity during fixation, prevents cell plasmolysis and improves preservation of structure. Change fixation solution 2 x during 5 hours. If samples float, place under gentle vacuum to improve penetration of fixative.

2. Wash samples in same P-S buffer for 4 x 15 min.

3. Soak samples in a preparation of Pronase B at 25 µg/ml in P-S buffer containing 5 mM EDTA for 15 minutes at 18°C.

4. Stop digestive (pronase) activity by incubating samples in 0.5 mM phenyl methylsulphonyl fluoride in P-S/EDTA buffer containing 4 mg/ml glycine to quench free aldehyde groups for 1 hour at 18°C.

c. Pre-hybridization and hybridization.

1.Place individual samples in siliconized Eppendorf tubes and equilibrate with 50 % formamide in 0.3 M NaCl/0.03 M Na citrate, 5 mM EDTA, 1 % Denhardt's, pH 7.0 (hybridization buffer) for 2 x 15 min - room temperature.

2 . Denature cDNA biotinylated probes prior to use by plunging in boiling water for 3 minutes, then place on ice.

3 . Incubate tissue samples in Eppendorf tubes containing 150 µl of 50 % formamide : 50 % hybridization buffer with 0.5 µg/ml of denatured biotinylated probe cDNA (either SSU probe, LSU probe or

control pBR322 probe) plus 250 µg/ml of salmon sperm DNA for 18-20 hours at 38°C. Another control consisted of incubation in the formamide-hybridization buffer without probe, followed by avidin-ferritin incubation.

d. Post-hybridization step.

1. Wash thoroughly with several changes of phosphate buffered saline. (PBS) over a 4-5 hour period.
2. Incubate in 5 µg/ml avidin-ferritin (Sigma) in PBS for 2 hours at 37°C
3. Wash tissue samples in PBS 4 x 15 minutes and leave overnight in PBS with gentle agitation.
4. Post-fix in 1 % osmium tetroxide in distilled H_2O for 1 hour, followed by several washes in H_2O.(optional)
5. Dehydrate in a graded series of alcohols, infiltrate and embed in Spurr resin. Since hybrid duplexes are stable in tissue blocks, conventional hydrophobic resins (Spurr, Epon) may be used.
6. Cut pale-gold ultrathin sections and place on 200-mesh copper grids. Do not post-stain as this can yield electron-dense cytoplasmic backgrounds which interfer with ferritin particle visualization.

We have used this protocol for protoplast preparations and isolated organelles. After pre-fixation step, samples were centrifuged at low speed to form a pellet and embedded in 1 % agar in H_2O. The agar blocks were then carried through steps as were tissue pieces.

Protocol for EM post-embedding method of ISH using biotinylated DNA probes

The same probes were used as described above.

a. Fixation and embedding.

1. Cut small samples (2 mm2) of leaf in a droplet of freshly made 4 % paraformaldehyde - 0.5 % glutaraldehyde in Na-phosphate-sucrose buffer (0.1 M phosphate buffer, 0.04 M sucrose, pH 7.2). Place samples in fixation solution for 5 h at 4°C using intermittent vacuum in a dessicateur until samples sink. Change fixation solution 3 X.
2. Wash samples in same P-S buffer for 4 x 15 min.
3. Dehydrate in a graded ethanol series (30 %, 50 %, 70 %, 80 %, 90 %, 95 %) for 15 min each, 100 % ethanol 2 x 10 min).
4. Orient leaf (or other tissue) samples as desired (eg. for X sections) in 1-2 % agar in H_2O, cut into cubes and continue dehydration (between 50 %, 70 % step).
5. Infiltrate LR white resin by transferring specimens to a 3:1 mixture of ethanol/LRW, a 1:1 mixture, 1:3 ethanol/LRW (2-h baths) and rotate overnight in pure LR white resins at room temperature. Change several times the following day prior to embedding in pure LR white. Use tightly-capped gelatin capsules to exclude oxygen during polymerization at 50°C for 10 h.
6. Prepare ultrathin light-gold sections and place on 200-mesh gold grids.
7. Place sectioned material in covered petri dish and store in a dessicator.

b. Pre-hybridization and hybridization.

1. Place grids in an oven at 60°C for 1 hour prior to hybridization.
2. Digest sections with 10 µg/ml proteinase K in 20mM Tris,2mM $CaCl_2$ for 15 min at 37°C (optional). Wash rapidly in PBS and fix with 4% paraformaldehyde for 10 min to stop the protease reaction. Wash again in PBS and in 2 X SSC for 5 min.
3. Prepare probe-hybridization buffer : 0.5 µg of biotinylated probe (LSU, SSU or pBR322) in 50% deionized formamide, 10% dextran sulfate, 6% Denhardts, 2X SSC, 50mM NaH2PO4/Na2HPO4 buffer, pH 7 and salmon sperm DNA at 250 µl/ml to block non-specific binding.Denature double-stranded probes prior to use (2 min in boiling water); immediately cool to 0°C in ice where kept until use.
4. Line a glass petri dish with parafilm and place into a larger recipient containing moist filter paper. Carefully place a number of 2-5 µl drops of probe-mixture on parafilm taking care not to spread them.
5. Place grids on droplets, seal the recipient with tape and hybridize overnight (16-18 h) at selected temperature (38°C-> 40°C).

5\. Rinse by placing grids on 4X SSC at 45° C, 2 X SCC at room T° and then on 1X SCC for 10 min.

c. Immuno-gold detection.

1. Pre-incubate grids in 0.1 M Tris-HCl, 0.1 M NaCl, 1 % BSA and 1 % Tween 20, pH 7.2 for 30 min. Drain off excess solution.

2. Incubate with rabbit anti-biotin (ENZO Biochemical Inc.) diluted 1/100 in pre-incubation buffer containing 0.1 % BSA and 0.2 % Tween 20.

3. Rinse in Tris-HCl (as above) without BSA for 4 x 5 min. Drain off excess buffer

4. Incubate in Protein A-gold conjugate (10 nm particles, Amersham Int) diluted to 1/50 in PBS for 1 hour. Wash with PBS 3 x 5 min.

5. Rinse by dipping grids several times in distilled H_2O.

6. Counterstain in saturated aqueous uranyl acetate for 15 min.

7. References

1. Binder, M. (1987) *In situ* hybridization at the electron microscope level. *Scan. Microscopy* **1**, 331..

2. Binder, M., Tourmente, S., Roth, J., Renaud, M., and Gehring, W.J. (1986) *In situ* hybridization at the electron microscope level: localization of transcripts on ultrathin sections of Lowicryl K4M-embedded tissue using biotinylated probes and protein-A gold complexes. *J. Cell Biol.*. **102**, 1986.

3. Brady M.A., and Finlan M.F.(1990) Radioactive labels: autoradiography and choice of emulsions for in situ hybridization, in *In situ hybridization principles and practice*. ed. Polak, J.M., and McGee, J.O'D. Oxford, New-York, Tokyo, chap. 3

4. Brangeon, J., Prioul, J.L., and Forchioni, A. (1988) Localization of mRNAs for the small and large subunits of rubisco using electron microscope *in situ* hybridization. *Plant Physiol.*. **86**, 990.

5. Brangeon, J., Nato, A., and Forchioni, A. (1989) Ultrastructural detection of ribulose-1,5-bisphosphate carboxylase protein and its subunit mRNAs in wild-type and holoenzyme-deficient *Nicotiana* using immunogold and *in situ* hybridization techniques. *Planta* **177**, 151.

6. Brangeon, J., and Forchioni, A. (1991) Détection ultrastructurale des ARNm de tissus foliaires par hybridation *in situ* à l'aide de sondes biotinylées, révélées à la ferritine, in *Techniques en Microscopie électronique, cyrométhodes, immunocytologie, autoradiographie, hybridation in situ,* Morel G., ed., Paris, chap. 52.

7. Bresser, J., and Evinger-Hodges, M.J. (1987) Comparison and optimization of in situ hybridization procedures yielding rapid, sensitive mRNA detections. *Gene Anal. Tech.* **4**, 89.

8. Budde, R. J.A. and Randall, D.D. (1990) Protein kinases in high plants, in *Inositol Metabolism in Plants,* Wilery-Liss eds. pp 351-367.

9. Chan,V.T.-W.and McGee, J.O'D. (1990) Non-radioactive probes : preparation, characterization and detection. in *In situ hybridization principles and practice*. Polak J.M. and McGee, J.O'D., eds., Oxford New York, Tokyo, chap. 4.

10. Cox, K.H. and Goldberg, R.B. (1988) Analysis of plant gene expression. in *Plant Molecular Biology: A Practical Approch,* Shaw,C.H.,ed.,Oxford, IRL Press, chap.1.

11. Giaid, A., Hamid, Q., Adams, C., Springall, D.R., Terenghi, G., and Polak, J.M. (1989) Non-isotopic RNA probes. Comparison between different labels and detection systems. *Histochemistry* **93**, 191.

12. Decroocq-Ferrant,V.,Decroocq,S.,VanWent,J.,Schmidt,E., Kreis,M.(1995) A homologue of the MAP/ERK family of protein kinase genes is expressed in vegetative and in female reproductive organs of *Petunia hybrida* . *Plant Mol. Biol* . 27, 339.

13. Decroocq-Ferrant, V.Van Went, J., Bianchi, M., de Vries, S.C. and Kreis,M. (1995) *Petunia hybrida* homologues of shaggy/zeste-white 3 expressed in female and male reproductive organs. Plant J. **7**, 897.

14. Deniyn, M., De Weger, R.A., Berends, M.J.H., Compier-Spies, P.I., Jansz, H., Van Unnik, J.A.M., and Lips, C.J.M.,(1990) Detection of calcitonin-encoding mRNA by radioactive and non-radioactive *in situ* hybridization: improved colorimetric detection and cellular localization of mRNA in thyroid sections. *J. Histochem. Cytochem.*. **38**, 351.

15. Ellis, R.J., (1981) Chloroplast proteins: synthesis, transport and assembly. *Annu. Rev. Plant Physiol.* **32**, 111.

16. Ferrant, V., Van Went, J., Kreis, M. (1994) Ovule cDNA clones of Petunia hybrida. encoding proteins homologous to MAP and shaggy/zeste-white 3 protein kinases. In *Molecular and Cellular Aspects of Plant Reproduction* ,ed. Scott,R. and Stead, A. Cambridge University Press. pp 159-172

17. Gall, J.G., and Pardue, M.L., (1969) Formation and detection of RNA-DNA hybrids molecules in cytological preaparations. *Proc. Natl. Acad. Sci. USA.* **63**, 378.

18. Geuskens, M., and May, E., (1974) Ultrastructural localization of SV 40 viral DNA in cells, during lytic infection, by *in situ* molecular hybridization. *Exper. Cell Res.* **87**, 175.
19. Gibson S.J. and Polak J.M.(1990) Principles and applications of complementary RNA probes in *In situ hybridization principles and practice.* ed. Polak, J.M., and McGee, J.O'D. Oxford University Press, New-York, chap. 6
20. Guitteny, A.F., and Bloch, B.,(1989) Ultrastructural detection of the vasopressin messenger RNA in the normal and Brattleboro rat. *Histochemistry* **92**, 277.
21. Guitteny, A.F., Fouque, B., Mougin, Ch., Teoule, R., and Bloch, B.,(1988) Histological detection of messenger RNAs with biotinylated synthetic oligonucleotide probes. *J. Histochem Cytochem..* **36**, 563.
22. Guitteny, A.F., Fouque, B., Teoule, R., and Bloch, B.,(1989) Vasopressin gene expression in the normal and brattleboro rat: a histological analysis in semi-thin sections with biotinylated oligonucleotides probes. *J. Histochem. Cytochem..* **37**, 1479.
23. Harris, N., and Croy, R.R.D.,(1986) Localization of mRNA for pea leguminin: *in situ* hybridization using a biotinylated cDNA probe. *Protoplasma* **130**, 57.
24. Harris, N., Grindley, H., Mulchrone, J., and Croy, R.R.D.,(1989) Correlated *in situ* hybridisation and immunochemical studies of leguminin storage protein deposition in pea (*Pisum sativum* L.). *Cell Biol. Intern. Reports* **13**, 23.
25. Hawes C., (1988) Subcellular localization of macromolecules by microscopy.in *Plant Molecular Biology : A Practical Approach,* Shaw C.H. ed. IRL Press Oxford , chap 5
26. Hoefler, H., Childers, H., Montminy, M.R., Lechan, R.M., Goodman, R.H., and Wolfe, H.J, (1986) *In situ* hybridization methods for the detection of somatostatin mRNA in tissue sections using antisense RNA probes. *Histochem. J.* **18**, 597.
27. Hutchinson, N.J., (1984) Hybridization histochemistry: *in situ* hybridization at the electron microscopie level. In *Immunolabelling for electron microscopy,* Polak, J.M. and Vardndell I.M., eds., Elsevier Sci. Publ. NY, chap. 23.
28. Jamrich, M., Mahon, K.A., Gavis, E.R. and Gall, J.G.,(1984) Histone mRNA in amphibian oocytes visualized by *in situ* hybridization to methacrylate-embedded tissue sections. *Embo* J. **3**, 1939.
29. John H.L., Bimstiel, M.L. and Jones, K.W. (1969) RNA-DNA hybrids at the cytological level. *Nature* **223**,912.
30. Kelly, A.J., Zagotta, M.T., White, R.A., Chang, C., and Meeks-Wagner, D.R.,(1990) Identification of genes expressed in the tobacco shoot apex during the floral transition. *Plant Cell* **2**, 963.
31. Langdale, J.A., Zelitch, I., Miller, E., and Nelson, T., (1988) Cell position and light influence C4 versus C3 patterns of photosynthetic gene expression in maize. *EMBO J.* **7**, 3643.
32. Lathe, R.,(1990) Oligonucleotide probes for in situ hybridization. In *In situ hybridization principles and practice.* ed. Polak, J.M., and McGee, J.O'D. Oxford, New-York, Tokyo, chap. 5.
33. Leitch, A.R., Mosgöller, W., Schwarzacher, T., Bennett, M.D., and Heslop-Harrison, J.S., (1990) Genomic *in situ* hybridization to sectioned nuclei shows chromosome domains in grass hybrids. *J. Cell Sci..* **95**, 335.
34. Le Guellec, D., Frappart, L., and Desprez, P.Y., (1991) Ultrastructural localization of mRNA encoding for the EGF receptor in human breast cell cancer line BT20 by *in situ* hybridization. *J. Histochem. Cytochem..* **39**, 1.
35. Mansfield, M.A., and Raikhel, N.V., (1990) Abscisic acid enhances the transcription of wheat-germ agglutinin mRNA without altering its tissue-specific expression. *Planta.* **180**, 548.
36. Manuelidis, L., and Ward, D.C., (1984) Chromosomal and nuclear distribution of the Hind III 1.9 kb human DNA repeat segment. *Chromosoma* **91**, 28.
37. Manuelidis, L.,(1985) Indications of centromere movement during interphase and differentiation. *Ann. Acad. Sci. NY* **450**, 205.
38. McFadden, G.I., Bönig, I., Cornish, E.C., and Clarke, A.E., (1988) A simple fixation and embedding method for use in hybridization histochemistry on plant tissues. *Histochem. J .* **20**, 575.
39. McFadden, G.I., (1989) *In situ* hybridization in plants: from macroscopic to ultrastructural resolution. *Cell Biol. Internat. Rep.* **13**, 3.
40. McFadden, G., Bönig, I., and Clarke, A.,(1990) Double label *in situ* hybridization for electron microscopy. *Trans. Roy. Micros. Soc.* **1**, 683,.
41. McFadden, G.I.,(1991) In situ hybridization techniques: molecular cytology goes ultrastructural. In *Electron microscopy of plant cells,* Hall J.L. and Hawes C., eds., Academic Press, London, chap.6
42. Narayanswami, S., and Hamkalo BA.,(1987) *Microscopy in molecular biology. A practical approach."* Rickwood, D and Hames, B.D.,eds, IRL Press. Oxford, chap. 10.
43. Okamuro, J.K., and Goldberg, R.B., (1989) Regulation of plant gene expression: general principles. In

The Biochemistry of Plants, Volume15, Marcus, M., ed., Academic Press, San Diego, pp.4-51.
44. Ozden, S., Aubert, C., Gonzales-Dunia, D., and Brahic, M., (1990) Simultaneous *in situ* detection of two mRNAs in the same cell using riboprobes labeled with biotin and ^{35}S. *J. Histochem. Cytochem..* **38,** 917.
45. Pelletier, G., Tong, Y., Simard, J., Zhao, H.F., and Labrie, F.,(1989) Localization of peptide gene expression by *in situ* hybridization at the electron microscope level. *Methods in Neurosci..* **1,** 197.
46. Penschow, J.D., Haralambidis, J., and Coghlan, J.P., (1991) Location of glandular kallikrein mRNAs in mouse submandibular gland at the cellular and ultrastructural level by hybridization histochemistry using ^{32}P and ^{3}H oligodeoxyribonucleotide probes. *J. Histochem. Cytochem..* **39,** 835.
47. Puvion-Dutilleul, F., and Puvion, E., (1989) Ultrastructural localization of viral DNA in thin sections of herpes simplex virus type 1 infected cells by *in situ* hybridization . *Eur. J. Cell Biol..* **49,** 99.
48. Raikhel, N.V., Bednarek, S.Y., and Lerner, D.R., (1989) *In situ* hybridization in plant tissues. in *Plant molecular biology manual*, section B9, Gelvin, S.B., and Schilprroort, R.A., eds., Kluwer, the Netherlands, chap. 1.
49. Shinozaki, K., Sugiura, M., (1982) The nucleotide sequence of tobacco chloroplast gene for the large subunits of ribulose-1,5-bisphosphate carboxylase/oxygenase. *Gene* **20,** 91.
50. Simmons, D.M., Arriza, J.L., and Swarson, L.W.,(1989) A complete protocol for *in situ* hybridization of messenger RNAs in brain and other tissues with radiolabelled single-stranded RNA probes. *J. Histotechnol..* **12,** 169.
51. Singer, R.H., Langevin, G.L., and Lawrence, J.B.,(1989) Ultrastructural visualization of cytoskeletal mRNAs ans their associated proteins using double-label *in situ* hybridization. *J. Cell Biol..* **108,** 2343.
52. Sossountzov, L., Ruiz-Avila, L., Vignols, F., Jolliot, A., Arondel, V., Tchang, F., Grosbois, M., Guerbette, F., Miginiac, E., Delseny, M., Puydomenech, P., and Kader, J.C.,(1991) Spatial and temporal expression of a maize lipid transfer protein gene. *Plant Cell* **3,** 923.
53. Thiry, M., Thiry-Blaise, L.,(1989) *In situ* hybridization at the electron microscopic level: an improved method for precise localization of ribosomal DNA and RNA. *Europ. J. Cell Biol..* **50,** 235.
54. Trembleau, A., Ferre-Montagne, M. and Calas, A., (1988) Ultrastructural visualization of oxytocin mRNA by *in situ* hybridization. A high reoslution radioautographic study using a tritiated oligonucleotide probe. *Comptes Rendus Acad. Sci..* Paris III, **307,** 869.
55. Troxler, M., Pasamontes, L., Egger, D., and Brenz, K., (1990) *In situ* hybridization for light and electron microscopy: a comparison of methods for the localization of viral RNA using biotinylated DNA and RNA probes. *J. Virol. Methods..* **30,** 1.
56. Vandenbosch K.A., (1991) Immunogold Labelling in *Electron Microscopy of Plant Cells* ,Hall J.L. and Hawes C., eds., Academic Press, London, chap.5.
57. Webster, H. de F., Lamperth, L., Favilla, J.T., Lemke, G., Tesin, D., and Manuelidis, L., (1987) Use of a biotinylated probe and *in situ* hybridization for light and electron microscopic localization of PomRNA in myelin-forming Schwann cells. *Histochemistry* **86,** 44.
58. Wenderoth, M.P., and Eisenberg, B.R., (1991) Ultrastructural distribution of myosin heavy chain mRNA in cardiac tissue: a comparison of frozen and LR White embedment. *J. Histochem. Cytochem..* **39,** 1025.
59. Wilkinson, D.G., (1989) mRNA *in situ* hybridzation and the study of development. in *In situ hybridization principles and practice.* eds. Polak, J.M., and McGee, J.O'D. Oxford, New-York, Tokyo, chap. 8.
60. Wolber, R.A., and Beals, T.F., (1989) Streptavidin-gold labeling for ultrastructural *in situ* nucleic acid hybridization. in *Colloïdal gold: principles, methods and applications,* Academic Press, London, chap. 19.
61. Wolber, R.A., Beals, T.F., and Maassab, H.F., (1989) Ultrastructural localization of herpes simplex virus RNA by *in situ* hybridization. *J. Histochem. Cytochem..* **37,** 97.
62. Yokoyama, M., Ogawa, M., Nozu, Y., and Hashimoto, J., (1990) Detection of specific RNAs by *in situ* hybridization in plant protoplasts. *Plant Cell Physiol..* **31,** 403.

IN SITU DETECTION OF POLYPHENOLS IN PLANT MICRO-ORGANISM INTERACTIONS

ANDARY, C.[1], MONDOLOT-COSSON, L.[1] and DAI, G.H.[1-2]
[1]: *Laboratoire de Botanique, Phytochimie et Mycologie, Faculté de Pharmacie, 34060 MONTPELLIER, France.*
[2]: *Laboratory of Plant Stress Physiology, Hebei Academy of Agricultural and Forestry Sciences, 050051 SHIJIAZHUANG, Hebei, China.*

1. Introduction

It is known that plant resistance may often be the result of multiple chemical constituents acting at specific sites.

Most studies on host resistance have focused on the presence of allelochemicals in homogenised whole organs and have not ascertained defence mechanisms at the site of infection. However, susceptible and resistant host cultivars may have similar concentrations of the chemicals in whole plant analysis but different levels in specific tissues or sites. To fully understand host defensive response to pathogenic fungi and other pests, one must first determine the plant's reaction to cellular invasion using histochemical investigation.

Various studies in this field have implicated polyphenols in numerous defence mechanisms against attack by microorganisms [20, 27]. These mechanisms can be considered as constitutive mechanisms : highly-lignified cell walls, thick cuticle [1], presence of fungitoxic polyphenols [16, 32] or as active mechanisms induced by the phytopathogen : phytoalexins [28, 22], hypersensitivity reactions [29, 14].

We set out to investigate these phenolics directly in the tissues, revealing their presence by means of specific chemical reactions that give rise to characteristic fluorescences or colorations.

Two plant-microorganism combinations were studied : *Helianthus / Sclerotinia sclerotiorum* and *Vitis / Plasmopara viticola*. The different techniques used and the results obtained before and after infection are presented in this paper.

For sunflower (*Helianthus annuus*) there is no cultivated genotype truly resistant to *Sclerotinia sclerotiorum* (white rot). *H. resinosus*, a wild species, is highly tolerant to numerous pathogens, in particular *S. sclerotiorum* [13], and represents a good model for study. For grapevine, three species were chosen for their varied susceptibility to *Plasmopara viticola* (downy mildew) : *Vitis vinifera* cv. Grenache (susceptible, S), *V. rupestris* cv. rupestris (intermediate resistant, M) and *V. rotundifolia* cv. Carlos (resistant, R).

M. Nicole and V. Gianinazzi-Pearson (eds.), Histology, Ultrastructure and Molecular Cytology of Plant-Microorganism Interactions, 43-53.

2. Materials and Methods

2.1. PLANT SPECIES

Specimens of sunflower were taken from the Melgueil experimental station (INRA, Mauguio, France, Table 1). Studies were performed on leaves and capitulum bracts taken from plants cultivated in the open field.

Grapevine plants were obtained from the grapevine collection of the Ecole Nationale Supérieure Agronomique de Montpellier (ENSAM).

Tests were conducted with *V. vinifera* cv. Grenache (susceptible, S), *V. rupestris* cv. rupestris (intermediate resistant, M) and *V. rotundifolia* cv. Carlos (resistant, R). Plants were propagated in the greenhouse from cuttings.

2.2. FUNGUS, INOCULATION AND SAMPLING

For *H. resinosus*, the test used to detect infection with mycelium of *S. sclerotiorum* (strain SS, on leaves) was that described by Bertrand and Tourvielle [3]. The "mycelium" test used on the capitulum was that of Vear and Guillaumin [35].

For grapevine, the pathogen was isolated from naturally-infected leaves of *V. vinifera* cv. Grenache in a vineyard of ENSAM. The fungus was maintained on *V. vinifera* cv. Grenache in the greenhouse.

Fully-expanded second and fourth leaves were inoculated by spraying the undersurface of the leaves with a sporangial suspension of *P. viticola* containing 5 x 10^5 sporangia ml^{-1} ; the third and fifth leaves were sprayed with deionised water for controls. Following inoculation, the leaves were enclosed in clear polyethylene bags for 48 h. Inoculated and control leaves were excised 1, 2, 5, 8 and 15 days after inoculation. The excised leaves were immediately placed in a freezer until required for analysis.

2.3. HISTOCHEMISTRY AND OBSERVATIONS

Discs (1 cm in diameter) were removed from the leaves and mounted on microscope slides with the abaxial surface uppermost. Sections (30 μm thickness) of test leaves were cut with a freezing-stage microtome. After staining with the different reagents, the discs or sections were mounted in the reagents or in glycerine : water (15 : 85, v/v) and examined using a light microscope (Nikon Optiphot) with two filter sets, a UV filter set with 365 nm excitation and a 400 nm barrier filter, and a blue filter set with 420 nm excitation and a 515-560 nm barrier filter.

For autofluorescence, discs or sections from test leaves were mounted in glycerine-water on glass slides and examined by epi-fluorescence microscopy.

Neu's reagent. Sections of leaf or discs were immersed for 2-5 min in 1% 2-amino-ethyldiphenyl-borinate (Fluka) in absolute methanol, then mounted in glycerine-water and examined by epi-fluorescence microscopy [26, 5].

Wilson's reagent. Sections were immersed for 15 min in citric acid : boric acid (Prolabo) (5 : 5, w/w) in 100 ml absolute methanol, mounted in glycerine-water and examined by epi-fluorescence microscopy [17].

Vanillin-HCl. Sections or leaf discs were immersed for 5 min in 10% (w/v) vanillin in 1 volume of absolute ethanol mixed with 1 volume of concentrated HCl [30], mounted in this reagent and observed with the light microscope.

Mirande's reagent. Sections were placed in sodium hypochlorite (48% active chlorine) for 5 min then rinsed with distilled water before immersing for 5 min in a solution of carmine (N° 40) and iodine green (Prolabo). They were then rinsed with distilled water and observed using a light microscope.

Phloroglucinol-HCl. Sections or leaf discs were fixed in absolute ethanol then immersed for 3 min in a solution containing 10 g phloroglucinol (Prolabo) dissolved in 95 ml absolute ethanol [12]. They were then washed in concentrated HCl and mounted in 75% glycerine. The sections or leaf discs were examined using a light microscope.

Sudan IV. Sections were stained for 15 min in a saturated solution of Sudan IV (Sigma) in 70% ethanol, then rinsed rapidly three times with 50% ethanol and examined using a light microscope [21].

Peroxidase (TMB reagent). Peroxidase (EC 1.11.1.1.7) activity in fresh sections was localised with 3,3',5,5'-tetramethylbenzidine (TMB, Sigma). Sections were collected in buffer (pH 4-5) at room temperature and then incubated for 10 min in the following medium : 10 mg TMB dissolved in 2-5 ml absolute ethanol, 47.5 ml distilled water, 50 ml buffer (pH 4-5) and 3 ml 3% hydrogen peroxide. The sections were then rinsed with distilled water and examined using a light microscope [19].

Calcofluor reagent. Leaf discs were placed on a microscope slide and covered with one drop of 0.01% (w/v) calcofluor solution (optical fluorescent brightener 28, Sigma) in sodium phosphate buffer (pH 8), then protected with a coverglass and examined using epi-fluorescence microscopy [4].

Identification and chromatographic analysis. Analysis of caffeoylquinic acid derivatives in sunflower was performed by HPLC using the technique of Mondolot-Cosson and Andary [25]. Identification of resveratrol was made following Dai *et al.* [8].

3. Results and discussion

3.1 *H. RESINOSUS / S. SCLEROTIORUM* MODEL.

Bazzalo *et al.* [2] and Hemery [18] suggest the implication of caffeic acid derivatives in the resistance of cultivated sunflower (*H. annuus*) to *Sclerotinia sclerotiorum*.

For this reason, we isolated the principal caffeoylquinic acid derivatives in the leaves of *H. resinosus* and analysed them using HPLC. The results (Table 1) show that these compounds are clearly predominant only in *H. resinosus*, a species resistant to *S. sclerotiorum*. This particularity was also found at cell level.

The observation *in situ* of the phenolic constituents of healthy leaves and bracts of *H. resinosus* revealed the presence of lignified cells giving strong blue autofluorescence under ultra-violet (UV) light, in particular in the cortical parenchyma of the leaf main veins. In the presence of Neu's reagent, this fluorescence became intensely greenish-white under UV light (Figure 1a) and greenish-yellow under blue light, which is characteristic of caffeic acid derivatives. Neu's reagent is a borate salt that forms complexes with certain groups of phenolic compounds giving them specific fluorescences (Table 2). These lignified cells, their walls inlaid with caffeic acid derivatives, form numerous cortical fibres in *H. resinosus*. This type of fibre was not observed in *H. annuus*. The lignified structures form effective defensive barriers to *S. sclerotiorum* which is unable to degrade ligneous fibres [33].

H. resinosus also contains flavonoids in the epidermal cells of the upper surface of bracts and leaves. In the presence of Neu's reagent, these flavonoids give a bright orange-yellow fluorescence under UV light and bright orange fluorescence under blue light. The cells form a continuous layer, the role of which is most certainly defensive. Indeed, the flavonoids act as protection against UV light and various attacks [31]. This type of epidermal flavonoid was not observed in cultivated sunflower.

After infection, there appeared in the stomatal cells a large accumulation of phenolic compounds giving whitish orange fluorescence in the presence of Neu's reagent under UV light. These compounds were absent prior to infection and seem to be flavonoid in nature. In *H. resinosus*, a strong concentration of flavonoids was noted in the many large secretory hairs on the lower surface of the epidermis giving an orange fluorescence in the presence of Neu's reagent under UV light and a bright orange fluorescence under blue light. In addition, a strong accumulation of caffeic acid derivatives was observed in the healthy areas adjacent to the necroses in both the infected leaves and bracts (Figure 1b). This increase in the levels of caffeic acid derivatives was confirmed by HPLC analysis of the principal caffeoylquinic acid derivatives isolated from *H. resinosus*. An increase in caffeoylquinic acid derivatives after infection was also observed in certain hybrids of *H. annuus*, which would confirm their involvement in the process of resistance to *S. sclerotiorum* (Figure 2).

TABLE 1. HPLC analysis of caffeoylquinic derivatives* from different *Helianthus* leaves (results in g per 100 g dry matter)

Genotypes	1,5-DCQ	3,5-DCQ	5-MCQ	Total CQ
Helianthus section				
MH 1.2 (*H.* x *annuus*)	0.10	0.27	0.21	0.58
NSH 45 (*H.* x *annuus*)	0	0.72	0.53	1.25
RHA 265 (*H.*x *annuus*)	0.07	0.76	0.31	1.14
H. anomalus 525	0	0.05	0.09	0.14
H. argophyllus 741	0.02	0.12	0.07	0.21
H. exilis 331	0.06	0.44	0.30	0.80
Divaricati section				
H. decapetalus 551	0.04	1.52	1.70	3.26
H. divaricatus 232	0.38	1.50	1.40	3.28
H. hirsutus 260	0.47	1.38	0.90	2.75
H. resinosus 243	3.22	1.44	2.13	6.79
H. resinosus 770	6.10	2.02	1.36	9.48
H. tuberosus 289	0.16	1.37	1.66	3.19

*: 1,5-DCQ = 1,5-dicaffeoylquinic acid ; 3,5-DCQ = 3,5-dicaffeoylquinic acid ; 5-MCQ = 5-monocaffeoylquinic acid (= chlorogenic acid) ; CQ = caffeoylquinic derivatives

TABLE 2. Coloration or fluorescence of phenolic compounds after treatment with different reagents and observation under UV light (F) or visible light (L)

Phenols *	Reagent	Maximum absorption	Colour
Caffeic acid d.	Neu	380 nm	Whitish green (F)
Gallic acid d.	Neu	325 nm	Deep blue (F)
Gallocatechin d.	Neu	325 nm	Deep blue (F)
Flavonoids	Neu	410 nm	Yellow to orange (F)
Flavonoids	Wilson	405 nm	Yellow (F)
Catechic d. (proanthocyanidins)	Vanillin-HCl	550 nm	Red (L)

* : d.: derivatives

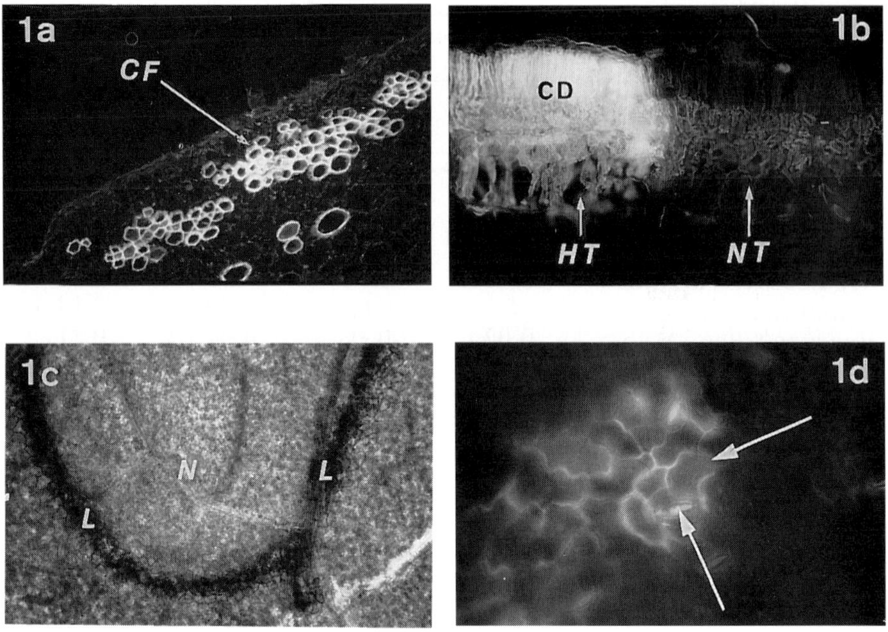

Figure 1

1a - Cross-section of *H. resinosus* healthy petiole (Neu's reagent, UV light, x 100) : cortical fibers (CF) are impregnated with caffeic acid derivatives (CD) [5].

1b - Cross-section of *H. resinosus* leaf infected by *S. sclerotiorum* (Neu's reagent, UV light x 100) : accumulation of caffeic acid derivatives (CD) in healthy tissue (HT) just near necrotic tissue (NT) [5].

1c - Lower leaf surface of *V. rupestris* cv. rupestris (intermediate resitant) 15 days after inoculation (treated with phloroglucinol-HCl, visible light, x 100) : notice lignin (L) formation around the necrotic area (N) [8].

1d - Lower leaf surface of *V. rotundifolia* cv. Carlos (résistant) 2 days after inoculation (Neu's reagent, UV light x 400) : arrows indicate yellow fluorescence of flavonoids in and around guard cells [9].

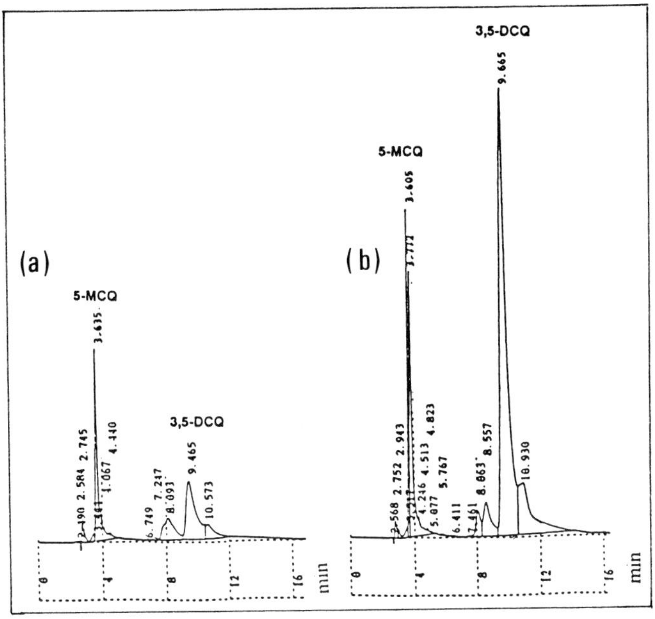

Figure 2. HPLC analysis of hybrid of *H. annuus* FLAMME [5]
(a) : healthy bract
(b) : infected bract, near necrosis
5-MCQ : 5-monocaffeoylquinic acid
3,5-DCQ : 3,5-dicaffeoylquinic acid

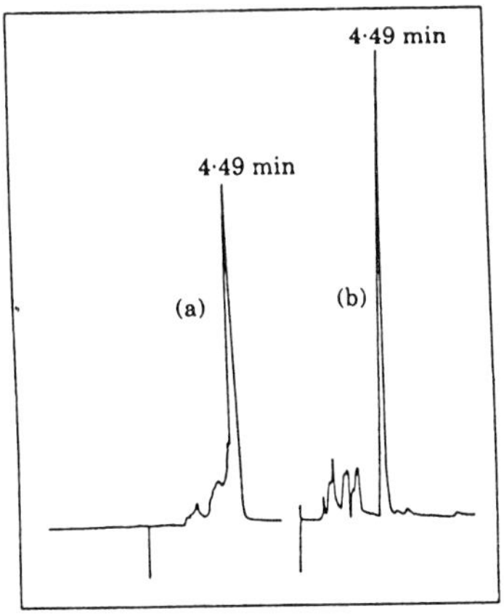

Figure 3. HPLC analysis of :
(a) : commercial resveratrol
(b) : a leaves extract of infected *Vitis rupestris* [8]

3.2. *VITIS* SPP / *P. VITICOLA* MODEL

In grapevine, the histolocalisation of the different phenolic compounds in healthy leaves of species R revealed a higher concentration of gallic acid derivatives (with strong deep blue fluorescence in the presence of Neu's reagent under UV light) as well as catechin tannins (coloured characteristically red when cold in the presence of vanillin-HCl) than the susceptible species S. The catechin tannins were found in the palisade tissue in all three species and in several layers of cells near the lower epidermis in species R only [6]. This localisation is an important factor and could be involved in limiting invasion by *P. viticola*, since this pathogen enters the leaves via the stomata on the under surface. Grapevine tannins are made up of gallocatechin derivatives (majority) and gallic acid derivatives. The catechin nucleus alone or total gallate nuclei (gallic acid derivatives + gallocatechin derivatives) can be analysed [10]. The species M presented levels of gallic acid derivatives and catechin tannins intermediate between those of species R and S. Statistical analysis of the concentrations of phenolic compounds in leaf extracts of species R, M and S revealed a strong correlation (90%) only between the levels of total gallate derivatives (gallic acid derivatives + gallocatechin derivatives) and the degree of resistance [6].

After infection, the use of calcofluor reagent, which gives a blue-tinged white fluorescence characteristic of fungal tissue, revealed the absence of zoospore germination in species R, weak sporangiophore development in species M and very abundant sporulation in species S [9].

The species M presented the largest number of reactions after infection. A blue autofluorescence under UV light was detected in the necrotic zones 5 days after infection. UV spectrophotometry and HPLC analysis of the extracted fluorescent molecule showed it to be trans-resveratrol (Figure 3), a molecule reported by Langcake and Pryce [23] to be a phytoalexin that appears in grapevine after infection with *P. viticola*. Dercks and Creasy [11] revealed the inhibitory activity of this phytoalexin on the germination of sporangia of *P. viticola*. In species M, we also observed peroxidase activity in the zone between the healthy and infected tissues 5 days after infection [8] and the formation of lignified tissue in this same zone 15 days after infection. This lignified zone around the necroses limited the spread of the pathogen (Figure 1c).

By inhibiting pathogen growth and enzyme activity, peroxidase-generated free radicals may provide additional time for the accumulation of stable end-products, such as lignin-like materials, that can probably induce host cell death by interrupting nutrient flow into and from cells, prevent intercellular hyphal growth and suppress sporulation [34, 15, 24].

In species R, a strong reaction in the stomata was observed from the second day after infection. A yellowish autofluorescence appeared in the guard cells of stomata situated in the necrotic zones as well as in the walls of the cells surrounding these same zones (Figure 1d). These compounds were identified as flavonoids since they produced characteristic yellow fluorescences in the presence of Neu's reagent and Wilson's reagent (Table 2). They appeared later in species M (5 days after infection) and later still and in much smaller quantities in species S [8].

These compounds, which appeared rapidly in the resistant species and highly localised in the infected zones, very certainly helped limit pathogen spread. However, it should be noted that we were not able to extract and identify the flavonoids due to their strong fixation to the cell walls which seems to rule out all non-denaturing extraction.

The pathogen developed weakly in species M, which must bring into play a whole set of mechanisms : accumulation of resveratrol and flavonoids in the necrotic zones and later lignification of these same zones. In species R, the very early appearance (from Day 2) of flavonoids in the infected tissues seemed to be sufficient to stop the spread of the pathogen. No lignification was observed in this species. In species S, the absence of an early reaction allowed the fungus to develop strongly and rapidly, the eventual appearance of flavonoids being too-little and too-late to contain the pathogen. The formation of resveratrol after infection is not constant in the grapevine. In this study, this compound appeared only in species M and not in species R or S. It seems therefore that the most effective mechanism to confine the pathogen is the very early appearance of flavonoids in the infected tissues. These flavonoids appeared at various speeds and intensities, depending on the level of resistance, in all the species and therefore constitute one of the key elements in the defence processes involved.

The resistance of the plants depends in part on the compounds present before infection as well as on the speed of induction of the multiple defence processes involving polyphenols. In sunflower, the caffeoylquinic acid derivatives play a leading role. In the grapevine, the flavonoids are strongly implicated and their presence is constant. These two types of molecule could constitute biochemical markers, being detected with simply set-up histochemical procedures and thus of potential use in improvement programmes to obtain higher, resistance varieties.

4. References

1. Akai, S. and Fukutomi, M. (1980) Performed internal physical defenses, in J.G. Horsfall and E.B. Colwling (eds.), *Plant Disease. An advanced Treatise,* Academic Press, London, New-york, vol. 5, pp. 139-159.
2. Bazzalo, M.E., Heber, E.M., Del Pero Martinez, M.A. and Caso, O.H. (1985) Phenolic compounds in stems of sunflower Plants inoculated with *Sclerotinia sclerotiorum* and their inhibitory effect on the fungus, *Phytopathol. Z.* **112,** 322-332.
3. Bertrand, F. and Tourvieille D. (1986) Recherche des tests de sélection pour la résistance à *Diaporthe helianthi* chez le tournesol, *Inf. Tech.* CETIOM **98,** 12-18.
4. Cohen, Y., Pe'er S., Balaas, O. and Coffey, M.D. (1987) A fluorescent technique for studing growth of *Peronospora tabacina* on leaf surfaces, *Phytopathology* **77,** 201-204.
5. Cosson, L. (1992) Composés caféoylquiniques et résistance du Tournesol cultivé et sauvage vis-à-vis de *Sclerotinia sclerotiorum*, Thèse Doctorat, Université Montpellier I, France, pp. 218.
6. Dai, G.H. (1994) Etude des facteurs biochimiques de résistance de la vigne (*Vitis spp.*) au mildiou (*Plasmopara vitocola*), Thèse Doctorat, Université Montpellier I, France, pp 144.

7. Dai, G.H., Andary, C., Mondolot-Cosson, L. and Boubals, D. (1994) Polyphenols and resistance of grapevines (*Vitis sp.*) to downy mildew (*Plasmopara viticola*), *Acta horticulturae* **381,** 763-766.
8. Dai, G.H., Andary, C., Mondolot-Cosson, L. and Boubals, D. (1995a) Histochemical studies on the interaction between three species of grapevine, *Vitis vinifera, V. rupestris* and *V. rotundifolia* and downy mildew (*Plasmopara viticola), Physiol. Mol. Plant Pathol.* **46,** 177-188.
9. Dai, G.H., Andary, C., Mondolot-Cosson, L. and Boubals, D. (1995b) Histochemical responses of leaves of *in vitro* plantlets of *Vitis spp.* to infection with *Plasmopara viticola, Phytopathology* **85,** 149-154.
10. Dai, G.H., Andary, C., Mondolot-Cosson, L. and Boubals, D. (1995c) Involvement of phenolic compounds in the resistance of grapevine callus to downy mildew (*Plasmopara viticola*), *E. J. Plant Pathol.* (in press).
11. Dercks, W. and Creasy, L.L. (1989) The significance of stilbene phytoalexins in the *Plasmopara viticola*-grapevine interaction, *Physiol. Mol. Plant Pathol.* **34,** 189-202.
12. Gahan, P.B. (1984) *Plant Histochemistry and Cytochemistry,* Academic Press, London, New-York, pp. 301.
13. Georgieva-Todorova , J. (1984) Interspecific hybridization in the genus *Helianthus* L., *Z. Pflanzenzüchtg.* **93,** 265-279.
14. Graham, T.L. and Graham, M.Y. (1991) Cellular coordination of molecular responses in plant defense, *Mol. Plant-Microbe Interact.* **4**, 415-422.
15. Hammond, K.E. and Lewis, B.D. (1986) Ultrastrural studies of the limitation of lesions caused by *Leptosphaeria maculans* in stems of *Brassica napus* var. *oleifera, Physiol. Mol. Plant Pathol.* **28**, 251-265
16. Harborne, J.B. (1989) Higher plant-lower plant interactions : phytoalexins and phytotoxins, in J.B. Harborne (ed.), *Introduction to Ecological Biochemistry,* Academic Press, London, New-York, pp 302-340.
17. Hariri, B., Sallé, G. and Andary, C. (1991) Involvement of flavonoids in the resistance of two poplar cultivars to mistletoe (*Viscum album* L.), *Protoplasma* **162,** 20-26.
18. Hemery, M.C. (1987) Recherche de marqueurs phénoliques pour la résistance du Tournesol au *Sclerotinia,* D.E.A. de phytomorphogénèse - Université Blaise Pascal, Clermont-Ferrand, France, pp 31.
19. Imberty, A., Goldberg, R. and Catesson, A.M. (1984) Tetramethylbenzidine and p-phenylenediamine-pyrocatechol for peroxidase histochemistry and biochemistry : two new, non-carcinogenic chromogens for investigating lignification process, *Plant. Sci. Lett.* **35**, 103-108.
20. Ingham, J.L. (1973) Disease resistance in higher plants. The concept of preinfectional and postinfectional resistance, *Phytopathol. Z.* **78**, 314-315.
21. Jensen, W.A. (1962) *Botanical histochemistry,* Froeman (ed.) San Francisco, California, pp 264.
22. Kim, J.J., Ben-Yehoshua, S., Shapiro, B., Henis, Y. and Carmeli, S. (1991) Accumulation of scoparone in heat-treated lemon fruit inoculated with *Penicillium digitatum, Plant Physiol.* **97**, 880-885.
23. Langcake, P. and Pryce, R.J. (1976) The production of resveratrol by *Vitis vinifera* and other members of the *Vitaceae* as a response to infection or injury, *Physiol. Plant Pathol.* **9**, 77-86.
24. Moersbacher, B.M., Flott, B.E., Noll, U. and Reisener, H.J. (1989) On the specificity of an elicitor preparation from stem rust which induces lignification in wheat leaves, *Plant Physiol. Biochem.* **27**, 305-314.

25. Mondolot-Cosson, L. and Andary, C. (1994) Resistance factors of a wild species of sunflower *Helianthus resinosus* to *Sclerotinia sclerotiorum, Acta Horticulturae* **381**, 642-645.
26. Neu, R., (1956) A new reagent for differentiating and determining flavones on paper chromatograms, *Naturwissenschaften* **43,** 82.
27. Nicholson, R.L. and Hammerschmidt, R. (1992) Phenolic compounds and their role in disease resistance, *Ann. Rev. Phytopathol.* **30**, 369-389.
28. Paxton, J., (1981) Phytoalexins. A working redefinition, *Phytopathol. Z.* **101**, 106-109.
29. Rouxel, T. (1989) Les phytoalexines et leur intervention dans la' résistance hypersensible aux champignons phytopathogènes, *Agronomie* **9**, 529-545.
30. Sarkar, S.K. and Howarth, R.E. (1976) Specificity of the vanillin test for flavanols, *J. Agric. Food Chem.* **24**, 317-320.
31. Tissut, M. and Ravanel, P. (1980) Répartition des flavonols dans l'épaisseur des feuilles de quelques végétaux vasculaires, *Phytochemistry* **19**, 2077-2081.
32. Tomas-Barberan, F.A., Msonthi, J.D. and Hostettmann, K. (1988) Antifungal epicuticular methylated flavonoids from *Helichrysum nitens, Phytochemistry* **27**, 753-755.
33. Tourvieille, D. and Vear, F. (1984) La sélection du Tournesol pour une meilleure résistance au *Sclerotinia sclerotiorum* i, *Inf. tech. CETIOM* **88**, 3-23.
34. Vance, C.P. (1980) Lignification as a mechanism of disease resistance, *Ann. Rev. Phytopathol.* **18**, 259-288.
35. Vear, F. and Guillaumin, J.J. (1977) Etude de méthodes d'inoculation du Tournesol par *Sclerotinia sclerotiorum* et application à la sélection, *Ann. Amélior. Plantes* **27**, 523-537.

GOLD CYTOCHEMISTRY APPLIED TO THE STUDY OF PLANT DEFENSE REACTIONS

Nicole BENHAMOU
Recherche en sciences de la vie et de la santé
Pavillon Charles-Eugène Marchand
Université Laval, Sainte-Foy, Québec, Canada, G1K 7P4

ABSTRACT. In the past decade, it has become increasingly apparent that *in situ* localization of plant molecules could lead to a better understanding of the functional activity of the plant cell during various biological processes. Recent advances in the purification of specific probes such as enzymes, lectins and antibodies and in the use of colloidal gold as a particulate marker of high electron density have provided opportunities for the development of cytochemical approaches which could not only allow an accurate localization of a wide range of plant molecules in their respective cell compartments but also help to elucidate their potential functions. Innovative developments in plant cytochemistry appear with increasing frequency and it is expected that improvements in both tissue processing and probe specificity will extend the applicability of this approach to more and more research areas in plant pathology.
Because polysaccharides are major plant and microbial cell surface components, their involvement in a number of functions including recognition, attachment, and adhesion has long been recognized of key importance in the outcome of a given interaction. Consequently, their precise in situ localization has been the subject of extensive studies and a number of probes such as lectins as well as enzymes including ß-1,4-exoglucanase, ß-glucosidase, xylanase and chitinase have proved useful for studying cell surface interactions during the infection process. Another facet that has been widely investigated in recent years concerns the drastic physiological changes induced in response to infection and thought to contribute to plant disease resistance. Among the molecules involved in this altered metabolism, callose, a structural polymer of ß-1,3-glucans, and phenolic compounds have been successfully targeted by means of two newly-introduced gold-complexed enzymes: a plant ß-1,3-glucanase and a fungal laccase.
The association of complementary technologies together with the refinement of existing methods and the introduction of modern approaches have undoubtedly contributed to open new avenues of investigation in the field of plant-pathogen interactions. In conjunction with biochemistry and molecular biology, cyto-and immunocytochemistry of plant tissues have proved useful for elucidating some aspects of the highly complex relationship between plants and pathogens. Thus, it is clear that both molecular and traditional approaches of plant pathology will continue to benefit from exciting, new findings generated by the *in situ* localization of plant and microbe molecules.

M. Nicole and V. Gianinazzi-Pearson (eds.), Histology, Ultrastructure and Molecular Cytology of Plant-Microorganism Interactions, 55-77.

1. Introduction

In recent years, cytological and biochemical investigations of plant tissues, as well as modern techniques of molecular biology have contributed to the unravelling of the highly complex strategy involved in plant disease resistance [8, 9, 28, 32]. Convincing evidence from a number of studies has shown that all plants, resistant or susceptible, respond to pathogen attack by the induction of an array of defense reactions designed to affect pathogen growth and viability [5]. However, in cases where a plant-pathogen interaction results in disease establishment, successful host colonization by the parasite is likely due to delayed plant defense expression, rather than to absence or inactivation of defense mechanisms [6]. Thus, the speed and extent of the plant response to microbial attack appear to be key determinants in the outcome of a given interaction and it is reasonable to assume that a faster response towards a pathogen may enhance the resistance of a previously susceptible plant.

Upon microbial attack, plants elaborate an array of inducible defense responses of both a structural and a biochemical nature [21, 36]. Plant pathogens have evolved special mechanisms to penetrate the barrier imposed by the cuticle and by the cellulose-containing cell walls [29]. It is, therefore, not surprising that, in turn, plants have developed the potential to prevent effective pathogen penetration by producing an array of substances that contribute to rapid modifications of the cell walls [42] and to the creation of a toxic environment [36]. Among the most important physical changes that affect the properties of the plant cell wall and contribute to the elaboration of permeability barriers preventing pathogen spread and enzymatic degradation, one may cite: 1) lignin and related compounds [46], 2) callose, a polymer of ß-1,3-glucans [20], 3) phenolics [36] and 4) hydroxyproline-rich glycoproteins (HRGPs) and glycine-rich glycoproteins [18, 38]. Major biochemical changes have been shown to be associated with the accumulation of phytoalexins, secondary metabolites known to be toxic for fungi and bacteria [36], the synthesis of pathogenesis-related (PR) proteins including enzymes with antimicrobial activities [23], and the accumulation of protease inhibitors [49]. In the past decade, significant advance has been made in understanding the sequential events taking place in the expression of such plant defense reactions [31, 37]. However, in spite of the significant move towards understanding how resistance mechanisms are regulated and coordinated at the plant cell level, the spatio-temporal localization of induced defense reactions in plants has not been deeply investigated. With regard to plant disease resistance, much research has been directed towards signal transduction and gene expression [32, 50], although growing evidence from an increasing number of studies demonstrates that visualizing the spatio-temporal distribution of molecular responses at the cellular level is an essential complement to biochemical and molecular analyses [9, 33].

Localization of induced defense reactions in plant tissues is of particular value when studying their mode of action in relation to pathogen growth inhibition *in planta*. Recent advances in the isolation and purification of lectins and enzymes and in the preparation of specific antibodies have provided opportunities for the development of cyto-and immunocytochemical approaches which could not only allow an accurate localization of various molecules in their respective cell compartments but also help to elucidate their potential functions [8, 9, 33]. Taken together, cyto- and immunocytochemical methods provide useful and often unique information on various topics including: 1) spatio-temporal distribution of newly-synthesized molecules during the course of infection; 2) vulnerability of plant cell wall polymers to microbial enzymes; 3) vulnerability of microbial cell walls to plant enzymes; 4) antimicrobial potential of some induced proteins; 5) accumulation of new macromolecules in the plant cell wall; 6) chemical composition of newly-formed physical barriers such as wall appositions; and 7) activation of the phenylpropanoid pathway during resistance. Innovative developments in plant cytochemistry appear with increasing frequency and it is realistic to believe that discovery of new probes and improvements in their specificity will extend the applicability of this approach to more and more research topics in the study of plant-pathogen interactions.

In recent years, several excellent reviews have been published that address the wide range of applications of immunocytochemistry in plant biology [8, 9, 47]. In this paper, emphasis will be more specifically given to recently emerging cytochemical approaches designed to investigate the spatio-temporal localization of structural and antimicrobial molecules involved in host-pathogen interactions.

2. Methodology: Principles and general practice

For several years, the major limitation of electron microscope studies used in conjonction with biochemical investigations was the lack of specific probes for the precise identification of the chemical nature of the material which was seen. In the past ten years, the use of well-characterized proteins has contributed to marked changes in this type of investigations so that it is now possible not only to identify major and minor compounds in a cell but also to accurately delineate the spatio-temporal location of a variety of molecules in intracellular compartments.

Cyto-and immunocytochemical approaches make use of the affinity properties existing between macromolecules[3, 43]. Once tagged to an appropriate electron-dense marker, several proteins and glycoproteins enable the ultrastructural localization of their target molecules. These techniques have been widely explored for localizing several compounds in higher plant cells where they play a number of key biological functions. Several excellent reviews have been published that address the breadth of applications of colloidal gold labeling and these should be consulted for additional information [8, 9, 47, 48]. The objective of this chapter is instead to describe and illustrate the most recent developments in plant cytochemistry with particular emphasis on the localization of molecules involved in plant defense responses. Since the author has recently reviewed the potential value of antibodies for the accurate localization of plant antigens [8, 9], the present paper will concentrate on the use of enzymes and lectins as powerful tools for studying the mechanisms underlying plant disease resistance.

2.1. TISSUE PROCESSING

Post-embedding techniques are the most flexible and widely used approaches in cytochemistry. Although there is no standard protocol for the preparation of tissue for on-grid cytochemistry, the main objective remains the preservation of the ultrastructure and the retention of substrate molecules. Thus, conditions for tissue fixation and embedding have to be worked out in order to obtain optimal degrees of structural preservation and immunoreactivity.

Tissue fixation is essential to restrict diffusion of compounds into and out of cells, and to strengthen the plant structures against the effect of other reagents during tissue processing. Fixation should therefore be efficient enough to retain good ultrastructure while maintening probe accessibility [22]. Glutaraldehyde, a dialdehyde that efficiently cross-links protein molecules, is probably the most widely used electron microscopical fixative. It can be used alone at a concentration ranging from 1-3%, or in combination with formaldehyde. Conventional fixation procedures usually recommend tissue post fixation with osmium tetroxide. However, this fixative which acts as an excellent membrane stabilizer and contrasting agent can also mask or destroy protein antigens. However, there is evidence that most lectin- and enzyme-binding sites are not altered by osmium tetroxide, allowing its use in lectin and enzyme cytochemistry [5]. Thus, for each system, conditions that yield optimal labeling and best ultrastructural preservation have to be worked out.

Several types of resins are available and have been successfully used for cytochemical purposes. Two broad types of resins are currently been used: the epoxy resins and the hydrophylic cross-linked acrylics. Epoxy resins such as Epon, Spurr and Araldites exhibit low water absorption and are hydrophobic. They offer the advantage of yielding good ultrastructural preservation and high beam stability. Successful results in retaining both tissue preservation and immunoreactivity have been obtained with Epon 812 [6, 13, 14, 15, 16].

2.2. SECTION PREPARATION

Grids of gold or other inoxidizable metals such as nickel are used to collect ultrathin tissue sections (70 nm in thickness). Copper grids should be avoided because copper is known to interact with buffer solutions during cytochemical procedures. In order to enhance stability of the sections, previous grid coating with Formvar or Collodion is recommended.

2.3. ENZYME CYTOCHEMISTRY

Since the first demonstration that gold-complexed nucleases were suitable probes for localizing nucleic acids in animal tissues [2], a number of enzymes have been applied to the detection of their corresponding substrate molecules in plant cells and tissues [13, 44, 48]. The potential value of the enzyme-gold technique relies in the use of highly purified enzymes. Parameters such as minimal amount of enzyme required for full stabilization of the colloidal gold solution, optimal pH for the absorption of the enzyme to the gold particles and optimal pH of labeling are of crucial importance for both reducing non specific binding and preserving the biological activity of the enzyme under study [3]. Nearly all enzymes can be directly complexed to colloidal gold under appropriate conditions and applied to tissue sections for localizing the corresponding substrate molecules. The most commonly used enzymes in plant biology and pathology are:

1) Exoglucanase, a ß-1,4-glucan cellobiohydrolase, specific for cellulosic ß-1,4-glucans [13]

2) ß-glucosidase, a ß-D-glucoside glucanohydrolase, specific for ß-D-glucosides [4]

3) Phospholipase A_2, an enzyme specific for phospholipids [30]

4) ß-1,3-glucanase, an enzyme purified from tobacco plants reacting hypersensitively to tobacco mosaic virus [7]

5) Laccase, an enzyme produced by the fungus *Rigidoporus lignosus* and specific for phenolic compounds [19]

Enzyme	Source	Substrate specificity	pH of colloidal gold	pH of optimal labeling
ß-1,4-exoglucanase	*Trichoderma*	ß-1,4-glucans	9.0	6.0
ß-glucosidase	Almond	ß-D-glucosides	9.3	6.0
Phospholipase A_2	Bee venom	Phospholipids	4.5	7.0
ß-1,3-glucanase	Tobacco	ß-1,3-glucans	5.5	6.0
Laccase	*Rigidoporus lignosus*	Phenolics	4.0	6.0
∞-amylase	Porcine pancreas	Starch	7.2	7.2
Chitinase	*Streptomyces griseus*	Chitin	7.0	6.0
Pectinase	*Aspergillus niger*	Polygalacturonic acids	7.3	7.2
Xylanase	*Polyporus tulipiferae*	Xylans	7.2	5.0

TABLE 1. Optimal conditions for preparation of some enzyme-gold complexes

Cytochemical labeling with enzyme-gold complexes is easy to perform. The most straightforward procedure is:

1) Sections are floated on a drop of phosphate buffered saline (PBS) containing 0.01% (w/v) polyethylene glycol (PEG 20000) for 5 min. The pH is adjusted according to the pH of optimal activity of the enzyme.

2) Sections are transferred onto a drop of the gold-complexed enzyme at the appropriate dilution in PBS-PEG for 30-60 min.

3) Sections are thoroughly washed with PBS, rinsed with distilled water, and air-dried.

4) Sections are contrasted with uranyl acetate and lead citrate.

Cytochemical controls performed to assess specificity of the labeling include: a) addition of the corresponding substrate to each enzyme-gold complex for a competition experiment [i.e. ß-1,4-glucans from barley for the ß-1,4-exoglucanase-gold complex; Salicin for the ß-glucosidase-gold complex; L-∂ phosphalidylcholine for the phospholipase A_2; laminarin for the ß-1,3-glucanase; and *p* -coumaric acid, ferulic acid or sinapinic acid for the laccase]; b) digestion of tissue sections with the uncomplexed enzyme prior to incubation with the enzyme-gold complex; c) substitution of the enzyme-gold complex under study by bovine serum albumin (BSA)-gold complex to assess the non-specific adsorption of the protein-gold complex to the tissue sections; d) incubation of the tissue sections with the enzyme-gold complex under non-optimal conditions for biological activity; e) incubation of the tissue sections with colloidal gold alone to assess non-specific adsorption of the gold particles to the tissue sections.

2.4. LECTIN CYTOCHEMISTRY

Lectins constitute a group of carbohydrate-binding proteins (usually glycoproteins) of nonimmune origin that occur predominantly in plants (mainly leguminosae) and invertebrates (molluscs). Because of their specific binding properties and their ability to recognize subtle differences in complex carbohydrate structures, lectins have provided versatile tools for topochemistry [5, 6]. In recent years, lectins have found wide application in the study of intracellular carbohydrate-containing molecules [14]. The list of available lectins has rapidly increased providing a large spectrum of accurate probes for sugar localization. The most currently used lectins in plant pathology are:

1) Wheat-germ agglutinin (WGA), specific for N-acetylglucosamine residues, and used for the localization of fungal chitin in fungus-infected plant tissues [6, 39]

2) *Helix pomatia* agglutinin (HpA), specific for N-acetylgalactosamine residues [5]

3) *Ricinus communis* agglutinin I (RcA), specific for galactose residues [14]

4) *Aplysia* gonad lectin (AGL), a lectin isolated from the sea mollusc *Aplysia depilans* and found to specifically bind to polygalacturonic acids (pectin) [16]

5) Concanavalin A (Con A), specific for terminal glucopyranosyl and mannopyranosyl residues [5]

Lectins with high MW (>15 KDa) can be directly complexed to colloidal gold and applied to tissue sections in a single-step procedure [6]. By contrast, lectins with low MW (<15 KDa) are not easy to conjugate to colloidal gold. In such cases, an indirect (two-step) labeling in which the marker is complexed to a secondary reagent that has affinity for the lectin is used [6]. Secondary reagents include glycoproteins such as ovomucoid (specific for WGA) or polysaccharides with appropriate sugar-binding sites.

Lectin	Sugar specificity	pH of colloidal gold	Optimal pH of labeling
Wheat germ agglutinin	*N*-Acetyl-D-glucosamine	—	—
Ovomucoid	—	5.3	6.0
Helix pomatia agglutinin	*N*-Acetyl-D-galactosamine	7.4	7.4
Ricinus communis agglutinin	D-galactose	8.0	8.0
Aplysia gonad lectin	Polygalacturonic acids	9.5	8.0
Concanavalin A	D-mannose and D-glucose	8.0	8.0

TABLE 2. Optimal conditions for preparation of some lectin-gold complexes

Both direct and indirect labeling with lectins can be applied to ultrathin tissue sections for localizing specific sugar residues [6]. As a general rule, all experiments are performed in a moist chamber to avoid dessication.
For direct labeling with lectins, the protocol is identical to that described for the enzyme-gold approach. For indirect labeling, the protocol is:

1) Sections are pre-incubated on a drop of PBS, pH 7.2.
2) Sections are transferred onto a drop of the uncomplexed lectin at the appropriate dilution in PBS for 30 min.
3) Sections are washed with PBS, pH 7.2. The excess of buffer is removed with filter paper.
4) Sections are incubated on the gold-complexed secondary reagent at the appropriate dilution.
5) Sections are washed with PBS, rinsed with distilled water and stained with uranyl acetate and lead citrate.

Specificity of the labeling obtained with lectin-gold complexes has to be assessed through several control tests. For direct labeling procedures, these controls include: a) incubation with the lectin-gold complex to which was previously added its corresponding sugar; b) incubation with the uncomplexed lectin, followed by incubation with the gold-complexed lectin. For indirect labeling procedures, the controls include: c) incubation with the lectin previously absorbed with is corresponding sugar, followed by incubation with the secondary reagent complexed to gold; and d) incubation with the gold-complexed secondary reagent alone.

3. Application of gold cytochemistry to the study of plant-pathogen interactions

The potential value of the enzyme- and lectin-gold approaches in elucidating some of the cellular and molecular events occurring during the infection process of plant tissues has been documented with increasing frequency in the past few years. In addition to provide key information on the extent of plant cell disorganization and host cell wall alteration upon microbial colonization, cytochemical investigations of plant tissues challenged by potential pathogens have brought new insights into the spatio-temporal distribution of newly-synthesized molecules involved in defense reactions and have contributed to a better understanding of their biological significance *in planta* [13, 14, 16, 24, 39]. The application of gold cytochemistry has also proved useful in delineating the mode of action of antagonistic fungi believed to be powerful biocontrol agents [10, 25, 34]. Collectively, the data generated by the use of enzyme- and lectin-gold complexes have greatly contributed to refine our knowledge on some aspects of the complex relationship established between partners during host-pathogen interactions.

3.1. ENZYME-GOLD CYTOCHEMISTRY

The potential value of the enzyme-gold approach to help understanding the various facets of the altered host metabolism following microbial attack has been clearly demonstrated by the successful localization of a number of substrate molecules, notably cellulose, ß-1,3-glucans and lignin-like compounds [7, 13, 19, 40]. Several other molecules such as glucosides, phospholipids, chitin and to a lesser extent pectin have also been identified *in situ*, allowing a better characterization of the chemical composition of diverse structures such as intercellular deposits and wall appositions formed in response to infection in resistant plant cultivars.

3.1.1. *Ultrastructural localization of cellulose subunits*

The plant cell wall is a complex structure that plays both structural and functional roles. In plant-microbe interactions, the plant cell wall plays a key role in that it provides a physical barrier that is difficult to breach without the hydrolytic action of microbial enzymes such as cellulases and pectinases. In recent years, particular attention has been paid to the structural changes that occur in host cell walls in response to pathogen attack [42]. Understandably, vulnerability of cellulose to microbial cellulases has been the subject of extensive studies. Traditionally, plant cell wall components have been identified by using extractive methods often lacking sufficient specificity and reliability. Benhamou *et al.* [13] were the first to introduce the enzyme-gold approach as a nondestructive method for localizing cellulose in thin sections of embedded plant tissues. The authors described the potential value of a ß-1,4-exoglucanase, purified from a cellulase produced by the fungus *Trichoderma harzianum* , for accurately labeling cellulose subunits in plant tissues. Since then, the exoglucanase-gold approach has been successfully applied to a number of host-pathogen interactions including among others tobacco-*Phytophthora parasitica* var.*nicotianae* [11] (Fig.1), cucumber-*Pythium ultimum* [26], rubber tree root-*Rigidoporus lignosus* [39], and rose-*Sphaerotheca pannosa* var.*rosae* [34]. In all cases, application of the gold-complexed exoglucanase proved useful for delineating the extent of cellulose degradation in time and space as well as the ability of some microbial cellulases to cause extensive damage, mainly because of their diffusion at a large distance from the point of fungal penetration.

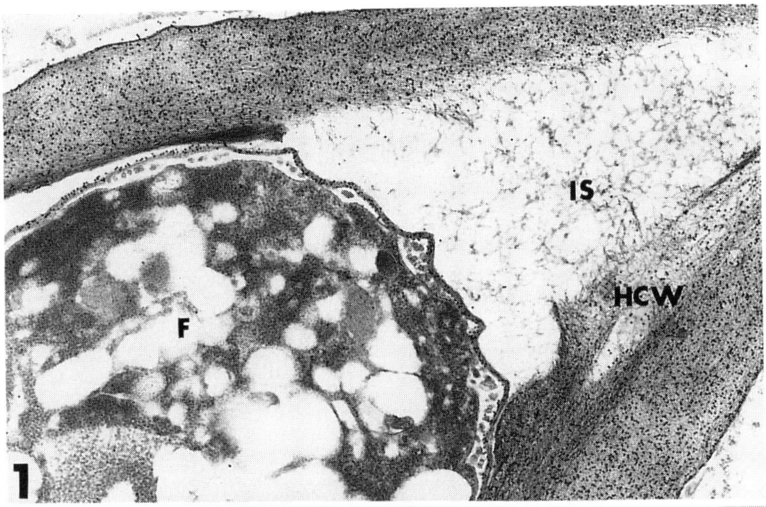

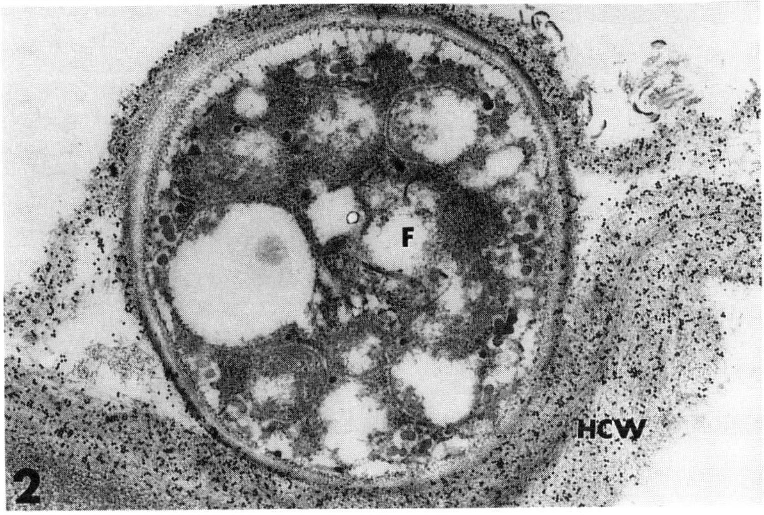

Figs. 1 and 2. Cytochemical labeling of cellulose with the exoglucanase-gold complex. **Fig. 1.** Tobacco root tissues infected by *Phytophthora parasitica* var. *nicotianae*. x 30 000. **Fig. 2.** Tomato root tissues infected by *Fusarium oxysporum* f. sp. *radicis-lycopersici* (FORL). x 30 000.

Abbreviations used in figures : F, fungus; HCW, host cell wall; IS, intercellular space.

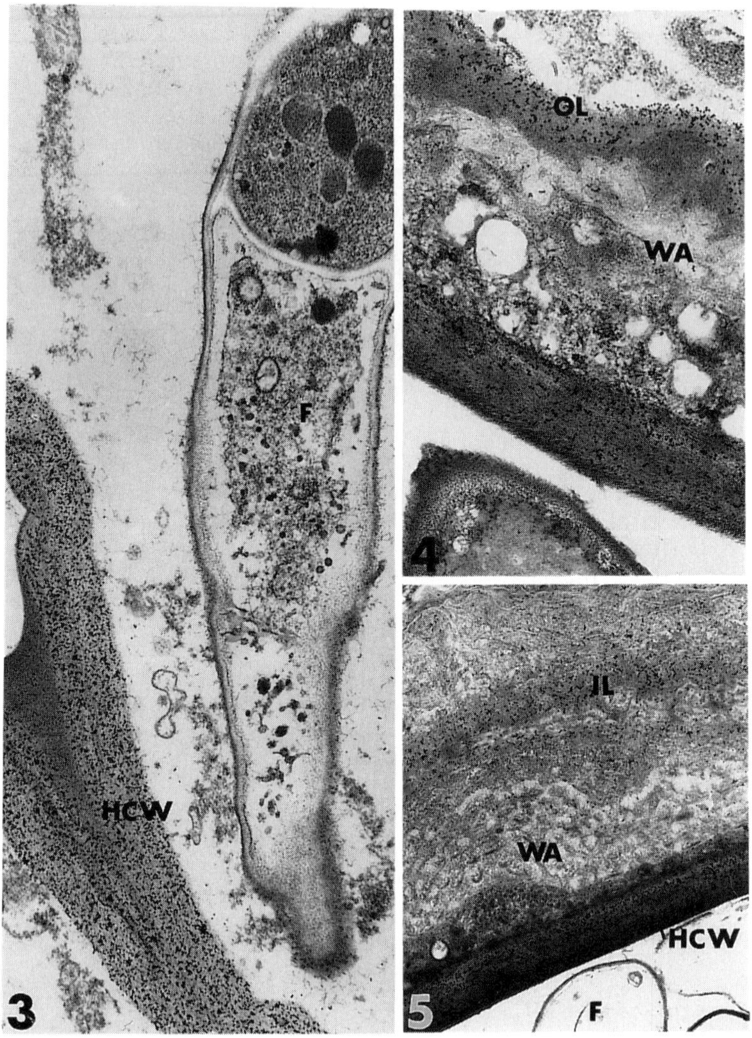

Figs. 3-5. Cytochemical labeling of cellulose with the exoglucanase-gold complex. **Fig. 3.** Tomato root tissues expressing resistance to FORL following elicitor treatment. x 20 000. **Figs. 4 and 5.** Formation of wall appositions in tomato root tissues expressing resistance to FORL following elicitor treatment. Fig.4. x 30 000; Fig.5. x 16 000.

Abbreviations used in figures : F, fungus; HCW, host cell wall; IL, intermediate layer; OL, outer layer; WA, wall apposition.

In a recent study, Benhamou and Lafontaine [12] reported that tomato plants, treated by elicitors such as fungal ß-1,3-glucans and chitosan, expressed increased resistance to *Fusarium oxysporum* f.sp. *radicis-lycopersici* (FORL). Ultrastructural investigations of the infected root tissues from water-treated (control) plants showed a rapid colonization of all tissues including the vascular stele. Fungal ingress was always associated with marked host cell disorganization and cell wall alteration (Fig.2). In root tissues from elicitor-treated plants, restriction of fungal growth to the epidermis and the outer cortex, decrease in pathogen viability (Fig.3), and formation of numerous wall appositions at sites of attempted penetration (Fig.4) were the main features of the host-pathogen interaction. The wall appositions were found to vary greatly in their appearance from multi-textured to multi-layered structures.

Labeling with the gold-complexed exoglucanase occurred over the wall appositions, but it was irregularly distributed (Figs.4 and 5). In multi-textured appositions composed of variously shaped zones, gold particles were predominantly associated with the electron-opaque layer bordering the regions enriched with vesicles and membrane fragments (Fig.4). Even over this layer, labeling appeared to be mainly distributed over the more external portion. In multi-layered appositions, an intense labeling occurred over a thick, compact layer, most often inserted between two stratified areas (Fig.5). Examination of a large number of sections revealed that cellulose was not widely distributed in the appositions. It was confined to the outer cover of the appositions or to an intermediate layer resembling the host cell wall in terms of architecture. In light of these observations, one may assume that the cellulosic layers correspond to retracted portions of the adjacent host cell wall. The early events leading to the development of complete appositions may involve a splitting of the host cell walls followed by a gradual deposition of polysaccharides such as callose and pectin between the splitted walls. Subsequent infiltration of phenolics and related substances may then contribute to compaction of the polysaccharidic matrix.

The few areas of application listed above indicate the powerful value of the exoglucanase-gold approach in studying some of the molecular events associated with plant tissue colonization by microorganisms. The technique is easy, reproducible and highly specific and it should acquire increasing applicability and relevance in the near future.

3.1.2. *Ultrastructural localization of ß-1,3-glucans*

Callose, a polymer of ß-1,3-glucans, is a minor plant cell wall compound that occurs mainly in sieve tube elements of phloem cells. However, it is known that plants respond to microbial injury by the rapid deposition of callose at sites where the pathogen is likely to be restricted [1]. In the past years, detection of callose in infected plant tissues has relied mainly on the use of aniline blue fluorescence in combination with the periodic acid-schiff reaction. Although valuable, this technique was not sensitive enough to allow accurate correlation between cytological details and fine localization of callose in cell compartments. Recently, Benhamou [7] reported that a purified tobacco ß-1,3-glucanase (PR protein), once complexed to colloidal gold, was a powerful probe for localizing ß-1,3-glucans in tobacco root tissues infected by the fungus *Phytophthora parasitica* var. *nicotianae*. The acidic enzyme was conjugated to gold at pH 5.5 and its specificity for ß-1,3-glucans was demonstrated by competition in binding using several polysaccharides as potential substrates.

In fungus-infected tobacco root tissues, large amounts of ß-1,3-glucans were found at sites of attempted pathogen penetration such as plasmodesmata, and intercellular spaces. They were also detected in papillae, mainly in the amorphous matrix (Fig.6). ß-1,3-glucans were found to occur in host cell wall areas adjacent to fungal cells. Since ß-1,3-glucans are also components of the fungal cell wall, the question remains to be answered whether the glucans found in the plant cell wall are newly-synthesized or derived from the fungus cell wall. The gold-complexed tobacco ß-1,3-glucanase was found to be a powerful probe although it could not differentiate between ß-1,3-glucans of plant and fungus origins.

In tomato plants expressing resistance to FORL following elicitor treatment, incubation with the tobacco ß-1,3-glucanase resulted in a strong labeling of wall appositions (Fig. 7). Labeling appeared to be more predominantly associated with the electron lucent regions, especially in the area bordering the apposition. Since success of a plant in warding off invading pathogens relies primarily on its capability of elaborating rapidly a first defensive line for protecting the cell walls against both penetration and enzymatic degradation, it is likely that deposition of callose is an early event probably followed by others including the accumulation of lignin and related compounds.

3.1.3. *Ultrastructural localization of phenolic-like compounds*

A purified laccase, produced by the white rot fungus *Rigidoporus lignosus* , was recently used to localize lignin-like compounds and polymerized phenols in tomato root tissues expressing resistance to FORL[19]. Upon incubation with the gold-complexed enzyme, labeling was found to be predominantly associated with electron-dense aggregates occurring in the lumen of reacting cortical cells and to a lesser extent with the host cell walls (Fig.8). By contrast, amorphous globules inserted between the dense deposits were unlabeled (Fig.9).

Accumulation of plant phenolic products including lignin in response to microbial attack and/or elicitor treatment has been abundantly documented on a biochemical basis [36] but seldom demonstrated *in planta*, mainly because of the limited availability of specific probes. Evidence is provided that laccase, a copper-containing polyphenol oxidase produced by wood rotting fungi, is a powerful probe for precisely localizing phenolic substances in plant tissues. Infiltration of host cell walls with phenolics and intracellular accumulation of similar compounds are likely to contribute to the elaboration of permeability barriers preventing pathogen spread and enzymatic degradation in addition to create a fungitoxic environment.

3.1.4. *Ultrastructural localization of other substrate molecules*

Various other enzymes, including ß-glucosidase [4], phospholipase and xylanase [48], have been applied for localizing their corresponding substrate molecules in plant tissues. The enzyme-gold approach has already a wide range of applications and it is expected that purification of more and more enzymes will open new avenues of investigations in plant pathology.

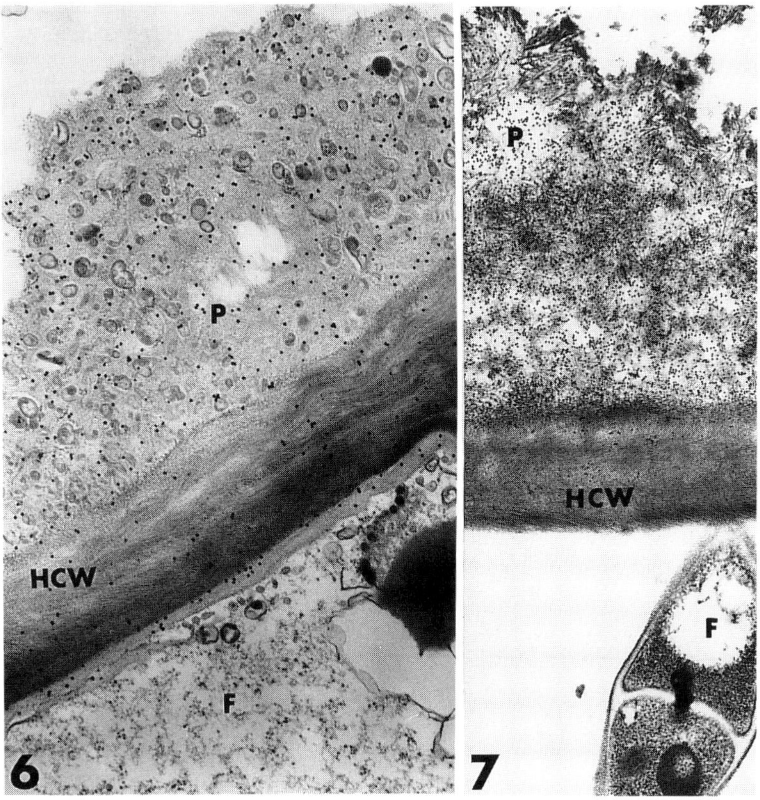

Figs. 6 and 7. Cytochemical labeling of callose with a gold-complexed ß-1,3-glucanase. **Fig. 6.** Formation of a papilla in tobacco root tissues infected by *Phytophthora parasitica* var.*nicotianae*. x 25 000. **Fig. 7.** Formation of a papilla in tomato root tissues expressing resistance to FORL following elicitor treatment. x 20 000.

Abbreviations used in figures : F, fungus; HCW, host cell wall; P, papilla.

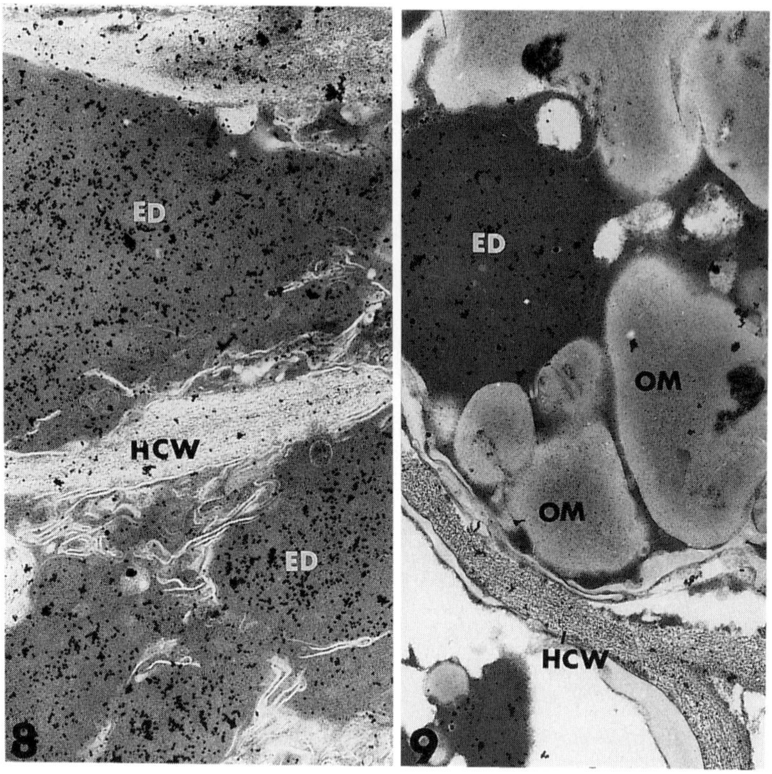

Figs. 8 and 9. Cytochemical labeling of phenolic-like compounds with a gold-complexed laccase. Accumulation of electron-dense deposits in tomato root tissues expressing resistance to FORL following elicitor treatment. The opaque masses inserted between the deposits are unlabeled. x 30 000.

Abbreviations used in figures : F, fungus; HCW, host cell wall; ED, electron-dense deposits; OM, opaque masses.

3.2. LECTIN-GOLD CYTOCHEMISTRY

The introduction of colloidal gold as a particulate marker of high electron-density has contributed to the extensive use of lectin-gold cytochemistry in a variety of biological systems including plant-pathogen interactions. Among the most widely used lectins, wheat germ agglutinin, WGA, a lectin with specific binding affinity for N-acetylglucosamine residues (chitin), and *Aplysia* gonad lectin (AGL), a lectin with polygalacturonic acid (pectin) binding affinity, have been widely applied to a number of host-pathogen interactions. Other lectins such as *Ricinus communis* agglutinin (RcA) [14], Concanavalin A [5], and *Helix pomatia* agglutinin (HpA) [6] have also been used, although less frequently.

3.2.1. *Ultrastructural localization of pectic compounds*

The *Aplysia* gonad lectin (AGL) was introduced in the field of plant cytochemistry by Benhamou *et al.* [16] and found very useful for localizing galacturonic acid-rich molecules. Studies dealing with the application of this lectin provided new information on both the vulnerabiliy of wall-bound pectin to fungal pectinases and the involvement of pectic fragments in stimulating plant defense reactions [11, 15, 41]. Fig. 9 illustrates the gradual alteration of pectin in bean leaf tissues infected by *Colletotrichum lindemuthianum* . Pectin breakdown often occurred at a distance from the point of fungal entry. Pectin, was in turn found to be involved in a number of host responses such as deposition in intercellular spaces (Fig.11) and accumulation in wall appositions (Fig.12). Labeling with the AGL-gold complex has greatly contributed to our understanding of the multifaceted role played by pectic compounds during the course of infection.

3.2.2. *Ultrastructural localization of chitin*

Application of the WGA-ovomucoid-gold complex for the localization of chitin has proved useful for delineating the mode of action of biocontrol agents such as *Trichoderma* spp. and visualizing the hydrolysis of fungal cell walls *in planta.*

In the last two decades, *Trichoderma* species have received considerable attention as potential biocontrol agents of a number of soil-borne pathogens [27]. Understandably, the mechanisms by which *Trichoderma* isolates could control pathogenic populations in the rhizosphere have been extensively studied. Recent progress in the purification and identification of *Trichoderma* metabolites has led to the consideration that, in most cases, the antagonistic process relied on the production of antibiotics and/or hydrolytic enzymes such as chitinases and ß-1,3-glucanases [45]. However, few studies have been able to correlate the production of hydrolases in media supplemented with chitin or laminarin with true antifungal activity *in vivo.* To this end, the use of the gold -complexed WGA-ovomucoid has offered a unique and powerful tool for visualizing any alteration of the pathogen cell walls in the presence of the antagonist. The reliability of this approach has been convincingly shown by Chérif and Benhamou [25], Hajlaoui *et al.* [35], and more recently Benhamou and Chet [10]. Using the lectin-gold method, the authors were able to provide evidence for the key role of either enzymatic hydrolysis [10], or antibiosis [35] in the antagonistic process. In the study dealing with the mycoparasitism of *Trichoderma harzianum* against *Rhizoctonia solani* [10], it was clearly shown that collapse and loss of the protoplasm preceded cell alteration as judged

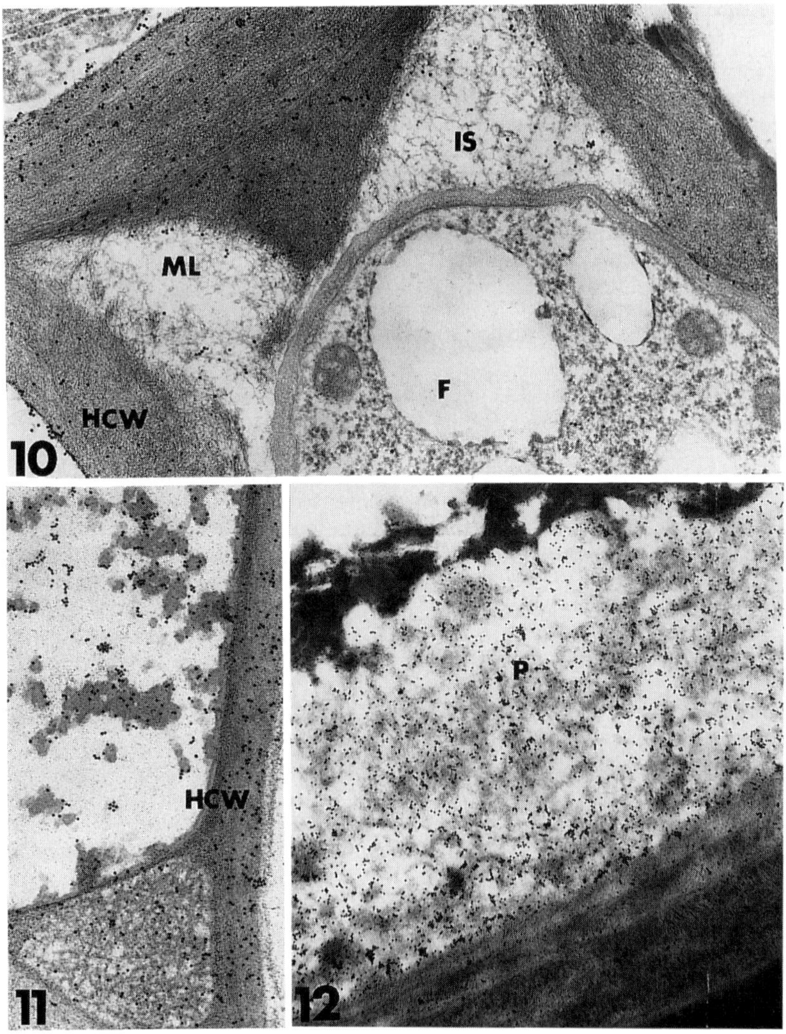

Figs. 10-12. Cytochemical labeling of pectin with the gold-complexed *Aplysia* gonad lectin. **Fig. 10.** Bean leaf tissues infected by *Colletotrichum lindemuthianum* . x 40 000. **Figs. 11 and 12.** Tomato root tissues expressing resistance to FORL following elicitor treatment. Fig.11. x 30 000; Fig. 12. x 25 000.

Abbreviations used in figures : F, fungus; HCW, host cell wall; IS, intercellular space; ML, middle lamella; P, papilla.

by the strong and regular labeling observed over the cell walls of highly altered hyphae of *R. solani* (Fig. 13). This provided indirect evidence that antibiosis rather than enzymatic process could be a major determinant of *Trichoderma* antagonism.

Extensive research conducted in several laboratories has demonstrated that plants react to pathogens by the activation of numerous genes leading to the synthesis and accumulation of various molecules [28, 31]. Proteins associated with defense reactions have been grouped into several categories such as: pathogenesis-related (PR) proteins [23], proteins acting as inhibitors of microbial enzymes [49], hydroxyproline-rich glycoproteins [38], and glycine-rich proteins [37]. The discovery that some PR proteins could be hydrolytic enzymes with chitinase and ß-1,3-glucanase activities has led to speculation that they could be involved in plant disease resistance [23]. Support for the hypothesis that such enzymes played an active role in the plant's defense response came largely from the observation that these enzymes were capable of degrading fungal cell walls *in vitro*. Although this finding was taken as an indication that chitinase and ß-1,3-glucanase were active determinants of plant disease resistance, a conclusive evidence was needed. Using the antibody-gold approach, Benhamou *et al.* [17] were able to study the spatio-temporal distribution of these hydrolases *in planta*. Application of the gold-complexed WGA-ovomucoid confirmed the production of plant chitinases as an important plant defense response. Indeed, in tomato plants expressing resistance to FORL, the amount of chitin in the cell walls of invading hyphae was greatly reduced as compared to normal, thus indicating the action of a hydrolytic process (Fig.14).

4. Concluding remarks

Thefew examples described above point out the interest of using ultrastructure and gold cytochemistry for a better understanding of the cellular and molecular events occurring *in planta*. Exciting progress has been made in the understanding of the cellular events induced in response to infection, mainly because of the introduction of new biological probes (i.e. laccase and ß-1,3-glucanase-gold complexes). In conjunction with biochemistry and molecular biology, cyto-and immunocytochemistry of plant tissues have proved useful for elucidating some aspects of the natural defense system that plants elaborate upon microbial attack. Thus, it is clear that both molecular and traditional approaches will continue to benefit from novel findings derived from the *in situ* localization of an increasing number of plant induced molecules. In addition, the advent of new complementary technologies such as Confocal Laser Microscopy and *in situ* hybridization will contribute to enhance our knowledge about the various genes that code for plant defense proteins.

5. Acknowledgments

The author wishes to thank Sylvain Noël for excellent technical assistance. Thanks are extended to Dr N. Gilboa-Garber (Bar-Ilan University, Israël) and Drs J.P. Geiger and Michel Nicole (ORSTOM, Montpellier, France) for providing the *Aplysia* gonad lectin and the laccase respectively. This work was supported by grants from FCAR (Fonds Québécois pour la Formation de Chercheurs et l'Aide à la Recherche) and from NSERC (Natural Sciences and engineering Research council of Canada).

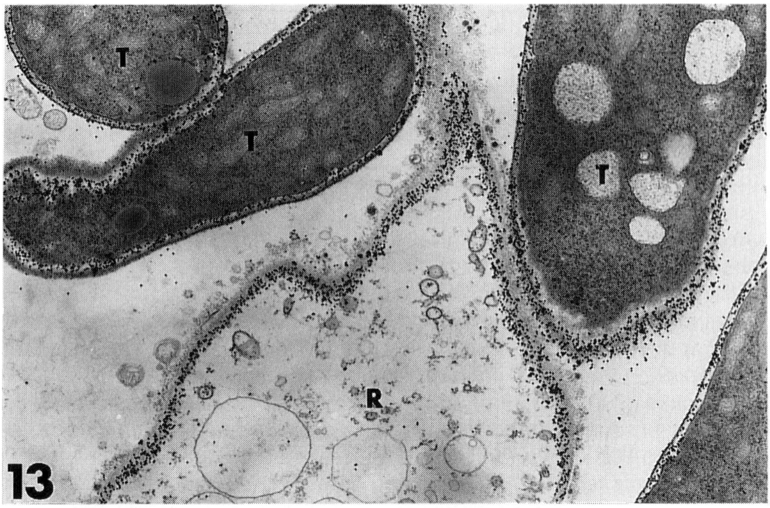

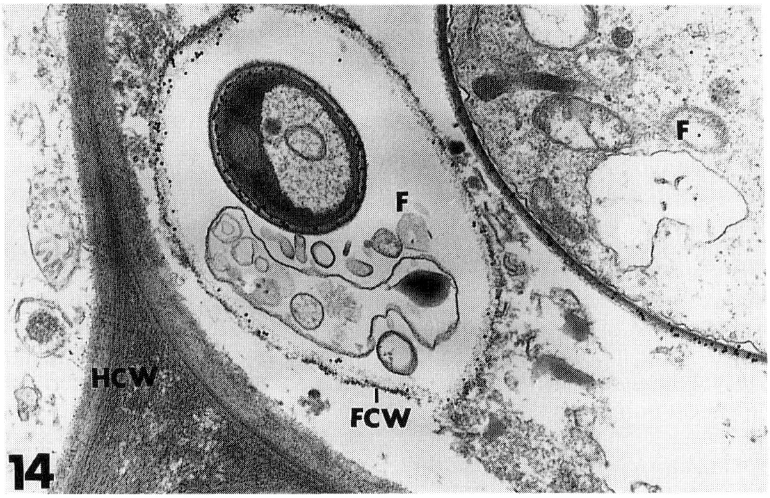

Figs. 13 and 14. Cytochemical labeling of chitin with the WGA-ovomucoid-gold complex. **Fig. 13.** Interaction between *Trichoderma harzianum* and *Rhizoctonia solani* in culture. x 14 000. **Fig. 14.** Tomato root tissues expressing resistance to FORL following elicitor treatment. x 50 000.

Abbreviations used in figures : F, fungus; FCW, fungus cell wall; HCW, host cell wall; R, cell of *Rhizoctonia solani* ; T, cell of *Trichoderma harzianum* .

6. References

1. Aist, J.R. (1976) Papillae and related wound plugs of plant cells, Annu. Rev. Phytopathol. **14**, 145-163.

2. Bendayan, M. (1981) Ultrastructural localization of nucleic acids by the use of enzyme-gold complexes. J. Histochem. Cytochem. **29,** 531-541.

3. Bendayan, M. (1984) Enzyme-gold electron microscopic cytochemistry: a new affinity approach for the ultrastructural localization of macromolecules. J. Elect. Microsc. Techn. **1,** 349-372.

4. Bendayan, M., and Benhamou, N. (1987) Ultrastructural localization of glucoside residues on tissue sections by applying the enzyme-gold approach. J. Histochem. Cytochem. **35**, 1149-1155.

5. Benhamou, N. (1988) Ultrastructural localization of carbohydrates in the cell walls of two pathogenic fungi: A comparative study. Mycologia, **80**, 324-337.

6. Benhamou, N. (1989) Preparation and application of lectin-gold complexes, in M.A. Hayat (ed.), *Colloidal Gold, Principles, Methods and Applications*, Vol. 1. Academic Press, New York. pp. 95-143.

7. Benhamou, N. (1992) Ultrastructural detection of ß-1,3-glucans in tobacco root tissues infected by *Phytophthora parasitica* var. *nicotianae* using a gold-complexed tobacco ß-1,3-glucanase.Physiol. Mol. Plant Pathol. **41**, 351-370.

8. Benhamou, N. (1993) Spatio-temporal regulation of defence genes: immunocytochemistry, in B. Fritig and M. Legrand, (eds.), *Mechanisms of Plant Defense Responses*, Kluwer Acad. Publ. Dordrecht, pp. 221-235.

9. Benhamou, N., and Asselin, A. (1993) *In situ* immunocytochemical localization of plant proteins, in J.E. Thompson and B.R. Glick (eds.), *Methods in Plant Molecular Biology and Biotechnology*, CRC Press , Boca Raton, London, pp. 225-241.

10. Benhamou, N., and Chet, I. (1993) Hyphal interactions between *Trichoderma harzianum* and *Rhizoctonia solani*: : ultrastructure and gold cytochemistry of the mycoparasitic process. Phytopathology, **83**, 1062-1071.

11. Benhamou, N., and Coté, F. 1992. Ultrastructure and cytochemistry of pectin and cellulose degradation in tobacco roots infected by *Phytophthora parasitica* var. *nicotianae*. Phytopathology, **82**, 468-478.

12. Benhamou, N. and Lafontaine, P.J. (1995) Ultrastructural and cytochemical characterization of elicitor-induced responses in tomato root tissues infected by *Fusarium oxysporum* f. sp. *radicis-lycopersici* . Planta (in press).

13. Benhamou, N., Chamberland, H., Ouellette, G.B., and Pauze, F.J. (1987) Ultrastructural localization of ß-1,4-D-glucans in two pathogenic fungi and in their host tissues by means of an exoglucanase-gold complex. Can. J. Microbiol. **33**, 405-417.

14. Benhamou N, Chamberland H, Ouellette GB, and Pauzé FJ. (1988) Detection of galactose in two fungi causing wilt diseases and in their host tissues by means of a gold-complexed *Ricinus communis* agglutinin. Physiol. Mol. Plant Pathol. **32**: 249-266.

15. Benhamou, N., Chamberland, H., and Pauzé, F.J. (1990) Implication of pectic components in cell surface interactions between tomato root cells and *Fusarium oxysporum* f. sp. *radicis-lycopersici*:. A cytochemical study by means of a lectin with polygalacturonic-acid binding specificity. Plant Physiology, **92**, 995-1003.

16. Benhamou, N., Gilboa-Garber, N., Trudel, J., and Asselin, A. (1988) Introduction of a new lectin-gold complex for the ultrastructural localization of galacturonic acids. J. Histochem.Cytochem. **36,** 1403-1411.

17. Benhamou, N., Joosten, M.H.A.J., and De Wit, P.J.G.M. (1990) Subcellular localization of chitinase and of its potential substrate in tomato root tissues infected by *Fusarium oxysporum* f. sp. *radicis-lycopersici*. Plant Physiol. **92**, 1108-1120.

18. Benhamou, N., Mazau, D., and Esquerré-Tugayé, M.T. (1990b) Immunocytochemical localization of hydroxyproline-rich glycoproteins in tomato root cells infected by *Fusarium oxysporum* f. sp. *radicis-lycopersici*: study of a compatible interaction. Phytopathology, **80**, 163-173.

19. Benhamou, N., Lafontaine, P.J. and Nicole, M. (1994) Seed treatment with chitosan induces systemic resistance to *Fusarium* crown and root rot in tomato plants. Phytopathology, **84**, 1432-1444.

20. Bonhoff, A., Reith, B., Golecki, J., and Grisebach, H. (1987) Race cultivar-specific differences in callose deposition in soybean roots following infection with *phytophthora megasperma* f.sp. *glycinea*. Planta, **172**, 101-105.

21. Bowles, D.J. (1990) Defense-related proteins in higher plants. Ann. Rev. Biochem. **59**, 873-907.

22. Brandtzaeg, P. (1982) Tissue preparation methods for immunocytochemistry, in G.R. Bullock, and P. Petruz (eds.), *Techniques in Immunocytochemistry*, Vol. 1, , Academic Press, New York. pp. 2-75.

23. Carr, J.P., and Klessig, D.F. (1989) The pathogenesis-related proteins of plants, in J.K. Setlow (ed.), *Genetic Engineering, Principles and Methods*, Plenum Press, New York, Vol. II. pp. 65-89.

24. Chamberland, H., Benhamou, N., Ouellette, G.B., and Pauzé, F.J. (1989) Cytochemical detection of saccharide residues in paramural bodies formed in tomato root cells infected by *Fusarium oxysporum* f. sp. *radicis-lycopersici.* Physiol. Mol. Plant Pathol. **34,** 131-141.

25. Chérif, M., Benhamou, N. (1990) Cytochemical aspect of chitin breakdown during the parasitic action of a *Trichoderma* sp. on *Fusarium oxysporum* f. sp. *radicis-lycopersici..* Phytopathology, **80**, 1406-1414.

26. Chérif, M., Benhamou, N., and Bélanger, R.R. (1991) Ultrastructural and cytochemical studies of fungal development and host reactions in cucumber plants infected by *Pythium ultimum..* Physiol. Mol. Plant Pathol. **39**, 353-375.

27. Chet, I. (1987) *Innovative Control to Plant Disease Control,* Wiley & Sons, New York. pp. 1-372.

28. Collinge, D.B., and Slusarenko, A.J. (1987) Plant gene expression in response to pathogens. Plant Mol. Biol. **9,** 389-410.

29. Collmer, A., and Keen, N.T. (1986) The role of pectic enzymes in plant pathogenesis. Annu. Rev. Phytopathol. **24,** 383-409.

30. Coulombe, P., Kan F.W.K., and Bendayan, M. (1988) Introduction of a high resolution cytochemical method for studying the distribution of phospholipidids in biological tissues. Eur J. Cell Biol. **46,** 564-576.

31. Dixon, R.A., and Lamb, C.J., (1990) Molecular communication in interactions between plants and microbial pathogens. Annu.Rev. Plant Physiol. Plant Mol. Biol. **41**, 339-367.

32. Dixon, R.A., Harrison, M.J., and Lamb, C.J. (1994) Early events in the activation of plant defense responses. Annu. Rev. Phytopathol. **32**, 479-501.

33. Graham, T.L., and Graham, M.Y. (1991) Cellular coordination of molecular responses in plant defense. Mol. Plant. Microbe Interact. **5,** 415-422.

34. Hajlaoui, M., Benhamou, N., and Bélanger, R.R. (1991) Cytochemical aspects of fungal penetration, haustorium formation, and interfacial material in rose leaves infected by *Sphaerotheca pannosa* var. rosae. Physiol. Mol Plant Pathol. **39**, 341-355.

35. Hajlaoui; M.R., Benhamou, N., and Bélanger, R.R. (1992) Cytochemical study of the antagonistic activity of *Sporothrix flocculosa* on rose powdery mildew, *Sphaerotheca pannosa* var. *rosae*. Phytopathology, **82,** 583-589.

36. Hahlbrock, K., and Scheel, D. (1987) Biochemical responses of plants to pathogens, in I. Chet (ed.) *Innovative Approaches to Plant Disease Control,* John Wiley & Sons, New York, pp. 229-254.

37. Lamb, C.J., Lawton, M.A., Dron, M., and Dixon, R.A., (1989) Signals and transduction mechanisms for activation of plant defense against microbial attack. Cell **56**, 215-224.

38. Mazau, D., and Esquerré-Tugayé, M.T. (1986) Hydroxyproline-rich glycoprotein accumulation in the cell walls of plants infected by various pathogens. Physiol. Mol. Plant Pathol. **29**, 147-157.

39. Nicole, M., and Benhamou, N. (1991) Ultrastructural localization of chitin in the cell walls of *Rigidoporus lignosus*, the white rot fungus of rubber tree roots. Physiol. Mol. Plant Pathol. **29**, 415-431.

40. Nicole, M., and Benhamou, N. (1991) Cytochemical aspects of cellulose breakdown during the infection process of rubber tree roots infected by *Rigidoporus lignosus*. Phytopathology, **81**, 1412-1420.

41. Nicole, M. and Benhamou, N. 1993. Pectin degradation during root decay of rubber trees by *Rigidoporus lignosus*. Can. J. Bot. **71**, 370-378.

42. Ride, J.P. (1983) Cell walls and other structural barriers in defence, in J.A. Callow (ed.), *Biochemical Plant Pathology*, John Wiley & Sons, New York, pp. 214-236.

43. Roth, J. (1983) The colloidal gold system for hight and electron microscopic cytochemistry, in G.R. Bullock and P. Petruz (eds.), *Techniques in Immunocytochemistry*, Vol. 2, Academic Press, New York, pp. 217-284.

44. Ruel, K., and Joseleau, J.P. (1984) Use of enzyme-gold complexes for the ultrastructural localization of hemicelluloses in the plant cell wall. Histochemistry, **81**, 573-580.

45. Sivan, A., and Chet, I. 1989. Degradation of fungal cell walls by lytic enzymes of *Trichoderma harzianum*. J. Gen. Microbiol. **135**, 675-682.

46. Vance, C.P., Kirk, T.K. and Sherwood, R.T. (1980) Lignification as a mechanism of disease resistance. Ann. Rev. Phytopathol. **18**, 259-288.

47. Vandesbosh, K.A. (1991) Immunogold labeling, in J.L. Hall and C. Hawes (eds.), *Electron Microscopy of Plant Cells*, Academic Press, New York, pp. 181-218.

48. Vian, B. (1986) Ultrastructural localization of carbohydrates. Recent developments in cytochemistry and affinity methods, in J. Bailey (ed.), *Biology and Molecular Biology of Plant-Pathogen Interactions*, NATO ASI Series, Vol.H1, Springer-Verlag, Berlin, Heidelberg, pp. 49-57.

49. Walker-Simmons, M., and Ryan, C.A. (1984) Proteinase inhibitor synthesis in tomato leaves. Induction by chitosan oligomers and chemically modified chitosan and chitin. Plant Pathol. **76,** 787-790.

50. Ward, E.R., Uknes, S.J., Williams, S.C., Dincher, S.S., Wiederhold, D.L., Alexander, D.C., Ahl-Goy, P., Metraux, J.P., and Ryals, J.A. (1991) Coordinate gene activity in response to agents that induce systemic acquired resistance. Plant Cell **3**, 1085-1094.

USE OF MONOCLONAL ANTIBODIES TO STUDY DIFFERENTIATION OF *COLLETOTRICHUM* INFECTION STRUCTURES

R.J. O'CONNELL[1], N.A. PAIN[2], J.A. BAILEY[1], K. MENDGEN[3] and J.R. GREEN[2].
[1]IACR-Long Ashton Research Station, Department of Agriculture, University of Bristol, Long Ashton, Bristol BS18 9AF, UK. [2]School of Biological Sciences, The University of Birmingham, P.O. Box 363, Birmingham B15 2TT, UK. [3]Fakultät für Biologie, Universität Konstanz, Postfach 5560, D-78434 Konstanz, Germany.

1. Introduction

Successful penetration and colonization of plant tissues by most fungal pathogens requires differentiation of specialised cell types or infection structures, e.g. germ tubes, appressoria, penetration hyphae, infection hyphae and haustoria. Each cell type is adapted to a particular role in the infection process, e.g. adhesion, contact-sensing, penetration and nutrient uptake [22,32]. Molecular genetic techniques, such as differential or subtractive hybridization and mutational analysis, are being used to identify genes involved in the morphogenesis and function of these infection structures. For example, many genes that are specifically expressed or up-regulated during the formation of appressoria by *Colletotrichum, Magnaporthe* and rust fungi have now been cloned [7,8,23,24,27,29,55,69,76]. Some of these genes have been sequenced and disrupted to determine their role in the infection process [24,69]. These approaches have so far been restricted to infection structures that can be obtained *in vitro*, such as appressoria. Identification of genes expressed by infection structures formed following host penetration is more difficult due to contamination with host mRNAs, although recent advances in the isolation of such structures from infected tissue may alleviate this problem [19,30,50].

An alternative approach is to use monoclonal antibodies (MAbs) to identify differentiation-related proteins and carbohydrates. MAbs can be raised against previously uncharacterised molecules that may only be minor components of a complex mixture [16]. Thus, following immunization with whole cells or crude cell extracts, MAbs binding to molecules of interest can be selected using suitable screening assays. This approach has been used to study cell surface components of the zoospores and cysts of *Phytophthora* and *Pythium* spp. [15,20], and the intracellular infection structures of *Erysiphe pisi* [30] and *Rhizobium* spp. [10].

We have used MAbs to study infection structures formed by the anthracnose fungus, *Colletotrichum lindemuthianum*, on tissues of *Phaseolus vulgaris*. In this chapter, we

M. Nicole and V. Gianinazzi-Pearson (eds.), Histology, Ultrastructure and Molecular Cytology of Plant-Microorganism Interactions, 79-97.

describe methods for the isolation of these structures from bean leaves and for the production of cell type-specific MAbs. Our research has focused on (a) the extracellular matrices (ECMs) surrounding conidia, germ-tubes and appressoria, (b) the plasma membrane of appressoria and (c) the biotrophic interface between intracellular hyphae and living host cells. Potential applications of the MAbs in the diagnosis and taxonomy of *Colletotrichum* are also discussed.

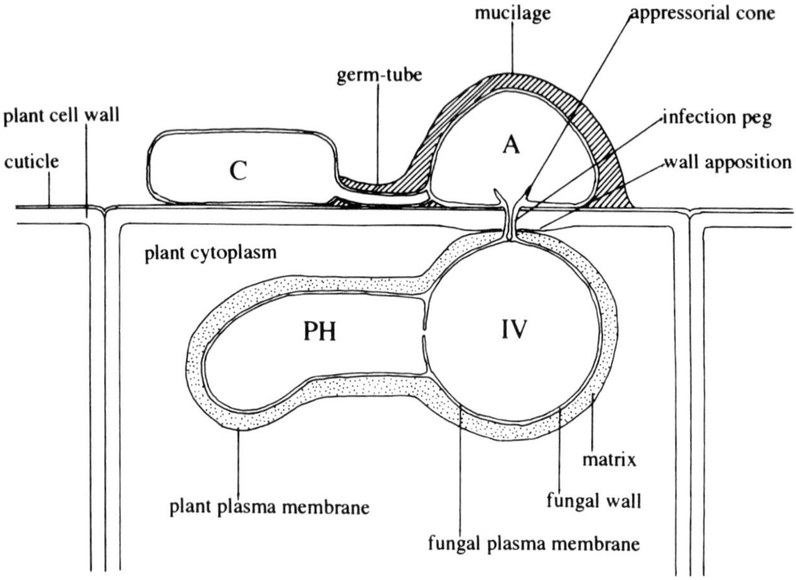

Figure 1. Diagram showing the infection structures produced by *C. lindemuthianum* during infection of a bean epidermal cell. Conidium (C), appressorium (A), infection vesicle (IV) and primary hypha (PH). Reprinted from Pain *et al.* [50] with permission of the Trustees of The New Phytologist.

2. Cytology of Infection Process

The cytology of infection of *P. vulgaris* by *C. lindemuthianum* is well-documented [5,28,35,40,42,43,48,73]. The infection structures produced by this pathogen are illustrated diagrammatically in Figure 1.

2.1. PRE-PENETRATION DEVELOPMENT

Freeze-substitution studies have shown that the conidia of *Colletotrichum* species are coated with a layer of short, densely-packed fibres, which are arranged perpendicular to the cell wall (Figure 2) [37,71]. In the case of *C. lindemuthianum*, this 'spore coat' contains irregularly-shaped pores, and is rich in glycoproteins containing vicinal hydroxyl groups and *N*-acetylgalactosamine residues [41,47]. The spore coat may function in the initial rapid attachment of conidia to substrata, which appears to involve hydrophobic proteins on the spore surface [36,61,77].

The conidia of *C. lindemuthianum* germinate in free water to form narrow germ-tubes, which are enveloped by a fibrillar sheath containing basic proteins and α-linked galactose residues [51]. Germ-tubes of several other *Colletotrichum* species are also surrounded by a fibrillar sheath [1,18,26,71]. The germ-tube sheath could perform several important functions, including adhesion to host surfaces, protection from dessication, maintenance of extracellular ion balances and deployment of fungal enzymes [38,39,59]. Negative-staining showed that the germ-tube sheath of *C. lindemuthianum* contains long, loosely-arranged, hair-like structures or 'fimbriae', which protrude from the cell surface (Figure 3) [51]. Fimbriae have been detected in a wide range of other fungi, including *C. graminicola* [12,57,75], and may be involved in adhesion or contact-sensing [12,58].

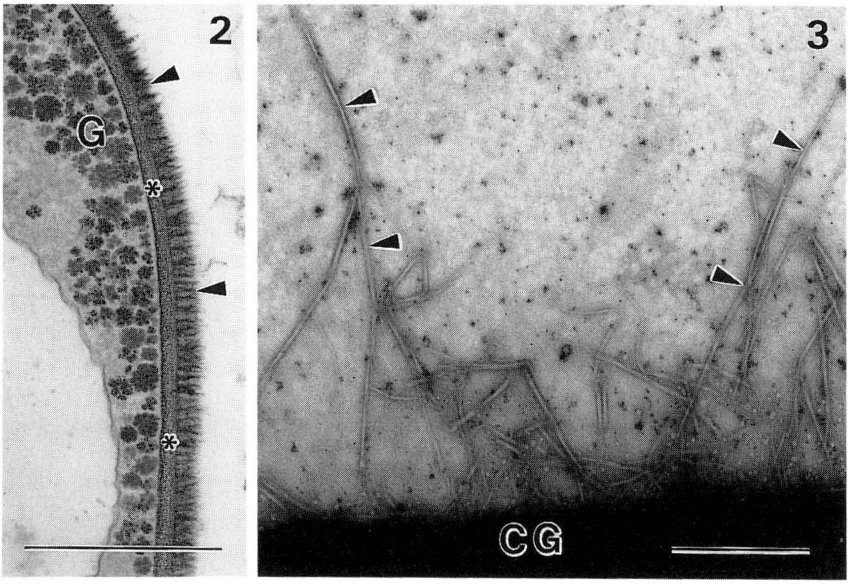

Figure 2. Transverse section through an ungerminated conidium of *C. lindemuthianum* prepared by propane jet-freezing and freeze-substitution and stained for carbohydrates using periodic acid-thiocarbohydrazide-silver proteinate. The spore coat (arrow heads), cell wall (asterisks) and glycogen granules (G) are stained. Bar = 1 μm. *Figure 3.* Conidial germ-tube (CG) of *C. lindemuthianum* growing on a Formvar-coated EM grid, stained with sodium phosphotungstate. Long fimbriae (arrowheads) protruding from the cell surface are negatively-stained. Bar = 1 μm.

Following contact with the plant cuticle, conidial germ-tubes cease apical growth and their tips swell and differentiate to form appressoria. Maturation of the appressorium involves the deposition of new wall layers containing melanin and the secretion of an ECM. The latter coats the domed region of the cell, forming a thickened ring around its base and extending outwards over the substratum as a thin film [5,51,70]. The ECM around *C. lindemuthianum* appressoria contains fimbriae, basic proteins, and residues of mannose, galactose and *N*-acetylglucosamine [41,51]. This material is likely to be involved in the firm adhesion of appressoria to the plant surface.

A pore develops in the ventral cell wall and becomes encircled by a funnel-shaped elaboration of the inner appressorial wall, called the appressorial cone [28]. A narrow infection peg or penetration hypha develops as an extension from the cone and penetrates host epidermal cells directly, through the cuticle and cell wall. Penetration of these barriers probably involves fungal enzymes, although appressoria can exert considerable mechanical pressure [5,32,34].

2.2. INTRACELLULAR BIOTROPHIC PHASE AND THE PLANT-FUNGAL INTERFACE

Following initial penetration, a large globular infection vesicle expands within the epidermal cell. Large-diameter primary hyphae later develop from the infection vesicle, forming an entirely intracellular mycelium, without true determinate haustoria. In a susceptible bean cultivar, host cells initially survive penetration; they plasmolyse normally, exclude permeability tracers and retain normal ultrastructure [44]. The host plasma membrane expands and invaginates around the developing infection vesicles and primary hyphae through the synthesis and incorporation of a large amount of new membrane material. However, in contrast to the extrahaustorial membranes around haustoria of obligate biotrophs, there is no evidence that the host plasma membrane invaginated around intracellular hyphae of *C. lindemuthianum* becomes structurally or physiologically specialised [40]. Moreover, there is no structure equivalent to a haustorial neckband binding the host plasma membrane to the cell wall of *Colletotrichum*.

In common with most intracellular biotrophs, the infection vesicles and primary hyphae of *C. lindemuthianum* are surrounded by an interfacial matrix layer, which separates the fungal wall from the host plasma membrane [40,44]. The liquid or gel-like matrix is rich in carbohydrates and appears to be a product of both plant and fungal activity. Immunogold cytochemistry has indicated the presence of fungal glycoproteins in the matrix [45], while the observation of numerous host Golgi and secretory vesicles around intracellular hyphae suggests that host products are also released into the matrix by exocytosis [32,40]. Exocytosis appears to be coupled to membrane re-cycling and possible endocytosis of material from the matrix into the host cell, since clathrin-coated pits are abundant in the invaginated region of the host plasma membrane [32,40]. Intracellular hyphae become enmeshed by large numbers of host microtubules that are closely appressed to the invaginated plasma membrane [32]. The interface between intracellular hyphae and living host cells is likely to have a key role in recognition phenomena and nutrient transfer [66].

2.3. NECROTROPHIC PHASE

About 24 hours after penetration, the host plasma membrane loses functional integrity and infected cells gradually senesce and die [44]. However, biotrophy is re-established in each newly-colonised cell, so that infected cells at the advancing margin of the primary mycelium are alive, while previously infected cells are dead. Approximately 4 days after initial penetration, narrow, thin-walled secondary hyphae develop as branches from the larger primary hyphae. This change in hyphal morphology is associated with an abrupt switch to a destructive, necrotrophic mode of nutrition. Secondary hyphae develop intramurally, as well as intracelluarly, rapidly dissolving host cell walls and killing host protoplasts ahead of infection [43,44]. The appearance of secondary hyphae coincides with an increase in the activity of endo-pectin lyase, an enzyme capable of killing and macerating bean tissues [72]. *C. lindemuthianum* is thus a facultative biotroph or hemibiotroph, initially feeding on living host cells before switching to necrotrophy [43]. *C. truncatum* and the cowpea anthracnose fungus also have an initial intracellular biotrophic phase and dimorphic primary and secondary mycelia [4,49]. However, in these fungi biotrophy is confined to a single host epidermal cell.

3. Production of MAbs

MAbs were prepared using mice inoculated with two types of immunogen; homogenates of conidia with short germ-tubes grown in liquid culture [52], or homogenates of infection structures isolated from bean leaves [50]. The latter were obtained by homogenizing infected bean leaves, removing plant debris by filtration through a 45 μm nylon mesh and collecting the fungal structures by isopycnic centrifugation (IPC) on Percoll, an inert suspension of colloidal silica [62]. The resulting preparation typically contained 2.5-4% intracellular hyphae (i.e. infection vesicles and primary hyphae), 3% conidia, 50% appressoria and 40% chloroplasts, together with some plant cell wall fragments and starch grains [50]. Approximately 2 x 10^5 intracellular hyphae can be obtained per gram of leaf tissue, of which over 70% are alive, as shown by their staining with fluorescein diacetate, normal ultrastructure and ability to grow in nutrient media [50].

A serious problem associated with immunization with complex mixtures of antigens is that the immune response can be dominated by highly antigenic determinants, such as the oligosaccharide side-chains of glycoproteins, which may be common to many molecules [10]. Various techniques have been developed to obtain antibodies to less abundant or less antigenic molecules, including chemical immunosuppression [74], cascade selection [10], neonatal tolerization [20] and co-immunization [6]. In the latter technique, polyclonal antisera generated against a mixture of antigens lacking the antigens of interest block the production of unwanted antibodies, possibly through the production of anti-idiotypic antibodies.

We used a co-immunization procedure in an attempt to generate MAbs specific for the intracellular hyphae of *C. lindemuthianum* [53]. First, a polyclonal antiserum was raised against an IPC preparation taken from bean leaves 40 hours after inoculation, which contained conidia, appressoria and plant contaminants but no intracellular hyphae. This antiserum was then incubated with an IPC preparation taken four days after inoculation, when intracellular hyphae were present. The mixture was used to immunise further mice, from which MAbs were prepared by conventional methods. The resulting hybridomas were screened for cell type-specific binding by indirect immunofluoresence (see below), using infection structures isolated from leaves.

From one such experiment we obtained MAbs specific for intracellular hyphae (three), conidia (one), appressoria (one), germ-tubes and appressoria (one) and chloroplasts (ten). Although intracellular hyphae comprised less than 4% of the immunogen, three out of the 16 MAbs obtained were specific for this cell type, suggesting that co-immunization was effective in increasing the immune response to such hyphae. The relatively small number of hybridomas produced appears to be a feature of the co-immunization procedure [6].

4. Characterisation of Antigens

Information on the molecular nature of the antigens recognised by the MAbs can be obtained using Western blotting and antigen modification techniques. The apparent molecular weight of the antigens is determined by separating proteins solubilised from cells or membranes using SDS-polyacrylamide gel electrophoresis, transfer of the proteins to nitrocellulose and Western blotting. However, some proteins are only labelled when separated under non-reducing conditions. This may indicate that the antibody recognises a conformational epitope.

The antigenic determinant or epitope recognised by an antibody is in the order of four to eight amino acids or sugar residues. Digestion of blotted proteins with a proteolytic enzyme, e.g. pronase or trypsin, will prevent binding of the MAb if a protein epitope is recognised. Similarly, if a MAb no longer binds after treatment of the blotted samples with sodium periodate, which oxidises vicinal hydroxyl groups, the antibody probably recognises a carbohydrate epitope in a glycoprotein antigen. However, carbohydrate epitopes lacking vicinal hydroxyl groups are not sensitive to periodate.

Antigens can be treated in a variety of ways before separation by SDS-PAGE and Western blotting. For example, digestion with peptide-*N*-glycosidase will remove *N*-linked carbohydrate side-chains from glycoproteins [31]. If the MAb recognises a protein epitope, the resulting bands in the Western blot will have a lower molecular weight when compared with untreated samples. Pre-treatment with trifluoromethane sulphonic acid removes both *N*- and *O*-linked side-chains, decreasing the apparent molecular weight of glycoprotein antigens [14]. Phase-partitioning in the detergent Triton X-114 allows the separation of hydrophobic proteins, which partition into the detergent phase, from hydrophilic proteins, which partition into the aqueous phase [56].

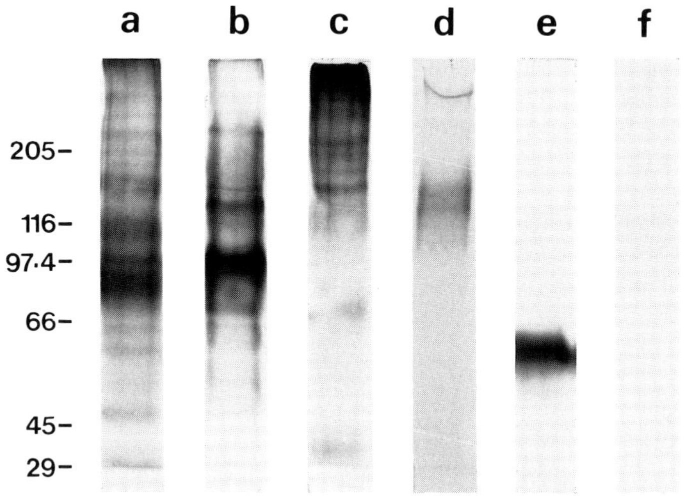

Figure 4. Western blots of proteins extracted from infection structures of *C. lindemuthianum* isolated form bean leaves. Proteins were separated by SDS-PAGE on 7.5 % acrylamide gels under non-reducing conditions, transferred to nitrocellulose and probed with (a) UB20, (b) UB22, (c) UB25, (d) UB26, (e) UB27 and (f) UBIM22 (a control antibody raised to rat bone cells). M_r (x 10^{-3}) are given to the left of the blots.

5. Localisation of Antigens

For indirect immunofluoresence labelling, infection structures were obtained *in vitro* by allowing conidia to germinate on multiwell slides or Formvar-coated EM grids, while infection structures isolated from leaves were air-dried onto gelatin-coated microscope slides. The cells were then labelled using conventional procedures, with or without pre-fixation with 4% formaldehyde [53].

For EM-immunogold labelling, a post-embedding method was used. The carbohydrate epitopes recognised by MAbs UB20 and UB22 (see below) were well-preserved by aldehyde fixation, dehydration by the Progressive Lowering of Temperature method [11] and low temperature embedding in LR White resin [52]. However, the protein epitopes recognised by MAbs UB25, UB26 and UB27 (see below) were poorly preserved by this method. Localisation of these antigens was achieved using cryo-fixation and freeze-substitution [51,53,54]. Cells on polycarbonate or cellophane membranes were cryo-fixed by plunging into liquid propane [21] or by propane jet-freezing [17]. For infected tissues, it was necessary to use high-pressure freezing, since this is the only method currently available for cryo-fixation of thick specimens (up to 0.5 mm) without the formation of damaging ice crystals [33]. After freeze-substitution in osmium-acetone, samples were embedded at -20°C in LR White resin [53]. Although osmium tetroxide is generally not recommended for immunocytochemistry of protein antigens, its inclusion in the freeze-substitution fluid did not prevent MAb binding.

To allow labelling of both sides of the section, grids were immersed in the immune reagents, while the use of small (5 nm) colloidal gold particles further increased labelling efficiency [64]. Silver-enhancement was necessary to visualise the gold probe at low magnifications [53].

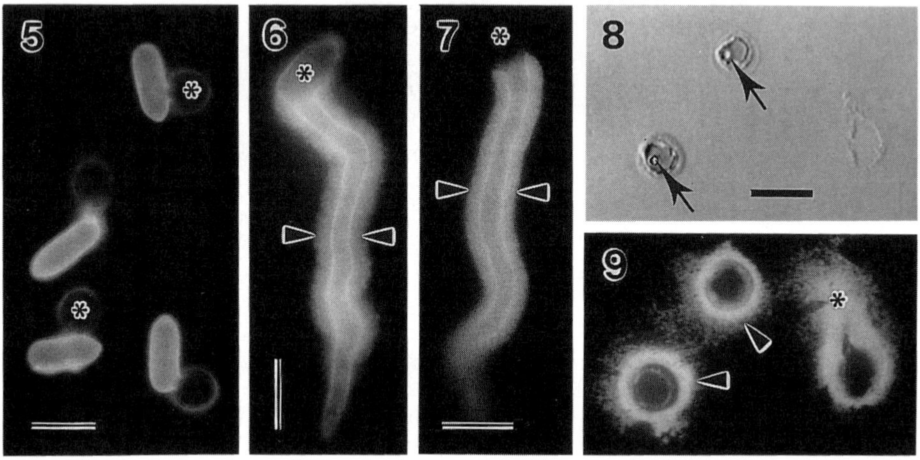

Figures 5-9. Immunofluorescence labelling of germlings of *C. lindemuthianum* grown on glass slides. Bars = 10 µm. *Figure 5.* UB20 labels conidia strongly but appressoria (asterisks) are weakly labelled. *Figure 6.* UB22 labels the conidium (asterisk) and a fibrillar sheath (arrowheads) around the germ-tube. *Figure 7.* UB26 labels a fibrillar sheath (arrowheads) around the germ-tube, but not the conidium (asterisk). *Figures 8 and 9.* Fragments of appressorial cell walls, containing penetration pores (arrows), remaining attached to a glass slide after ultrasonic disruption, viewed with differential interference contrast (*8*) or epi-fluorescence (*9*). UB26 labels haloes (arrowheads) around the disrupted appressoria and a track marking the previous position of a germ-tube (asterisk).

6. MAbs UB20, UB22 and UB26 Recognise Extracellular Glycoproteins on Conidia, Germ-tubes and Appressoria

MAbs UB20 and UB22, which had been raised to germlings grown in liquid culture, bound to different carbohydrate epitopes carried on two distinct sets of glycoproteins with a wide range of molecular weights (Figure 4) [52]. Immunofluorescence of germlings grown on glass slides showed that UB20 strongly labelled the spore coat, with only weak labelling of germ-tubes and appressoria (Figure 5). In contrast, UB22 labelled the fibrillar sheath around germ-tubes strongly, while appressoria and the apices of conidia were labelled less strongly (Figure 6). Fimbriae were not labelled by this antibody [O'Connell, unpublished data]. EM-immunogold labelling of infected bean tissue showed that both MAbs labelled the cell walls of biotrophic intracellular hyphae and the interfacial matrix that surrounds them, supporting earlier conclusions that the interfacial matrix contains fungal glycoproteins [45].

MAb UB26, which had been produced by co-immunization with fungal structures isolated from leaves, recognised a protein epitope on two high molecular weight glycoproteins (133 and 146 kDa) which contain both *O*- and *N*-linked carbohydrate side-chains (Figure 4) [51]. Immunofluorescence of fungal structures isolated from bean leaves and immunogold labelling of infected tissues showed that UB26 bound to germ-tubes and appressoria but, unlike UB22, did not bind to conidia or intracellular hyphae [51]. Labelling of germlings grown *in vitro* showed that UB26 bound to the germ-tube sheath (Figure 7), but the antibody did not label fimbriae in negatively-stained preparations [51].

UB26 also labelled a halo of material surrounding appressoria [51]. When appressoria which had formed on glass were disrupted by sonication, the upper portion of the cell was removed, leaving a small fragment of appressorial wall around the penetration pore attached to the glass (Figure 8) [70]. The halo surrounding these appressorial fragments was still labelled by UB26 (Figure 9), suggesting that the glycoproteins recognised by this antibody adhered firmly to the glass substratum. The glycoproteins also appear to be physically attached to the fungal cell wall, since they were not released into liquid culture media and were not removed by repeated washing during the isolation of germ-tubes and appressoria from infected leaves [51]. These results suggest that the glycoproteins recognised by UB26 are involved in the adhesion of germ-tubes and appressoria to substrata.

7. UB27 Recognises a Glycoprotein Unique to the Plasma Membrane of Appressoria

MAb UB27 was obtained by co-immunization with fungal structures isolated from bean leaves [54]. Immunofluorescence of such isolated cells showed that the antibody bound only to appressoria and not to any other fungal cell type. This specificity for appressoria was confirmed by immunogold labelling of freeze-substituted fungal germlings, grown either on plastic membranes or on plant surfaces (Figures 10 and 11). Within appressoria, the antigen was concentrated along the plasma membrane-wall interface, although the cell wall and ECM were also weakly labelled. Labelling was absent from the basal penetration pore and a region surrounding the pore, including the appressorial cone (Figure 10). An identical labelling pattern was observed with immunofluorescence, using appressoria mechanically removed from leaves or disrupted by ultrasonic vibration or freeze-fracture. In such appressoria, the internal surface of the cell wall was intensely labelled, except for a circular region around the penetration

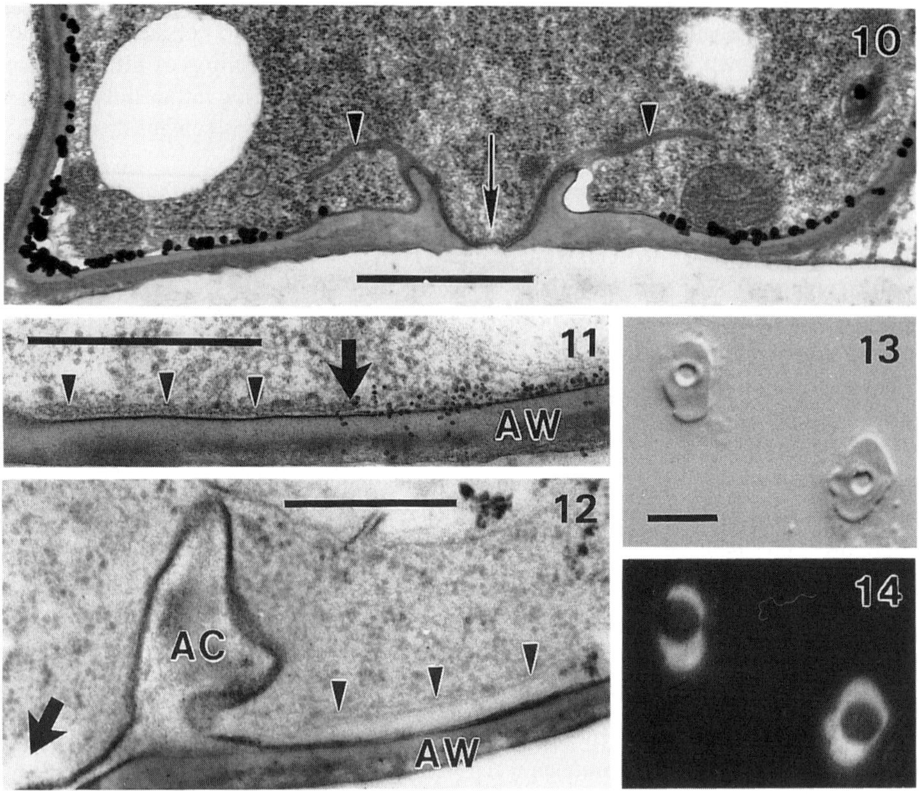

Figure 10. Section through the base of an appressorium of *C. lindemuthianum* formed on a polycarbonate membrane, cryo-fixed by plunge-freezing in liquid propane and immunogold labelled with UB27 (5 nm colloidal gold with silver enhancement). The plasma membrane is labelled, except in the region of the penetration pore (arrow) and appressorial cone (arrowheads). Bar = 1 μm. *Figure 11.* Section through the base of an appressorium formed on a bean hypocotyl, cryo-fixed by high-pressure freezing and immunogold labelled with UB27. Labelling of the plasma membrane and appressorial wall (AW) stops at the point indicated by the arrow. A flattened cisterna of smooth membrane (arrowheads) is associated with the unlabelled region of the plasma membrane. Bar = 0.5 μm. *Figure 12.* Section through the base of an appressorium formed on a bean hypocotyl (chemical fixation and dehydration), stained with PACP. The plasma membrane lining the appressorial wall (AW), appressorial cone (AC) and penetration pore (arrow) is uniformly stained. A cisterna of smooth membrane (arrowheads) encircling the pore is not stained. Bar = 0.5 μm. *Figures 13 and 14.* Fragments of appressorial cell walls, containing penetration pores, remaining attached to a glass slide after freeze-fracture (crushing under liquid nitrogen), labelled with UB27 and viewed with differential interference contrast (*13*) or epi-fluorescence (*14*). The internal surface of the wall is labelled, except for a circular region around the pore. Bar = 10 μm. Figures 10 and 11 reprinted from Pain *et al.* [54] with the permission of Springer-Verlag.

pore (Figures 13 and 14). Intact appressoria, on the other hand, were poorly labelled by UB27 in immunofluorescence experiments, presumably because the antibodies could not penetrate the melanised cell wall.

In Western blots of appressorial proteins (Figure 4), UB27 recognised a protein epitope in a single glycoprotein (Mr 48-50 kDa) with large, *O*-linked carbohydrate side-chains. In phase-separation experiments, the glycoprotein partitioned into the detergent phase, indicating that it is a hydrophobic protein and possibly an integral membrane component [54].

The region of appressorial plasma membrane not labelled by UB27 was sometimes closely associated with a flattened cisterna of smooth membrane, which appeared to form a ring around the penetration pore (Figure 11). This membrane ring, which was also visible in chemically-fixed appressoria, was not stained by periodic acid-chromic acid-phosphotungstic acid (PACP, Figure 12), suggesting that it is not derived from the plasma membrane. We therefore speculate that this structure is a component of the fungal endomembrane system.

Our data provide the first evidence that the plasma membrane of *Colletotrichum* appressoria is differentiated into two distinct domains; one containing the glycoprotein recognised by UB27 and another, around the penetration pore, from which it is absent. Physical links between this glycoprotein and components of the cell wall and/or cytoskeleton may be involved in maintaining separation between the two membrane domains [54]. Using PACP staining, Smereka *et al.* [65] found that the plasma membrane of *Venturia inaequalis* appressoria is also specialised around the penetration pore. However, the plasma membrane of *C. lindemuthianum* appressoria was uniformly stained by PACP (Figure 12). Since the appressorial plasma membrane is continuous with that of the infection peg and infection vesicle, specialization of the plasma membrane in the pore region might be a necessary preparation for subsequent penetration and intracellular biotrophic development within host cells.

The distribution of the glycoprotein recognised by UB27 within appressoria suggests that it may play a role in establishing cell polarity, perhaps by determining the position of the penetration pore. Alternatively, it could be involved in the synthesisis or assembly of wall polymers specific to the appressorium.

8. UB25 Recognises Glycoproteins Restricted to the Surface of Biotrophic Intracellular Hyphae

MAb UB25, which had been obtained by co-immunization with infection structures isolated from bean leaves [53], bound specifically to infection vesicles and primary hyphae in immunofluorescence experiments (Figures 15 and 16). In Western blots, this antibody recognised a protein epitope carried on a set of N-linked glycoproteins (Figure 4).

Immunogold labelling of bean tissues prepared by high-pressure freezing and freeze-substitution showed that these glycoproteins were present in the cell walls of intracellular infection vesicles and primary hyphae and in the interfacial matrix layer that surrounds them (Figure 17). In contrast, the glycoproteins were not detected around primary hyphae growing in intercellular spaces. They were also absent from hyphae

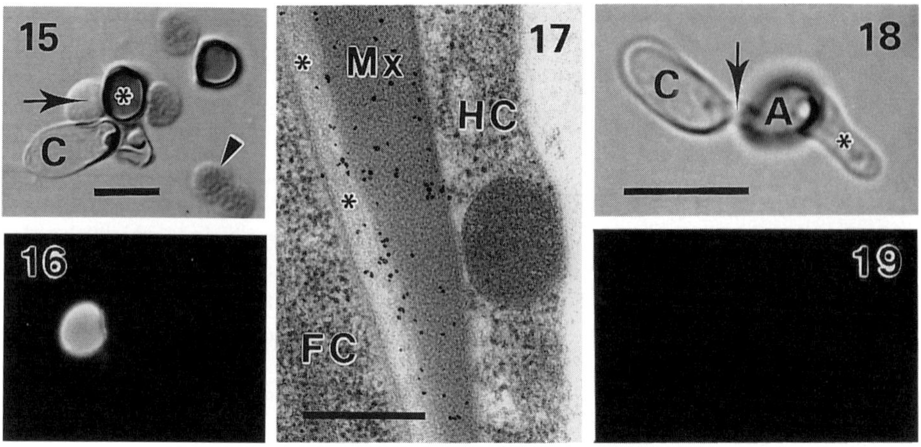

Figures 15 and 16. Immunofluorescence labelling with UB25 of infection structures of *C. lindemuthianum* isolated from bean leaves, viewed with differential interference contrast (*15*) or epi-fluorescence (*16*). Infection vesicle (arrow) is labelled but appressoria (asterisk), conidium (C) and chloroplasts (arrowhead) are not. Bar = 10 μm. *Figure 17.* Part of a primary hypha in a bean epidermal cell, cryo-fixed by high-pressure freezing and immunogold labelled with UB25 (5 nm colloidal gold with silver enhancement). The fungal wall (asterisks) and interfacial matrix (Mx) are labelled, but the host cytoplasm (HC) and fungal cytoplasm (FC) are not. Bar = 0.5 μm. *Figures 18 and 19.* Immunofluorescence labelling with UB25 of germling penetrating a Formvar plastic membrane (100 nm thick), viewed with bright field (*18*) or epi-fluoresence (*19*). The conidium (C), conidial germ-tube (arrow) and appressorium (A) above the membrane and the appressorial germ-tube (asterisk) below the membrane are not labelled. Bar = 10 μm. Figures 17-19 are reprinted from Pain *et al.* [53] with permission of the Trustees of The New Phytologist.

growing on the plant surface and from appressorial germ tubes (Figures 18 and 19), which are formed *in vitro* after penetration of plastic membranes and are developmentally equivalent to infection vesicles. The glycoproteins recognised by UB25 are thus entirely confined to the biotrophic interface between intracellular hyphae and living host protoplasts.

While the glycoproteins identified by UB25 are only expressed on the surface of intracellular hyphae, the expression of other glycoproteins, e.g. those recognised by MAb UB26 and the lectin GSI-B4, appears to be suppressed (see above) [51]. Thus, the fungal cell wall becomes specialised during growth within living host cells, in contrast to the host plasma membrane, which appears to remain largely unmodified [40]. Such changes in the composition of the fungal cell surface presented to the host plasma membrane may be important in the establishment of a compatible interaction and maintenance of host cell viability. This view is supported by the finding that glycoproteins recognised by UB25 are present very soon after initial penetration of epidermal cells, when the fungus first contacts the host plasma membrane.

9. Use of MAbs for Selective Cell Enrichment

Although intracellular hyphae were purified 3000-fold from leaf homogenates after IPC on Percoll, such preparations still contained only 2.5-4% hyphae, due to major contamination with other fungal cell types and plant components (see above). The specificity of MAb UB25 for these hyphae was exploited to purify them further using immunomagnetic separation. IPC preparations were first incubated with UB25, washed by centrifugation and then incubated with magnetic beads coated with goat anti-mouse IgG antibodies. Magnetic separation enriched the intracellular hyphae 10-fold, yielding a sample which contained 30-40% hyphae, of which 60% were viable [50]. More recently, it has been possible to obtain preparations containing up to 95% intracellular hyphae with yields of 1-3 x 10^5 per gram of leaf tissue. This was achieved by washing hyphae attached to the magnetic beads with buffer and repeating the magnetic separation step (N.A. Pain, unpublished results). Immunomagnetic separation has been widely used in animal cell biology and microbiology for the purification of cells, bacteria and viruses from mixed populations [62], but this is the first report of its use for the purification of fungal cells. The purified intracellular hyphae are now being used for further immunizations to generate specific antibodies.

10. Interspecific Cross-reactions of MAbs and their Relevance to the Taxonomy and Diagnosis of Colletotrichum

Until recently, the taxonomy of the genus *Colletotrichum* was very confused and in some cases inaccurate, being based largely upon host origin and descriptive criteria such as conidial size and morphology. Many isolates with indistinguishable morphologies were given different names simply because they were obtained from different host plants. As a result, more than 1000 different species were recognised by von Arx in 1957 [3]. When host origin was given less emphasis, the number of species was reduced to less than 40 [68].

C. lindemuthianum was initially viewed as a host-specific form of *C. gloeosporioides* [3] but was later regarded as a distinct species [67,68]. In a recent molecular analysis, based on sequences of rDNA, Sherriff *et al.* [63] compared isolates of *C. lindemuthianum* from bean (*P. vulgaris*) and cowpea (*Vigna unguiculata*) with other species of *Colletotrichum*. The results revealed several important findings (Figure 20). First, *C. lindemuthianum* from bean is clearly not a form of *C. gloeosporioides*. Second, the cowpea anthracnose fungus is neither a form of *C. lindemuthianum* nor of *C. gloeosporioides*; its present taxonomic status remains unclear [4]. Finally, these data clearly show that *C. lindemuthianum* is closely related to three other species, namely *C. orbiculare* (= *C. lagenarium*), *C. malvarum* and *C. trifolii*, which are distinct from all other members of the genus. It was thus proposed that *C. lindemuthianum* is a

Phaseolus-specific form of a species complex which should be correctly regarded as *C. orbiculare* (Berk. et Mont) [63].

Associated morphological and cytochemical studies on conidia revealed new criteria that were valuable in identifying *C. lindemuthianum* and other forms of *C. orbiculare* [46]. When produced, the conidia of all *Colletotrichum* species are aseptate, and in species of the *C. orbiculare* complex they remain aseptate after germination. However, in all other species a septum is produced in conidia during germination. Conidia of the *C. orbiculare* complex also differ from those of all other species in being coated with glycoproteins that are recognised by the lectin BPA [41,46].

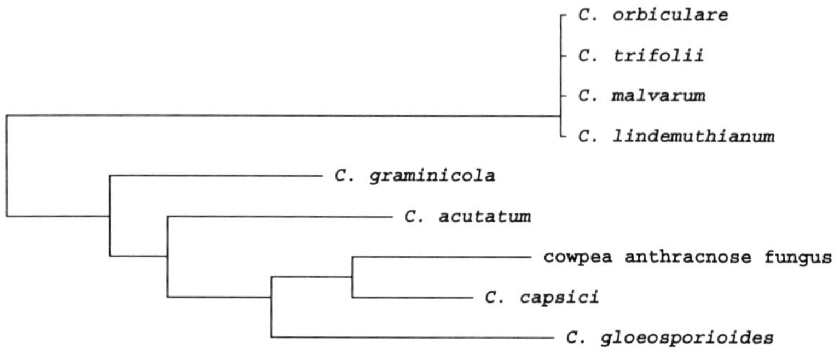

Figure 20. Dendrogram, based on analysis of ITS2 and D2 rDNA sequences, illustrating the relatedness of the *C. orbiculare* complex (*C. orbiculare*, *C. trifolii*, *C. malvarum* and *C. lindemuthianum*) and their distinction from other members of the genus. Scale: 2 mm horizontal distance is equivalent to 0.1% difference in nucleotides. Data taken from Sherriff *et al.* [63].

The patterns of cross-reaction shown by MAbs raised to cells of *C. lindemuthianum* correlate well with these molecular and morphological data and fully support the distinction made between the *C. orbiculare* complex and other species. For example, when the MAbs UB20 and UB22 were screened by ELISA for cross-reaction with 12 *Colletotrichum* species, UB20 bound only to members of the *C. orbiculare* complex, while UB22 also cross-reacted weakly with *C. acutatum*, *C. graminicola* and *C. gloeosporioides*. Neither MAb cross-reacted with the cowpea anthracnose fungus [52]. Similarly, MAb UB26 cross-reacted with the germ-tubes and appressoria of all members of the *C. orbiculare* complex in immunofluorescence experiments but did not bind to those of six other species, which included the cowpea anthracnose fungus [51].

Some of the MAbs may be valuable for disease diagnosis and epidemiological studies. For example, UB20 and UB26 could be used to distinguish between the *C. orbiculare* complex and morphologically similar species, such as *C. gloeosporioides*, that affect the same crop [52]. Preliminary results indicate that UB20 can be used in rapid spore agglutination and dot-blot tests [O'Connell, unpublished data]. These and other applications of MAbs in the diagnosis of fungal diseases have been reviewed recently by Dewey and Oliver [13].

TABLE 1. Binding characteristics of MAbs for cell surfaces of *C. lindemuthianum* conidia and infection structures, and nature of the antigens recognised

MAb	Molecular nature of glycoproteins recognised (kDa)	Epitope recognised	Conidia	Germ-tubes	Appressoria	Intracellular Hyphae
UB20	(29-205)	CHO	++	+	+	+
UB22	(50-205)	CHO	+	++	+	+
UB26	(133, 146) *N*- and *O*-linked	Protein	-	++	++	-
UB27	(48)	CHO	-	-	++	-
UB25	(multimers of 44) *N*-linked	Protein	-	-	-	++

CHO carbohydrate, ++ strong, + weak, - unlabelled. Data from Pain *et al.* [51-54].

11. Conclusions and Future Prospects

We have used MAbs to identify and partially characterise a number of developmentally-regulated glycoproteins present in *C. lindemuthianum* infection structures, which may be relevant to adhesion, morphogenesis, recognition and biotrophy. The properties of the MAbs and the glycoproteins that they bind to are summarised in Table 1.

The antibodies could be used in several ways to provide further information on the nature of these glycoproteins and their role in the infection process. For example, MAbs could be included in functional assays to study their possible inhibitory effect on adhesion, penetration or intracellular development. They could also be used for immunoaffinity chromatography to purify the glycoproteins for carbohydrate analysis and amino acid sequencing [2]. Sequence information could be used to generate oligonucleotide probes for screening a genomic or cDNA library [60]. Alternatively, polyclonal antibodies raised to the purified proteins could be used to immunoscreen cDNA expression libraries [60]. MAbs which recognise protein epitopes, such as UB25, UB26 and UB27, could be used to screen expression libraries directly. Gene disruption or replacement [9] or antisense techniques [25] would provide proof of the function of genes cloned in these ways. Compared to differential or subtractive hybridization, MAbs offer a more targeted, albeit slower, approach to the identification of genes involved in the development and function of fungal infection structures and biotrophic interfaces.

12. Acknowledgements

This work was supported by a BBSRC LINK award (Grant No. P01439). IACR-Long Ashton receives grant-aided support from the Biotechnology and Biological Sciences Research Council of the United Kingdom. We would like to thank Philip Keen and Caroline Nash for providing Figures 5 and 20, respectively, and Ellen Cooke for assistance with the preparation of the manuscript.

13. References

1. Akai, S. and Ishida, N. (1968) An electron microscopic observation on the germination of conidia of *Colletotrichum lagenarium, Mycopath. et Mycol. Appl.* **34**, 337-345.
2. Arvieux, J. and Williams, A.F. (1988) Immunoaffinity chromatography, in D. Catty (ed.), *Antibodies, Vol. I, A Practical Approach*, IRL Press, pp.113-136.
3. Arx, J.A. von (1957) Die Arten der Gattung *Colletotrichum* Cda, *Phytopath. Z.* **29**, 413-468.
4. Bailey, J.A., Nash, C., O'Connell, R.J., and Skipp, R.A. (1990) Infection process, host specificity and taxonomic relationships of a *Colletotrichum* species causing anthracnose disease of cowpea, *Vigna unguiculata, Mycol. Res.* **94**, 810-814.
5. Bailey, J.A., O'Connell, R.J., Pring, R.J., and Nash, C. (1992) Infection strategies of *Colletotrichum* species, in J.A. Bailey and M.J. Jeger (eds.), *Colletotrichum. Biology, Pathology and Control*, CAB International, Wallingford, pp. 88-120.
6. Barclay, S.L. and Smith, A.M. (1986) Rapid isolation of monoclonal antibodies specific for cell surface differentiation antigens, *Proc. Nat. Acad. Sci. USA* **83**, 4336-4340.
7. Bhairi, S., Buckley, E.H., and Staples, R.C. (1990) Protein synthesis and gene expression during appressorium formation in *Glomerella magna, Exp. Mycol.* **14**, 207-217.
8. Bhairi, S.M., Staples, R.C., Freve, P., and Yoder, O.C. (1989) Characterization of an infection structure-specific gene from the rust fungus *Uromyces appendiculatus, Gene* **81**, 237-243.
9. Bowen, J.K., Templeton, M.D., Sharrock, K.R., Crowhurst, R.N., and Rikkerink, E.H.A. (1995) Gene inactivation in the plant pathogen *Glomerella cingulata*: three strategies for the disruption of the pectin lyase gene *pnlA, Mol. Gen. Genet.* **246**, 196-205.
10. Bradley, D.J., Wood, E.A., Larkins, A.P., Galfre, G., Butcher, G.W., and Brewin, N.J. (1988) Isolation of monoclonal antibodies reacting with peribacteroid membranes and other components of pea root nodules containing *Rhizobium leguminosarum, Planta* **173**, 149-160.
11. Carlemalm, E., Villiger, W., Hobot, J.A., Acetarin, J-D., and Kellenberger, E. (1985) Low temperature embedding with Lowicryl resins: two new formulations and some applications, *J. Microsc.* **140**, 55-63.
12. Day, A.W. and Gardiner, R.B. (1987) Fungal fimbriae, *Studies in Mycology* **30**, 333-349.
13. Dewey, F.M. and Oliver, R. (eds.) (1994) *Modern Assays for Plant Pathogenic Fungi: Identification, Detection and Quantification*, CAB International.
14. Edge, A.S.B., Faltynek, C.R., Hof, L., Reichert, L.E., and Weber, P. (1981) Deglycosylation of glycoproteins by trifluoromethane sulfonic acid, *Anal. Biochem.* **118**, 131-137.
15. Estrada-Garcia, M.T., Green, J.R., Booth, J.M., White, J.G., and Callow, J.A. (1989) Monoclonal antibodies to cell surface components of zoospores and cysts of the fungus *Pythium aphanidermatum* reveal species-specific antigens, *Exp. Mycol.* **13**, 348-355.
16. Galfre, G. and Milstein, C. (1981) Preparation of monoclonal antibodies: strategies and procedures, *Methods in Enzymology* **73**, 3-46.
17. Gilkey, J.C. and Staehelin, L.A. (1986) Advances in ultrarapid freezing for the preservation of cellular ultrastructure, *J. Electron Microsc. Techn.* **3**, 177-210.

18. Griffiths, D.A. and Campbell, W.P. (1973) Fine structure of conidial germination and appressorial development in *Colletotrichum atramentarium, Trans. Br. Mycol. Soc.* **61**, 529-536.
19. Hahn, M. and Mendgen, M. (1992) Isolation by Con A binding of haustoria from different rust fungi and comparison of their surface qualities, *Protoplasma* **170**, 95-103.
20. Hardham, A.R., Gubler, F., Duniec, J., and Elliott, J. (1991) A review of methods for the production and use of monoclonal antibodies to study zoosporic plant pathogens, *J. Microsc.* **162**, 305-318.
21. Hoch, H.C. (1986) Freeze-substitution in fungi, in H.C. Aldrich and W.J. Todd (eds.), *Electron Microscopy of Microorganisms*, Plenum, New York, pp.183-212.
22. Hoch, H.C. and Staples, R.C. (1991) Signaling for infection structure formation in fungi, in G.T. Cole and H.C. Hoch (eds.), *The Fungal Spore and Disease Initiation in Plants and Animals*, Plenum Press, New York, pp. 25-46
23. Hwang, C.-S. and Kolattukudy, P.E. (1995) Isolation and characterization of genes expressed uniquely during appressorium formation by *Colletotrichum gloeosporioides* conidia induced by the host surface wax, *Mol. Gen. Genet.* **247**, 282-294.
24. Hwang, C.-S., Flaishman, M.E., and Kolattukudy, P.E. (1995) Cloning of a gene expressed during appressorium formation by *Colletotrichum gloeosporioides* and a marked decrease in virulence by disruption of this gene, *Plant Cell* **7**, 183-193.
25. Judelson, H.S., Dudler, R., Pieterse, C.M.J., Unkles, S.E., and Michelmore, R.W. (1993) Expression and antisense inhibition of transgenes in *Phytophthora infestans* is modulated by choice of promoter and position effects, *Gene* **133**, 63-69.
26. Kozar, F., and Netolitzky, H.J. (1978) Studies on hyphal development and appressorium formation of *Colletotrichum graminicola, Can. J. Bot.* **56**, 2234-2242.
27. Kubo, Y., Nakamura, H., Kobayashi, K., Okuno, T., and Furusawa, I. (1991) Cloning of a melanin biosynthetic gene essential for appressorial penetration of *Colletotrichum lagenarium, Mol. Plant-Microbe Interact.* **4**, 440-445.
28. Landes, M. and Hoffman, G.M. (1979) Zum Keimungs- und Infektionsverlauf bei *Colletotrichum lindemuthianum* auf *Phaseolus vulgaris, Phytopath. Z.* **95**, 259-273.
29. Lee, Y-H. and Dean, R.A. (1993) Stage-specific gene expression during appressorium formation of *Magnaporthe grisea, Exp. Mycol.* **17**, 215-222.
30. Mackie, A.J., Roberts, A.M., Callow, J.A., and Green, J.R. (1991) Molecular differentiation in pea powdery mildew haustoria -idenfication of a 62kDa *N*-linked glycoprotein unique to the haustorial plasma membrane, *Planta* **183**, 399-408.
31. Maley, F., Trimble, R.B., Tarentino, A.L., and Plummer, T.H. (1989) Characterization of glycoproteins and their associated oligosaccharides through the use of endoglycosidases, *Anal. Biochem.* **180**, 195-204.
32. Mendgen, K. and Deising, H. (1993) Infection structures of fungal plant pathogens - a cytological and physiological evaluation, *New Phytol.* **124**, 193-213.
33. Mendgen, K., Welter, K., Scheffold, F., and Knauf-Beiter, G. (1991) High pressure freezing of rust infected plant leaves, in K. Mendgen and D.E. Lesemann (eds.), *Electron Microscopy of Plant Pathogens*, Heidelberg, Springer-Verlag, pp. 31-42.
34. Mercer, P.C., Wood, R.K.S., and Greenwood, A.D. (1971) Initial infection of *Phaseolus vulgaris* by *Colletotrichum lindemuthianum*, in T.F. Preece and C.H. Dickinson (eds.), *Ecology of Leaf Surface Microorganisms*, Academic Press, London, pp. 381-390.
35. Mercer, P.C., Wood, R.K.S., and Greenwood, A.D. (1975) Ultrastructure of the parasitism of *Phaseolus vulgaris* by *Colletotrichum lindemuthianum, Physiol. Plant Pathol.* **5**, 203-214.
36. Mercure, E.W., Leite, B., and Nicholson, R.L. (1994) Adhesion of ungerminated conidia of *Colletotrichum graminicola* to artificial hydrophobic surfaces, *Physiol. Mol. Plant Pathol.* **45**, 421-440.
37. Mims, C.W., Richardson, E.A., Clay, R.P., and Nicholson, R.L. (1995) Ultrastructure of conidia and the conidium aging process in the plant pathogenic fungus *Colletotrichum graminicola, Int. J. Plant Sci.* **156**, 9-18.
38. Moloshok, T.D., Leinhos, G.M.E., Staples, R.C., and Hoch, H.C. (1993) The autogenic extracellular environment of *Uromyces appendiculatus* urediospore germlings, *Mycologia* **85**, 392-400.

39. Nicholson, R.L. and Epstein, L. (1991) Adhesion of fungi to the plant surface. Prerequisite for pathogenesis, in G.T. Cole and H.C. Hoch (eds.), *The Fungal Spore and Disease Initiation in Plants and Animals*, Plenum, New York pp 3-23.
40. O'Connell, R.J. (1987) Absence of a specialized interface between intracellular hyphae of *Colletotrichum lindemuthianum* and cells of *Phaseolus vulgaris*, *New Phytol.* **107**, 725-734.
41. O'Connell, R.J. (1991) Cytochemical analysis of infection structures of *Colletotrichum lindemuthianum* using fluorochrome-labelled lectins, *Physiol. Mol. Plant Pathol.* **39**, 189-200.
42. O'Connell, R.J. and Bailey, J.A. (1986) Cellular interactions between *Phaseolus vulgaris* and the hemibiotrophic fungus *Colletotrichum lindemuthianum*, in J.A. Bailey (ed.), *Biology and Molecular Biology of Plant-Pathogen Interactions*, NATO ASI Series H, Vol.1, Springer-Verlag, Berlin, pp. 39-49.
43. O'Connell, R.J. and Bailey, J.A. (1991) Hemibiotrophy in *Colletotrichum lindemuthianum*, in K. Mendgen and D.-E. Lesemann (eds.), *Electron Microscopy of Plant Pathogens*, Springer-Verlag, Berlin, pp. 211-222.
44. O'Connell, R.J., Bailey, J.A., and Richmond, D.V. (1985) Cytology and physiology of infection of *Phaseolus vulgaris* infected by *Colletotrichum lindemuthianum*, *Physiol. Plant Pathol.* **27**, 75-98.
45. O'Connell, R.J., Bailey, J.A., Vose, I.R., and Lamb, C.J. (1986) Immunogold labelling of fungal antigens in cells of *Phaseolus vulgaris* infected by *Colletotrichum lindemuthianum*, *Physiol. Mol. Plant Pathol.* **28**, 99-105.
46. O'Connell, R.J., Nash, C., and Bailey, J.A. (1992) Lectin cytochemistry: a new approach to understanding cell differentiation, pathogenesis and taxonomy in *Colletotrichum*, in J.A.Bailey and M.J. Jeger (eds.), *Colletotrichum. Biology, Pathology and Control*, CAB International, Wallingford, pp 67-87.
47. O'Connell, R.J., Pain, N.A., Hutchison, K.A., Jones, G.L., and Green, J.R. (1996) Ultrastructure and composition of the cell surfaces of infection structures formed by the fungal plant pathogen *Colletotrichum lindemuthianum*, *J. Microsc.* (in press).
48. O'Connell, R.J. and Ride, J.P. (1990) Chemical detection and ultrastructural localization of chitin in cell walls of *Colletotrichum lindemuthianum*, *Physiol. Mol. Plant Pathol.* **37**, 39-53.
49. O'Connell, R.J., Uronu, A.B., Waksman, G., Nash, C., Keon, J.P.R., and Bailey, J.A. (1993) Hemibiotrophic infection of *Pisum sativum* by *Colletotrichum truncatum*, *Plant Pathol.* **42**, 774-783
50. Pain, N.A., Green, J.R., Gammie, F., and O'Connell, R.J. (1994) Immunomagnetic isolation of viable intracellular hyphae of *Colletotrichum lindemuthianum* (Sacc. & Magn.) Briosi & Cav. from infected bean leaves using a monoclonal antibody. *New Phytol.* **127**, 223-232.
51. Pain, N.A., Green, J.R., Jones, G.L., and O'Connell, R.J. (1996) Composition and organisation of extracellular matrices around germ-tubes and appressoria of *Colletotrichum lindemuthianum*, *Protoplasma* (in press).
52. Pain, N.A., O'Connell, R.J., Bailey, J.A., and Green, J.R. (1992) Monoclonal antibodies which show restricted binding to four *Colletotrichum* species: *C. lindemuthianum*, *C. malvarum*, *C. orbiculare* and *C. trifolii*, *Physiol. Mol. Plant Pathol.* **41**, 111-126.
53. Pain, N.A., O'Connell, R.J., Mendgen, K., and Green, J.R. (1994) Identification of glycoproteins specific to biotrophic intracellular hyphae formed in the *Colletotrichum lindemuthianum*-bean interaction, *New Phytol.* **127**, 233-242.
54. Pain, N.A., O'Connell, R.J., and Green, J.R. (1995) A plasma membrane-associated protein is a marker for differentiation and polarisation of *Colletotrichum lindemuthianum* appressoria, *Protoplasma* (in press).
55. Perpetua, N.S., Kubo, Y., Okuno, T., and Furusawa, I. (1994) Restoration of pathogenicity of a penetration-deficient mutant of *Colletotrichum lagenarium* by DNA complementation, *Curr. Genet.* **25**, 41-46.
56. Pryde, J.G. (1986) Triton X-114: a detergent that has come in from the cold, *Trends in Biochem. Sci.* **11**, 160-163.
57. Rghei, N.A., Castle, A.J., and Manocha, M.S. (1992) Involvement of fimbriae in fungal host-mycoparasite interaction, *Physiol. Mol. Plant Pathol.* **41**, 139-148.

58. Romantschuk, M. (1992) Attachment of plant pathogenic bacteria to plant surfaces, *Ann. Rev. Phytopathol.* **30**, 225-243.
59. Ruel, K. and Joseleau, J-P. (1991) Involvement of an extracellular glucan sheath during degradation of *Populus* wood by *Phanerochaete chrysosporium, Appl. Environ. Microbiol.* **57**, 374-384.
60. Sambrook, J., Fritsch, E.F., and Maniatis, T. (1989) *Molecular Cloning: a Laboratory Manual*, Cold Spring Harbor Laboratory Press.
61. Sela-Buurlage, M.B., Epstein, L., and Rodriguez, R.J. (1991) Adhesion of ungerminated *Colletotrichum musae* conidia, *Physiol. Mol. Plant Pathol.* **39**, 345-352.
62. Sharpe, P.T. (1988) *Methods of Cell Separation: Laboratory Techniques in Biochemistry and Molecular Biology, Vol. 18*, R.H. Burdon and P.H. van Knippenberg (Series eds.), Elsevier, Amsterdam.
63. Sherriff, C., Whelan, M.J., Arnold, G.M., Lafay, J-F., Brygoo, Y., and Bailey, J.A. (1994) Ribosomal DNA sequence analysis reveals new species groupings in the genus *Colletotrichum*, *Exp. Mycol.* **18**, 121-138.
64. Slot, J.W. and Geuze, H.J. (1981) Sizing of Protein A - colloidal gold probes for immunoelectron microscopy, *J. Cell Biol.* **90**, 533-536.
65. Smereka, K.J., Machardy, W.E, and Kausch, A.P. (1987) Cellular differentiation in *Venturia inaequalis* ascospores during germination and penetration of apple leaves, *Can. J. Bot.* **65**, 2549-2561.
66. Smith, S.E. and Smith, F.A. (1990) Structure and function of the interfaces in biotrophic symbioses as they relate to nutrient transport, *New Phytol.* **114**, 1-38.
67. Sutton, B.C. (1980) *The Coelomycetes*, Commonwealth Mycological Institute, Kew, London.
68. Sutton, B.C. (1992) The genus *Glomerella* and its anamorph *Colletotrichum*, in J.A. Bailey and M.J. Jeger (eds.), *Colletotrichum: Biology, Pathology and Control*, CAB International, Wallingford, pp. 1-26.
69. Talbot, N.J., Ebbole, D.J., and Hamer, J.E. (1993) Identification and characterization of *MPG1*, a gene involved in pathogenicity from the rice blast fungus *Magnaporthe grisea, Plant Cell* **5**, 1575-1590.
70. Uchiyama, T., Ogasawara, Y., Nanba, Y., and Ito, H. (1979) Conidial germination and appressorial formation of the plant pathogenic fungi on the coverglass or cellophane coated with various lipid components of plant leaf waxes, *Agric. Biol. Chem.* **43**, 383-384.
71. Van Dyke, C.G. and Mims, C.W. (1991) Ultrastructure of conidia, conidium germination, and appressorium development in the plant pathogenic fungus *Colletotrichum truncatum*, *Can. J. Bot.* **69**, 2455-2467.
72. Wijesundera, R.L.C., Bailey, J.A., and Byrde, R.J.W. (1984) Production of pectin lyase by *Colletotrichum lindemuthianum* in culture and in infected bean *(Phaseolus vulgaris)* tissue, *J. Gen. Microbiol.* **130**, 285-290.
73. Wolkow, P.M., Sisler, H.D., and Vigil, E.L. (1983) Effect of inhibitors of melanin biosynthesis on structure and function of appressoria of *Colletotrichum lindemuthianum, Physiol. Plant Pathol.* **22**, 55-71.
74. Wycoff, K.L and Ayres, A.R. (1990) Monoclonal antibodies to surface and extracellular antigens of a fungal plant pathogen, *Phytophthora megasperma* f.sp. *glycinea*, recognize specific carbohydrate epitopes, *Physiol. Mol. Plant Pathol.* **37**, 55-79.
75. Xu, J. and Day, A.W. (1992) Multiple forms of fimbriae on the sporidia of corn smut, *Ustilago maydis*, *Int. J. Plant Sci.* **153**, 531-540.
76. Xuei, X., Bhairi, S., Staples, R.C., and Yoder, O.C. (1992) Characterisation of INF 56, a gene expressed during infection structure development of *Uromyces appendiculatus, Gene* **110**, 49-55.
77. Young, D.H. and Kauss, H. (1984) Adhesion of *Colletotrichum lindemuthianum* spores to *Phaseolus vulgaris* hypocotyls and to polystyrene, *Appl. Environ. Microbiol.* **47**, 616-619.

THE PLANT CELL WALL, FIRST BARRIER OR INTERFACE FOR MICROORGANISMS: IN SITU APPROACHES TO UNDERSTANDING INTERACTIONS

Brigitte VIAN, Danièle REIS, Laura GEA and Valérie GRIMAULT
Laboratoire de Pathologie Végétale, INA P-G, 16 rue Claude Bernard, 75 231 PARIS Cedex 05, FRANCE

1. Introduction

Microbial colonization within host plants involves a cascade of events in which cell surface contact between both partners plays a determinant role. The two protagonists have to enter in close contact and identify each other, whatever the interaction may be, i.e. pathologic, saprophytic or symbiotic.
The objective of this chapter is to show how *in situ* approaches allowing to detect cell wall organization and modifications can help to understand plant/microorganism interactions. First, we will emphasize that the wall is highly organized, forming a domain-specific apoplastic barrier around the cell and leading to the necessity for the invading microorganism to develop adapted colonizing strategies. Then, we will analyze some cell wall changes occurring in two apparently unrelated situations, parasitic and symbiotic, which however share common features.

2. The cell wall is basically ordered and multidomain

The cell wall constitutes a continuous and dynamic network through the whole body of the plant. Great advances have been realized at the molecular level in our kowledge of the constituting subunits of the matrix and of their relation with the cellulose microfibrils. New structural models have been provided that take into account the great amount of information on the various wall polymers, their specific cross-links and the dynamic changes occurring during growth and wall consolidation (6,12, 25, 40,43).
More controversial is the discussion concerning the three-dimensional architecture of the plant cell wall at the supramolecular level. From the literature it becomes clear that the cell wall -whatever it is elongating or not - is basically constructed with a certain degree of order which is both cell- and tissue-specific (26, 32, 35, 36). The highest degree of order is expressed in the helicoidal pattern described from the analysis of many types of cell walls in the recent years(14, 34, 35, 38, 39, 53). This helicoidal pattern is not unique. It corresponds to the basic construction of extracellular fibrous composites which are widespread in biological systems from plant and animal kingdom and have been recently extensively reviewed (27). Figure 1 shows an example of helicoidal organization in the tangential wall from an epidermal cell of mung bean hypocotyl in the course of elongation. The helicoidal order is clearly seen in the internal part of the wall, i.e. the most recently deposited, and appears as a series of typical superimposed bow-shaped arcs.
The helicoidal organization is not visible at the first glance and if the occurrence and the geometry of helicoidal constructions have been often controversial, this is due to the

M. Nicole and V. Gianinazzi-Pearson (eds.), Histology, Ultrastructure and Molecular Cytology of Plant-Microorganism Interactions, 99-115.

difficulty of finding the appropriate methodologies allowing to visualize the actual texture (35, 36, 53). Thus the choice of the most appropriate technique for specimen preparation

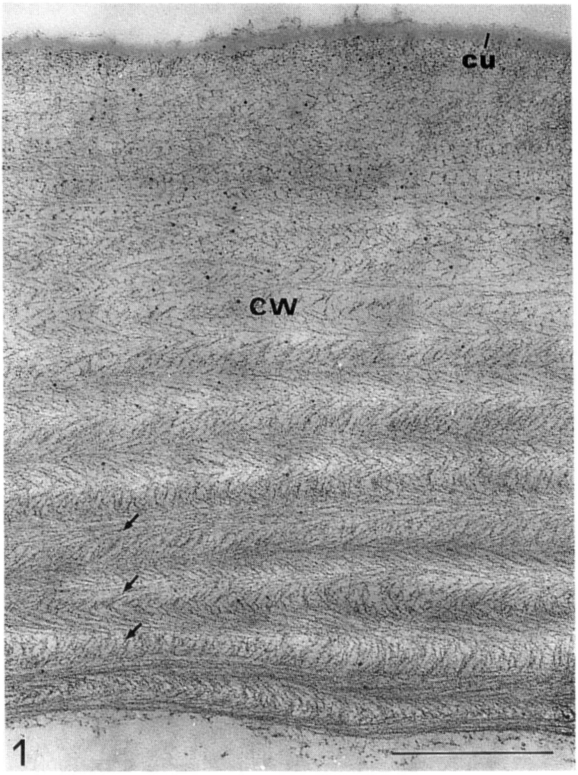

Figure 1. **Helicoidal order in the plant cell wall.**
Epidermal cell of the elongating zone of *Vigna radiata* hypocotyl. Methylamine partial extraction of matrix components, in order to expose the fibrillar framework (see 33). Transverse section. PATAg staining. Outer tangential wall. Thick cell wall (cw) in which the helicoidal order (arrows) is particularly visible in the innermost part, recently deposited; cu: cuticle. Bar represents 0.5 µm.

and analysis, i.e. subtractive cytochemistry with mild extraction of the matrix, delignification, freeze-fracture/deep-etch replicas, stereoreconstruction - is very important when one wants to have access to the three-dimensional organization and to its possible changes with time (33, 35, 38, 52).

Helicoids have been described in a wide variety of cell types and may occur in both primary and secondary walls, although in primary walls they may be transient (fig.1 and 34, 53). In higher plants, helicoids are not found in all tissues. They are encountered in the external tissues that are strategically placed for strengthening the organs, such as epidermis outside leaves and stems, outer cortical parenchyma, bundles of collenchyma (27). They are also found in conducting tissues and associated bundles of sclerenchyma. In terms of microorganism penetration, it is clear that helicoids are generally found in the tissues that are engaged in the colonization of plants and that the microorganisms have to negociate their entry into a host tissue through these complex barriers.

Another point that should be emphasized is that for a single cell the spatial organization of the wall is also uneven. It is becoming clear that the cell wall has to be viewed as a mosaic of specific domains, the composition and organization of which change with time and differentiation steps. The uneveness is visible at different levels. At the supramolecular level, the three-dimensional arrangement of the wall is not uniform all around the cell. In thickened expanding walls for example, the helicoids are unevenly distributed and form independent ribs at the cell edges. In such cells, the longitudinal walls are dramatically different from the transverse walls. The latter are thin, without any apparent order, whereas the former are thick and organized in helicoids, at least along longitudinal facets. Collenchyma is an extreme case in which the neighbouring thickenings can fuse laterally and cover the whole surface forming coaxial cylinders (53).

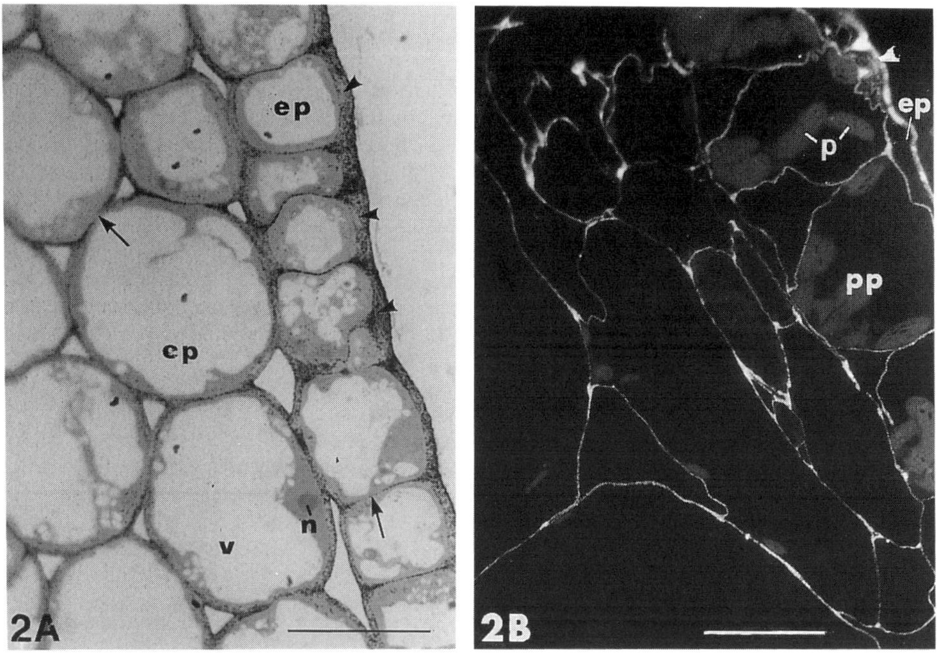

Figure 2. **Domain-specific distribution of wall polymers.**
Homogalacturonan detected by a monoclonal antibody JIM 5. 2A: thick (0.1µm) transverse section of *Vigna radiata* hypocotyl. Immunolabelling and silver intensification. Toluidine blue counterstaining. 2B: thick transverse section of *Saintpaulia ionantha* leaf. Immunofluorescence, FITC. In both cases, the labelling seen on cell walls is intense on tangential outer wall (arrowheads) of epidermis (ep) and less intense on radial walls. On the latter, homogalacturonan-deprived domains are seen (arrows); cp: cortical parenchyma, pp: palisade parenchyma rich in plastids (p), v: vacuole, n: nucleus. Bars represent 25 µm.

The uneveness is also visible in the distribution of the wall polymers themselves which appear domain-specific. An example is provided that shows an immunolocalization of homogalacturonans, i.e. a pectic polymer, in two different organs, an hypocotyl and a leaf, respectively (fig. 2). In both systems, the distribution of the homogalacturonans is not uniform. It roughly follows the gradient of thickness of the cell walls but it is irregular in the different regions of the wall of each cell. In mung bean hypocotyl (fig.2A), the silver deposits are more intense on the tangential walls than on the radial ones and even pectin-deprived domains can be seen. In the saintpaulia leaf (fig. 2B), in the spongy parenchyma, highly reactive junctions are seen close to wall areas where the immunolabelling is weaker. This uneven polymer distribution (it could be also shown for other polymers) is important since such reinforced intersects may play a role of bolt in host/microorganism interactions and heterolysis, whereas other less reinforced regions may facilitate their access. The fact that different wall regions or domains, with different composition and architecture can coexist and constitute an apoplastic mosaic similar to a patchwork quilt (32) is regularly emphasized in the literature because of the development of a panoply of polymer probes (4, 11, 13, 21, 24, 37, 49, 52, 54).

3. Towards destruction of the host cell wall: the parasitic situation

The description of the modalities of penetration of pathogens accross the mural barrier of host plants is out of the scope of the present chapter. It has been the object of many descriptions and reviews (see 3, 28, 31, 41, 42 and bibliography therein). Informations on wall depolymerases produced by plant pathogens and their role in plant disease have been also extensively provided (see 55 for a recent review). We will only illustrate a few examples of wall attack by two different pathogenic microorganisms.

The first example concerns a fungal attack of *Cucumis melo* hypocotyl by *Colletotrichum lindemuthianum. C. lindemuthianum* is responsible for anthracnose disease in a variety of host plants (5, 29, 46, 56). It was chosen because it degrades the wall progressively after penetration into the outer cells of hypocotyl. The use of the PATAg (periodic acid-thicarbohydrazide-silver proteinate) test for visualization of polysaccharides (33), allows to precisely follow the the wall deconstructive process. Clearly visible is a front of attack creating a dramatic discontinuity between very compact and highly PATAg-reactive unattacked zones and their attacked counterparts becoming less compact and poorly reactive. The degradation of host cell wall matrix by fungal enzymes exposes the fibrillar framework. This fungal extraction is mild and the native ordered construction, non detectable at once in the absence of the fungus, is therefore visualized. The wall organization appears layered and strikingly similar to what is exposed following subtractive cytochemistry (compare with fig. 1). Beyond the attack front, the wall deconstruction is more irregular, showing microdomains with degraded matrix polymers adjacent to still not degraded areas.

Such images allow to get insight into the colonization process. In the present case, they indicate that the pathogen first progresses by depolymerizing specifically some wall polymers without disturbing the general ordered construction of the wall. To go further, we need to localize the fungal enzymes and to identify the remaining substrates, which is now a reasonable challenge due to the availability of many new affinity probes.

Figure 3. **Progressive host cell wall deconstruction by a fungus.**
Host: hypocotyl of *Cucumis melo*; fungal pathogen: *Colletotrichum lindemuthianum*. Transverse section of cortical parenchyma. PATAg staining. Near the fungus (3A, f), a gradual lysis of the cell wall (hcw) is observed, with more or less clear domains. Beyond the attack front (3B, arrows), the helicoidal architecture of the wall is exposed ; fcw: fungus cell wall. Bars represent 1 μm.

Figure 4. **Differential host cell wall deconstruction by bacteria.**
Host: leaf of *Saintpaulia ionantha*; bacterial pathogen: *Erwinia chrysanthemi, strain 3937*. Transverse section of leaves. PATAg staining. On fig. 4A, an intercellular pocket containing bacteria (b) is seen. The wall maceration is heterogeneous and shows PATAg reactive areas (hcw) and swollen and poorly reactive areas (**). Between both states, the transition may be abrupt (arrows). Fig. 4B shows a detail of a region that is strongly macerated. The host cell wall (hcw) is weakly reactive. Note the highly reactive bacterial cell membrane (bm), pressed against the cell wall ghost (arrowhead). On the right, a bacteria is close to a plastid (p) of a destroyed palisade cell. Bars represent 1 μm.

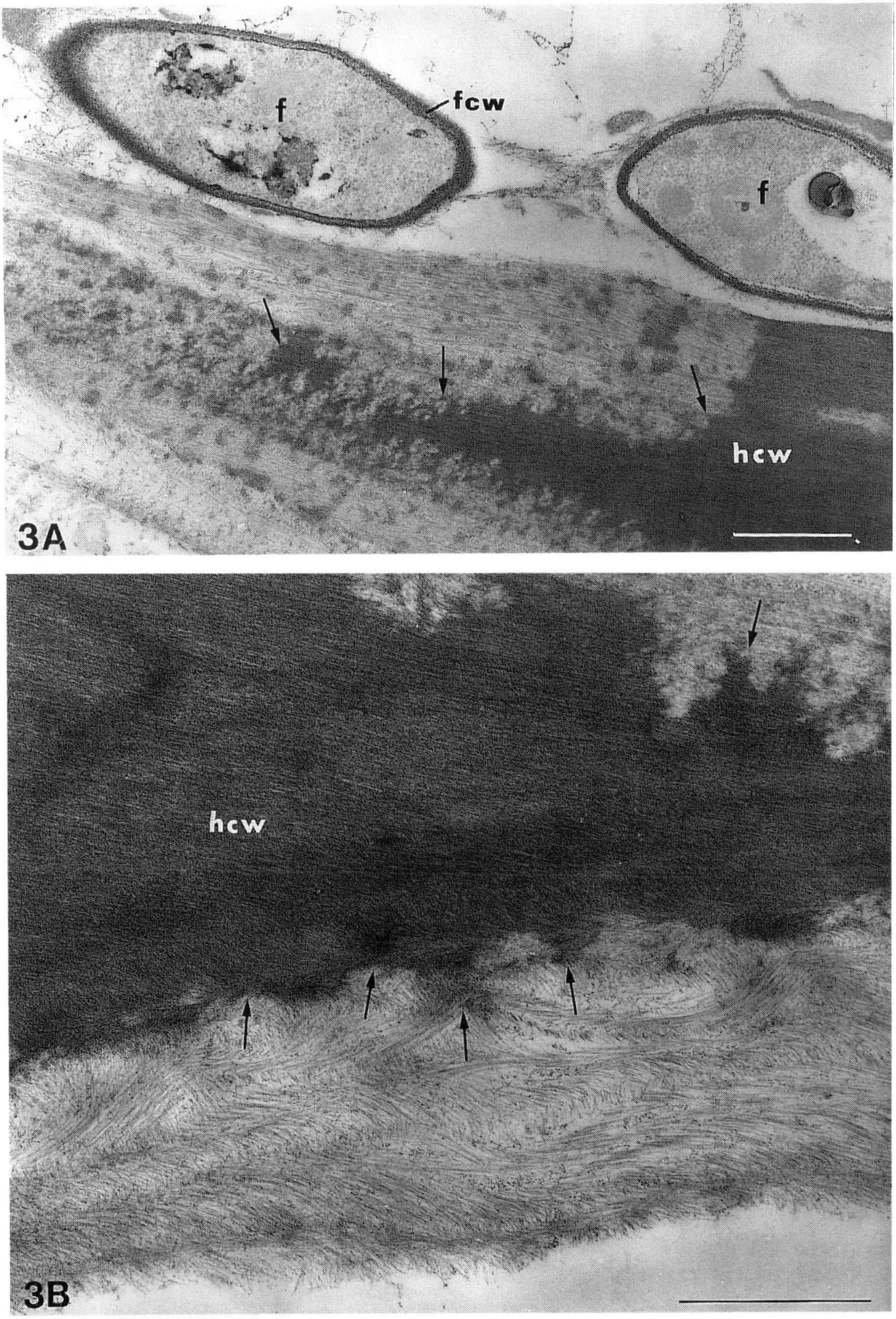
fcw
f
f
hcw
3A
hcw
3B

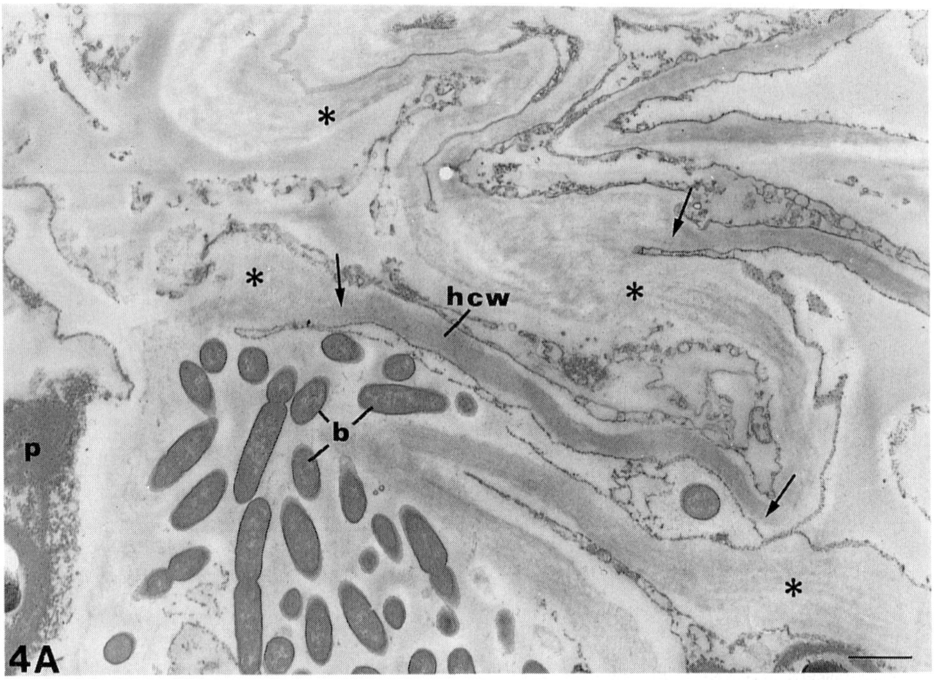
*
*
hcw
*
b
p
*
4A

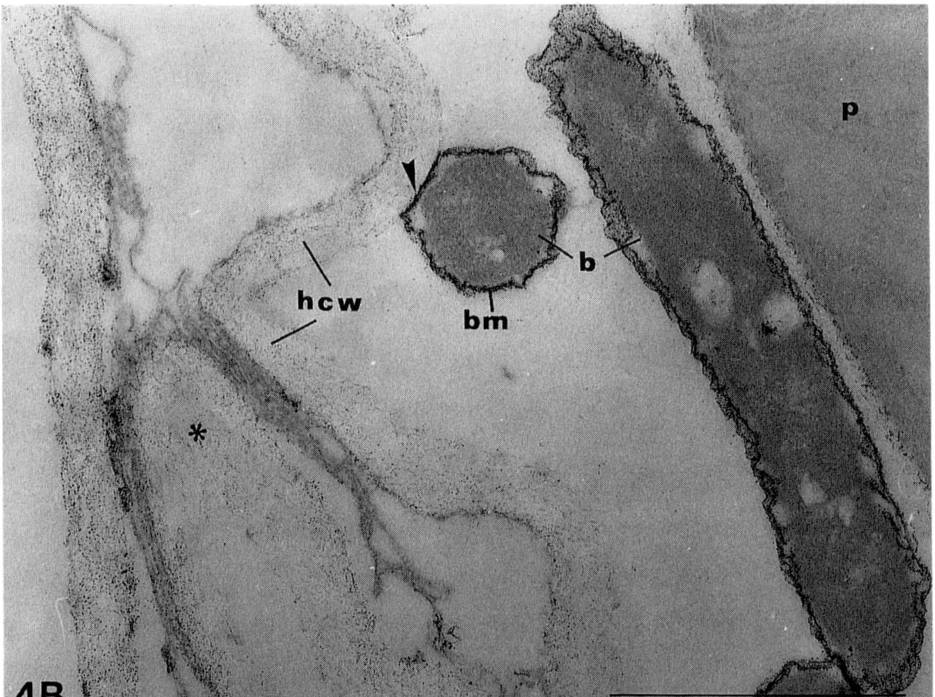
p
b
hcw
bm
*
4B

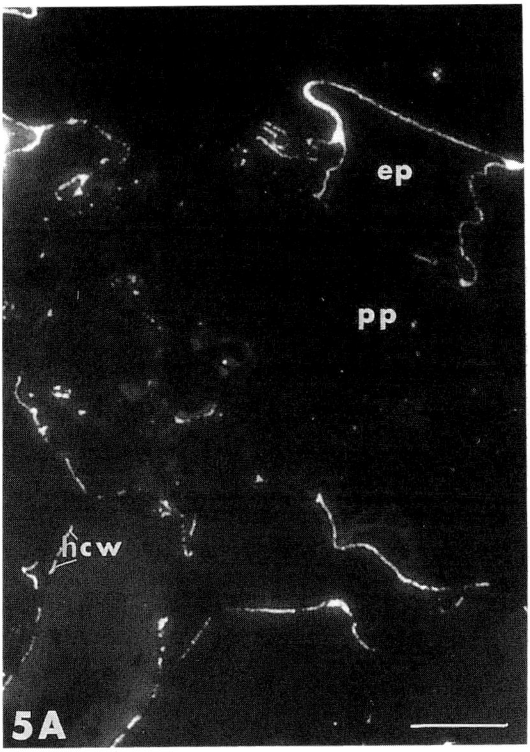

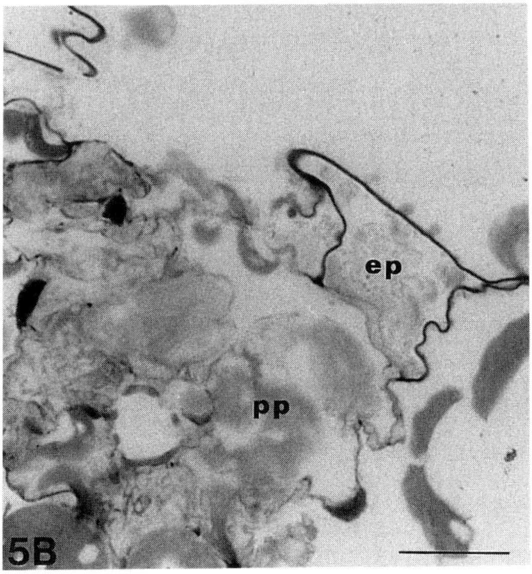

Figure 5. **Disappearance of wall substrate following bacterial attack.**
Host: leaf of *Saintpaulia ionantha*, bacterial pathogen: *Erwinia chrysanthemi*, *strain 3937*. Serial and thick transverse sections in an infected zone of leaf. 5A: monoclonal antibody JIM 5-FITC immunolocalization. 5B: toluidine blue. In the macerated area, cellular contours are hardly recognizable, just an epidermal cell (ep) is still identifiable. In the center of infected zone, pectic polymers are almost completely degraded, except some narrow points that correspond to resistant junctions. Peripheral cells are partially degraded; hcw: host cell wall; pp: palisade parenchyma. Bars represent 50 μm.

The model *Erwinia chrysanthemi/Saintpaulia ionantha* provides another interesting example of cell surface interaction. Pectinolytic Erwinia species cause various diseases including soft rot symptoms in a wide variety of plants and isolated organs, and a great amount of literature has been devoted to this pathogenic bacteria (23, 44, 45; see 2 for a recent and complete review). The high degree of virulence of *E. chrysanthemi* holds from the capacity of bacteria to secrete a set of depolymerizing enzymes, the repertoire of which becomes quite complex (2). A puzzling question concerns the occurrence of multiple isoforms of pectate-lyases (2, 23).

In the present model, the destruction of the host tissues is rapid in the vicinity of the point of inoculation, though it is different according to tissues and cells (44, 45). As described above for the fungal infection, ultrastructural cytochemistry allows to locally detect the degree of cell lysis. In particular, the PATAg test reveals that the cell wall deconstruction occurs differentially (fig. 4A). Within a single area one can observe the different degrees of wall disorganization: simple swelling, complete loss of reactivity, occurrence of a patchwork of reactive domains adjacent to unreactive ones. In other macerated regions, the host cell walls are strongly depolymerized and appear as poorly reactive cell wall ghosts that keep a certain coherence (cellulose can be still detected, results not shown) and against which bacteria are pressed and sticked (fig. 4B).

The mechanisms of interaction can be further investigated using a battery of potential tools: probes that target the different wall polymers, cellulose and/or matrix, and their progressive disappearance, probes that target the wall active enzymes and especially those of microbial origin. Here again, the model Erwinia/Saintpaulia provides a good system to illustrate that particular aspect . Figure 5 shows the disappearance of host pectic polymers

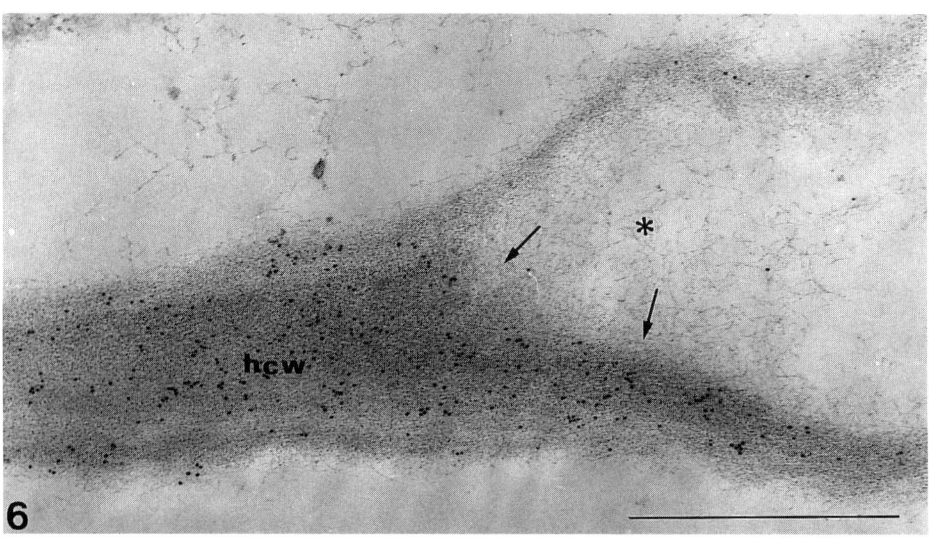

Figure 6. **Local wall substrate disparition following bacterial attack.**
Host: leaf of *Saintpaulia ionantha*; bacterial pathogen: *Erwinia chrysanthemi, strain 3937*. Transverse section in infected leaf. Monoclonal antibody JIM 7 - colloidal gold immunolabelling, PATAg staining. In the attacked wall areas *, there is a correspondence between the loss of PATAg reactivity and the disappearance of JIM7 labelling. Bar represents 1 µm.

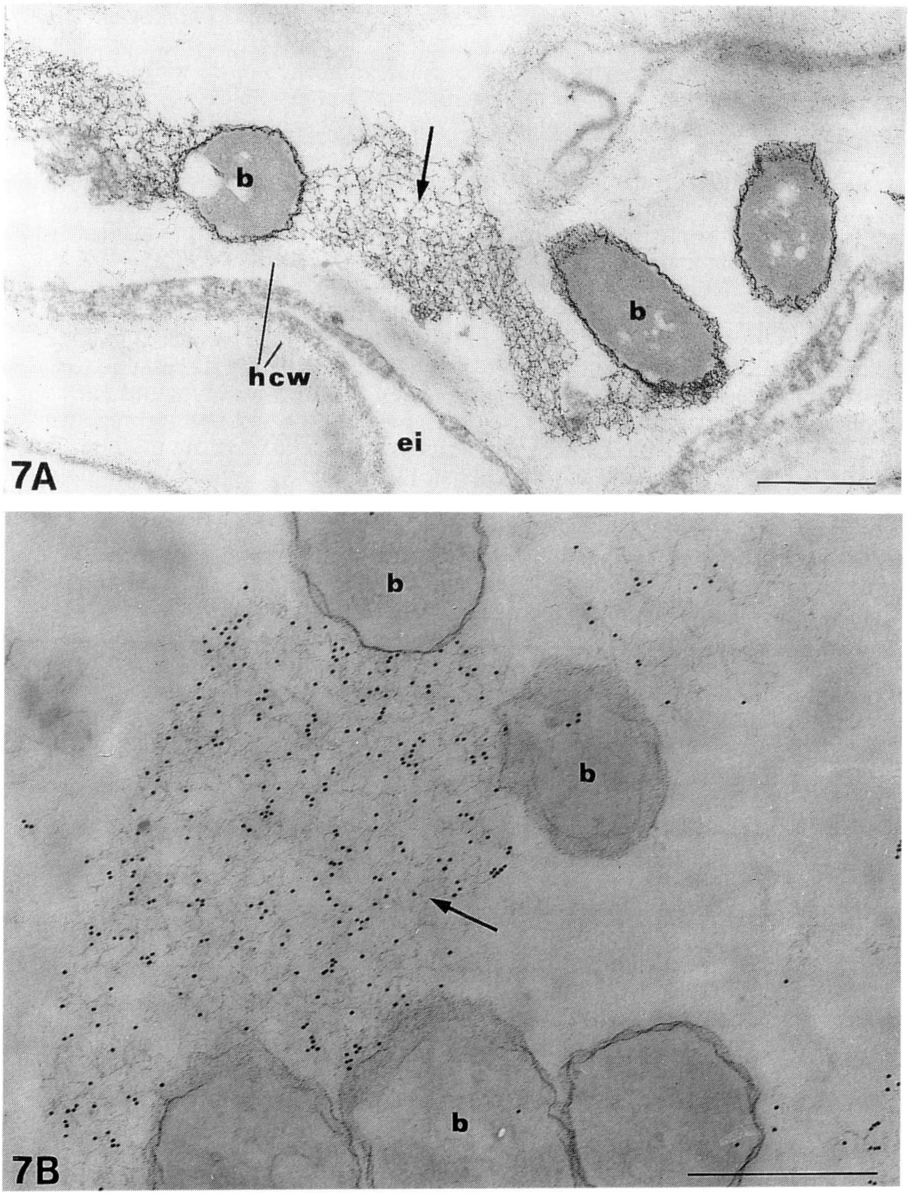

Figure 7. **Occurrence of pectic-rich network around invading bacteria.**
Host: leaf of *Saintpaulia ionantha*; bacterial pathogen: *Erwinia chrysanthemi, strain 3937*. Transverse section in infected leaf. 7A: PATAg staining. 7B: monoclonal antibody JIM 5-colloidal gold immunolabelling, PATAg staining. The host cell wall (hcw) is weakly reactive. Around the bacteria, a polysaccharidic network (arrow) strongly reactive with JIM 5 is observed; ei, intercellular space. Bars represent 0.5 μm.

in the vicinity of the inoculation point by means of immunofluorescence. The probe is JIM 5, a monoclonal antibody that recognizes unesterified sequences of homogalacturonans (21, 48). In the macerated area, one can precisely locate the regions where pectates have been depolymerized by observing the loss of fluorescence (compare to fig. 2B). Again, the deconstruction appears cell-specific, with a more complete disorganization in the spongy parenchyma than in the epidermis. The wall deconstruction is also domain-specific as shown by a combination of immunolabelling and PATAg test for polysaccharides (fig.6). The immunolabelling probe is JIM 7, a monoclonal antibody that recognizes esterified homogalacturonans (21, 48). It shows a correspondence between strongly PATAg-positive and JIM7-labelled areas as well as between swollen, poorly PATAg- and JIM7-reactive areas from which linear polysaccharides escape.
Several informations are provided on figure 7. It shows the occurrence of a network that develops around the invading bacteria after a relatively prolonged state (18h) of infection. Following the PATAg test, the constrast of this network is high compared to that of the neighbouring deconstructed cell walls of the host that are hardly discernible (fig. 7A). Affinity techniques reveal that this network is highly JIM 5-positive, thus enriched in unesterified pectins (fig. 7B) whereas it is not labelled when probes targeting other wall polymers, cellulose, esterified homogalacturonans, hemicelluloses, are used (results not shown). Therefore, from these data, it becomes possible to propose hypotheses on the origin of this network material. The possibility that it results from the rearrangement of a part of the host wall matrix cannot be ruled out.
Other examples showing the relationship between the heterogeneity in composition and organization of cell walls of hosts and that of the diverse arms involved by pathogen are available. Moreover, cell wall degrading enzymes provide inducers of a great variety of plant resistance responses which need also to be considered (2, 22, 42, 50, 55). It is obvious that the local and minute changes occurring at the onset of the infective process can remain undetected upon biochemical investigations. Certainly, the new possibilities offered by structural approaches using modern cytochemistry, affinity cytochemistry and *in situ* hybridization constitute a new and promising research avenue in plant pathology.

4. Towards a remodeling of the cell wall: the mycorrhizal situation

Mycorrhizas are widespread symbiotic associations established between roots of the great majority of land plants and many soil fungi. Compared to the pathogenic situation, the strategies of colonization by mycorrhizal fungi are generally non-aggressive and do not lead to intense degradation of the wall. The success of the mycorrhizal status requires a good compatibility and complementarity at the surface of both partners (7).
The study of mycorrhizas has been the object of many books and reviews (1, 7, 8, 18, 19, 30). We will only focus on some aspects of wall changes with a special attention to the newly interactive interface created during the mycorrhization process.
An ectomycorrhizal model is first briefly presented. It corresponds to the symbiosis established between *Pinus pinaster* and an auxin-overproducer mutant of the fungus *Hebeloma cylindrosporum* that is more invasive than the wild type (15, 16, 17). The mutant forms a hypertrophic Hartig net which takes the appearance of a multilayered pseudoparenchymatous sheet of fungal tissue separating cortical cells from each other (fig. 8).
The highly successful settlement of the mutant fungus implies splitting of the cementing middle lamella of the host cell walls and controlled alterations of both cell surfaces, i.e. modifications which may be necessary to facilitate the nutritional transfer between the partners. These alterations strongly depend on the state of mycorrhization. In particular,

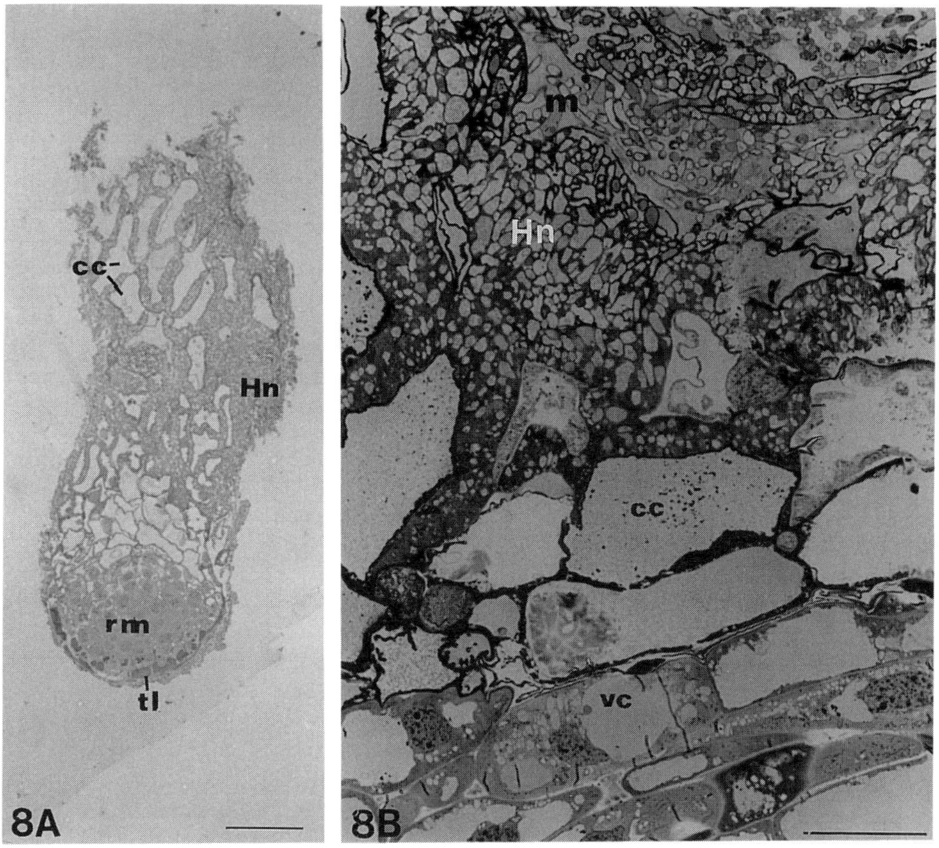

Figure 8. **Overdevelopment of Hartig net in an ectomycorrhiza.**
Host: root of *Pinus pinaster*; symbiotic fungus: *Hebeloma cylindrosporum* mutant type 331 that is auxin overproducer. Thick longitudinal section. Toluidine blue. 8A: tangential section. General view showing the important development of the Hartig net (Hn) at a certain distance of root apex (rm). Bar represents 100 µm. 8B: internal section. Detail of the peripheral zone. Multiseriate Hartig net (HN) developed between cortical cells (cc); m: mantle, tl: tanin layer, vc: vascular cylinder. Bar represents 50 µm.

the respective walls of both partners have well defined limits at the early steps while they become indiscernible at the later steps. This corresponds to modifications in composition and distribution of wall polymers and associated elements (results not shown, see 15). Changes in distribution of Ca^{2+}, known to be directly involved in the stabilization of the wall structure (20, 24, 52) seem particularly significant and will be illustrated here as an example. Calcium can be mapped by secondary ion mass spectrometry (SIMS) (16, 37, 47). The mapping of calcium on mutant mycorrhizas shows that the ion distribution is highly modified because of the presence of the fungus. The calcium signal becomes associated to the overall area of the Hartig net (fig. 9) whereas it is mainly associated to the apoplastic compartment in the absence of fungus (16). Thus an important calcium delocalization seems to intervene when the fungus develops between the cortical cells. This point is important regarding the role of calcium in mediating cell-cell adhesion/separation and in the message exchange and communication that may take place during the establishment of the symbiotic state (see 16 for a more complete discussion).

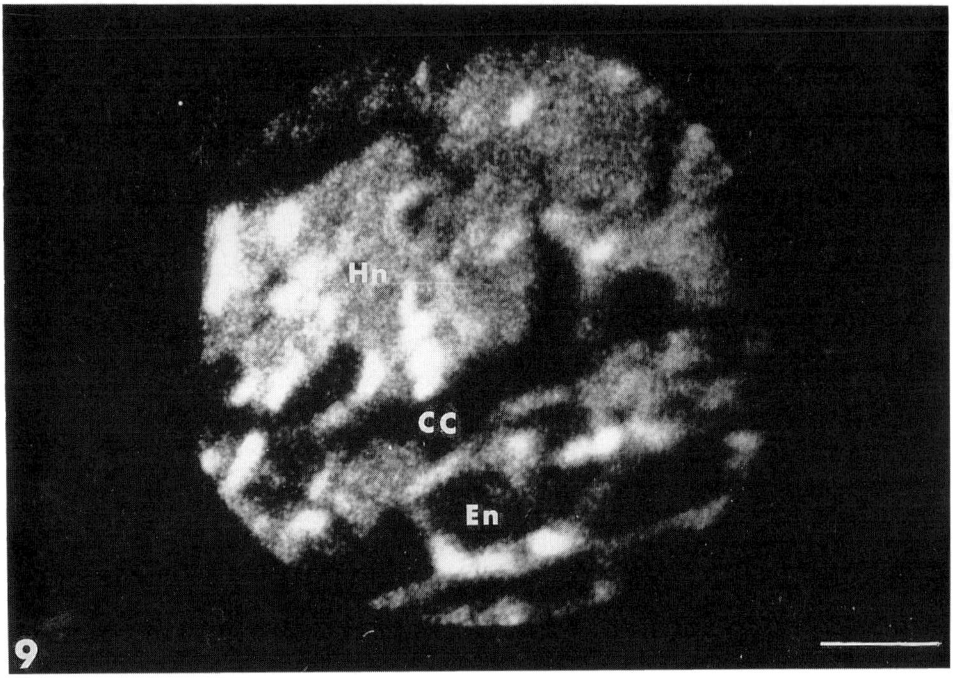

Figure 9. **Calcium distribution in an ectomycorrhiza.**
Host: root of *Pinus pinaster*, symbiotic fungus: *Hebeloma cylindrosporum* auxin overproducer mutant type 331. Secondary Ion Mass Spectrometry (SIMS). Two different regions in terms of calcium distribution are distinguished: above left, the calcium signal is intensively bright and dispersed on the overall area of the Hartig net (Hn), bottom right; the signal is less intense and located on cell walls of cortical cells (cc) and endoderm (En). Bar represents 50 µm.

The occurrence of a new interfacial material which is deposited between the host and fungal walls is another point that has retained much attention in the literature (7, 8). In ectomycorrhizas it is described as a cementing material (the so-called involving layer) that obscures the apoplastic plant /fungus contact, both symbionts loosing their wall integrity. The cytochemical characterization of this interface is still too partial and controversial for definite conclusions in terms of its origin and role (7). For endomycorrhizas, and more specially for the arbuscular mycorrhizas, the situation is much clearer. The cyto-molecular dissection of this new compartment indicates that the created interface is an apoplastic compartment where many newly synthesized polymers are the same as those of the host

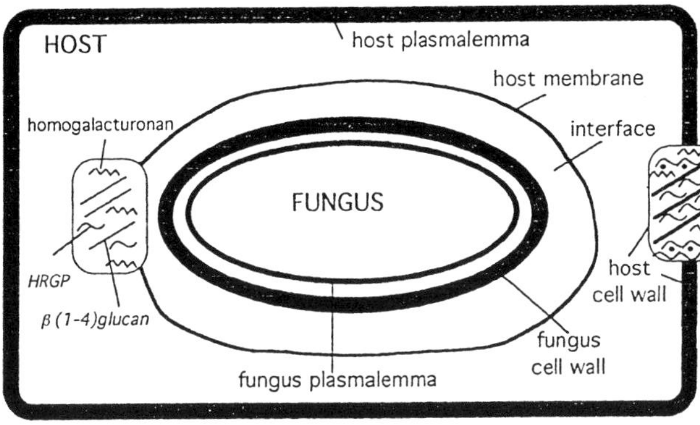

Figure 10. **Wall-like interfacial material deposited between fungus and host.**
The same polymers can be detected in host cell wall and in interface, but the organization is different. Adapted from (11) and (7).

cell wall (fig. 10; see 7, 9, 10, 11). However, the morphological construction of this cell-wall like compartment is different from the host cell wall which means that the presence of the fungus can affect the assembly but not the expression of the molecules of the cell surface.The presence of this interface compartment is important in terms of mutualistic symbiosis since it likely prevents a direct contact between the fungal wall and the host cytoplasm (7).

5. Concluding remarks

In this chapter we have chosen to illustrate different models. This is not without shortcomings since the analysis of each model cannot be advanced. However some points have to be emphasized from the foregoing data: (i) it is clear that as many models exist, as many interactive mechanisms are encountered; (ii) cell surface events appear determinant in all studied examples; (iii) the variety of the chosen models has allowed to illustrate the variety of possible structural approaches (subtractive cytochemistry, affinity cytochemistry, analytical methods for elements..) and we have tried to highlight the type of information provided by each. Indeed, this remains still a partial approach. At the present time the new probes developed in molecular biology appear highly promising tools in terms of molecular cytology. They should contribute to a renewal in plant pathology in the next future.

6. Acknowledgments

The authors are grateful to D. Expert (Laboratoire de Pathologie végétale, INA P-G, Paris) for critical reading of the manuscript. They are also grateful to K. Roberts and J.P. Knox (John Innes Institute, Norwich, UK) for providing the monoclonal antibodies JIM 5 and JIM 7 and to D. Mazau (Université Paul Sabatier, Toulouse, France) for providing infected melons. They want to thank A. Jauneau for his contribution in SIMS analysis.

7. References

1. Allen, M.S. (1992). *An integrative plant-fungal process,* Chapman et al., New York

2. Barras, F., Van Gijsegem, F. and Chatterjee, A.K. (1994) Extracellular enzymes and pathogenesis of soft-rot Erwinia, *Ann. Rev. Phytopathol.* **32**, 201-234.

3. Bateman, D.F. and Basham, H.G. (1976) Degradation of plant cell walls and membranes by microbial enzymes, in R. Heitefuss and P.H. Williams (eds.), *Encyclopedia of Plant physiology*, Vol. 4, Springer-Verlag, New York, pp 316-355.

4. Benhamou, N. (1991) Cell surface interactions between tomato and *Clavibacter michiganse*: localization of some polysaccharides and hydroxyproline-rich glycoproteins in infected host leaf tissues, *Physiol. Mol. Plant Pathol.* **41**, 351-370.

5. Benhamou, N., Lafitte , C., Barthe, J.P., and Esquerré-Tugayé, M.T. (1991) Cell surface interactions between bean leaf cells and *Colletotrichum lindemuthianum.* Cytochemical aspects of pectin breakdown and fungal endopolygalacturonase accumulation, *Plant physiol* **97**, 234-244.

6. Bolwell, G.P. (1993) Dynamic aspects of the plant extracellular matrix, *Int. Rev. Cytol.* **146**, 261-325.

7. Bonfante, P. (1994) Alteration of host cell surfaces by mycorrizal fungi, in O. Petrini and B. Ouellette. (eds), *Host Wall Alterations by Parasitic Fungi*, APS Press, St Paul Minnesota. pp 103-114.

8. Bonfante-Fasolo, P and Scannerini, S. (1992) The cellular basis of plant-fungus interchanges in mycorrhizal associations, in M.S. Allen (ed.), *An integrative plant-fungal process*, Chapman et al., New York, pp.67-101.

9. Bonfante-Fasolo, P. and Vian, B. (1989) Cell wall architecture in mycorrhizal poots of *Allium porrum* L. *Ann Sci. Nat.Paris*, **10**, 97-109.

10. Bonfante, P., Vian, B., Perotto, S., Faccio, A. and Knox, J.P. (1990) Cellulose and pectin localization in roots of mycorrhizal *Allium porrum*: labelling continuity between host cell wall and interfacial material, *Planta* **180**, 537-547.

11. Bonfante-Fasolo, P., Tamagnone, L., Esquerré-Tugayé, M.T., Mazau,D., Mosiniak, M. and Vian, B. (1991) Immunocytochemical location of hydroxyproline-rich glycoproteins at the interface between a mycorrhizal fungus and its host plants, *Protoplasma* **165**, 127-138.

12. Carpita, N.C. and Gibeaut, D.M. (1993) Structural models of primary cell walls in flowering plants: consistency of molecular structure with the physical properties of the walls during growth, *Plant J.* **3**, 1-30.

13. Chamberland, H. (1994) Gold labeling methods for the ultrastructural localization of host wall and pathogen components, in O. Petrini and B. Ouellette (eds), *Host Wall Alterations by Parasitic Fungi*, APS Press, St Paul Minnesota. pp 1-12.

14. Emons, A.M.C. and Van Maaren N. (1987) Helicoidal cell-wall texture in root hairs, *Planta* **170**, 145-151.

15. Gay, G., Sotta B., Tranvan H., Gea, L. and Vian, B. (1995) Fungal auxin is involved in ectomycorrhiza formation: genetical, biochemical and ultrastructural studies with IAA overproducer mutants of *Hebeloma cylindrosporum*, in Eurosilva Contribution to Forest Tree Physiology, in press.

16. Gea, L., Jauneau, A. and Vian, B. (1994) Preliminary SIMS imaging of calcium distribution in ectomycorrhizas of *Pinus pinaster* and *Hebeloma cylindrosporum*, *J. Trace and Microprobe Techniques* **12**, 323-329.

17. Gea, L., Normand, L., Vian, B. and Gay, G. (1994) Structural aspects of ectomycorrhiza of *Pinus pinaster* (Ait) Sol. formed by an IAA-overproducer mutant of *Hebeloma cylindrosporum* Romagnesi, *New Phytol.* **128**, 659-670.

18. Gianinazzi-Pearson, V. and Gianinazzi, S. (1989) Cellular and genetical aspects of interactions betweenhosts and fungal symbionts in mycorrhizae, *Genome* **31**, 336.

19. Harley, J.L. and Smith, S.E. (1983) *Mycorrhizal symbiosis*, Academic Press, London.

20. Jarvis, M.C. (1984) Structure and properties of pectin gels in plant cell walls, *Plant Cell Environm.* **7**, 153-164.

21. Knox, J.P., Linstead P.J., King, J., Cooper C. and Roberts K. 1990. Pectin esterification is spatially regulated both within cell walls and between developing tissues of root apices, *Planta* **181**, 512-521.

22. Kolattukudy, P.E., Kämper, J., Kämper, U., Gonzalez-Candelas, L. and Guo, W. (1994) Fungus-induced degradation and reinforcement of defensive barriers of plants, in O. Petrini and B. Ouellette (eds), *Host Wall Alterations by Parasitic Fungi*, APS Press, St Paul Minnesota. pp. 67-80.

23. Kotoujansky, A. (1987) Molecular genetics of pathogenesis by soft rot Erwinias, *Ann. Rev. Phytopathol.* **25**, 405-430.

24. Liners, F. and Van Cutsem, P. (1991) Immunocytochemical localization of homopolygalacturonic acid on plant cell walls, *Micron Microsc. Acta* **22**, 265-266.

25. McCann, M.C. and Roberts, K. (1991) Architecture of the primary cell wall, in C.W. Lloyd (ed.), *The cytoskeletal basis of plant growth and form*, Acad. Press, London. pp. 109-129.

26. McCann, M.C., Stacey, N.J., Wilson, R. and Roberts, K. (1993). Orientation of macromolecules in the walls of elongating carrot cells, *J. Cell Sci.* **106** 1347-1356.

27. Neville, A.C. (1993) *Biology of fibrous composites. Development beyond the cell membrane*, Cambridge University Press.

28. Nicole,M., Ruel, K. and Ouellette G.B. (1994) Fine morphology of fungal structures involved in host wall alteration, in O. Petrini and B. Ouellette (eds), *Host Wall Alterations by Parasitic Fungi*, APS Press, St Paul Minnesota. pp. 13-30.

29. O'Connell, R.J., Bailey, J.A. and Richmond, D.V. (1985) Cytology and physiology of infection of *Phaseolus vulgaris* by *Colletotrichum lindemuthianum*, *Physiol. Plant Pathol.* **27**, 75-98.

30. Ried, D.G., Lewis, D.H., Fitter, A.H. and Alexander, I.G. (1992) *Mycorrhizas in ecosystems*, C.A.B. Int., Wallingford, U.K.

31. Rioux, D. and Biggs A.R. (1994) Cell wall changes in host and nonhost systems: microscopic aspects, in O. Petrini and B. Ouellette (eds), *Host Wall Alterations by Parasitic Fungi*, APS Press, St Paul Minnesota. pp. 31-44.

32. Roberts, K. (1994) The plant extracellular matrix: in a new expansive mood, *Curr. Opinions Cell Biol.* **6**, 688-694.

33. Roland, J.C. and Vian, B. (1991) General preparation and staining of thin sections, in J.H.Hall and C. Hawes (eds), *Electron Microscopy of Plant Cells*, Academic Press, London. pp. 1-66.

34. Roland, J.C., Reis, D. and Vian, B. (1992) Liquid crystal order and turbulence in the planar twist of the growing plant cell walls, *Tissue and Cell* **24**, 335-345.

35. Roland, J.C., Reis, D., Vian, B. and Roy, S. (1989) The helicoidal plant cell wall as a performing cellulose-based composite, *Biology of the Cell* **67**, 209-220.

36. Roland, J.C., Reis, D.,Vian, B., Satiat-Jeunemaitre, B. and Mosiniak, M. (1987) Morphogenesis of cell walls at the supramolecular level: internal geometry and versatility of helicoidal expression, *Protoplasma* **140**, 75-91.

37. Roy, S. Jauneau, A. and Vian, B. (1994) Analytical detection of calcium ions and immunocytochemical visualization of homogalacturonic sequences in the ripe cherry tomato, *Plant Physiol. Biochem.* **32**, 633-640

38. Satiat-Jeunemaitre, B., Martin, B. and Hawes, C. (1992) Plant cell wall architecture is revealed by rapid-freezing and deep-etching, *Protoplasma* **167**, 33-42.

39. Satiat-Jeunemaitre, B. (1992) Spatial and temporal regulations in helicoidal extracellular matrices: comparison between plant and animal systems, *Tissue and Cell* **24**, 315-334.

40. Stone, B.A., Liyama, K. and Lam, T.B.T (1994) Covalent cross-links in the cell wall, *Plant Physiol.* **104** 315-320.

41. Stone, J.K., Viret, O., Petrini, O. and Chapela, I.H. (1994) Histological studies of host penetration and colonization by endophytic fungi, in O. Petrini and B. Ouellette (eds), *Host Wall Alterations by Parasitic Fungi*, APS Press, St Paul Minnesota. pp. 115-128.

42. Sutic, D.D. and Sinclair J.B. (1991) *Anatomy and physiology of diseased plants*, CRC press, Boca Raton.pp 1-79.

43. Talbott L.D. and Ray P.M. (1992) Molecular size and separability features of pea cell wall polysaccharides, *Plant Physiol* **98**, 357-368.

44. Temsah-Makkouk, M. (1992) Approche ultrastructurale et immunocytochimique du pouvoir pathogène d'*Erwinia chrysanthemi* : un nouvel outil pour la connaissance de la paroi de l'hôte. Thèse Université P. et M. Curie, Paris.

45. Temsah, M., Bertheau, Y. and Vian, B. (1991) Pectate-lyase fixation and pectate disorganization visualized by immunocytochemistry in *Saintpaulia ionantha* infected by *Erwinia chrysanthemi*, *Cell Biol.Intern. Reports* **15**, 611-620.

46. Tepper, C.S. and Anderson, A.J. (1990) Interaction between pectic fragments and extracellular components from the fungal pathogen *Colletotrichum lindemuthianum*, *Phys. Mol. Plant Pathol.* **36** 147-158.

47. Thellier, M., Ripoll, C. and Berry, J.P. (1991) Biological applications of secondary ion mass spectrometry, *Eur. Micros. Anal.* **11**, 9-11.

48. Vandenbosch K.A., Bradley, D.J., Knox, J.P., Perotto, S., Butcher, G.W. and Brewin, N. (1989) Common components of the infection thread matrix and the intercellular space identified by immunocytochemical analysis of pea nodules and uninfected roots, *EMBO J.* **8**, 335-342.

49. Vandenbosch, K.A. 1991 Immunogold labelling, in J.L. Hall and C. Hawes (eds.), *Electron microscopy of plant cells*, Academic Press, London, pp.181-219.

50. Verma, D.P.S. (1992) *Molecular signals in plant-microbe communications*, CRC Press, Boca Raton.

51. Van Buren J.P. (1991) Function of pectin in plant tissue structure and firmness, in R.H. Walter (ed.), *The chemistry and technology of pectin*, Academic Press, London, pp.1-22

52. Vian, B. and Roland, J.C. (1991) Affinodetection of the sites of formation and the further distribution of polygalacturonans and native cellulose in growing plant cells, *Biol. Cell* **71**, 43-55.

53. Vian, B., Roland, J.C. and Reis, D. (1993) Primary cell wall texture and its relation to surface expansion, *Int. J. Plant Sci.* **154**, 1-9.

54. Vian, B., Temsah, M., Reis, D. and Roland, J.C. (1992) Colocalization of the cellulose framework and cell-wall matrix in helicoidal constructions, *J. Microsc.* **166**, 111-122.

55. Walton, J.D. (1994) Deconstructing the cell wall, *Plant Physiol.* **104** 113-1118.

56. Wijesundera, R.L.C., Bailey, J.A., Byrde, R.J.W. and Fielding, A.H. (1989) Cell wall degrading enzymes of *Colletotrichum lindemuthianum*: their role in the development of bean anthracnose, *Physiol. Mol. Plant Pathol.* **34**, 403-413.

ADHESION OF FUNGAL PROPAGULES

Significance to the Success of the Fungal Infection Process

RALPH L. NICHOLSON
Department of Botany and Plant Pathology
Purdue University
West Lafayette, IN 47907, USA

1. Introduction - A Review of Recent Literature

Adhesion is a well established phenomenon that occurs during the normal growth of organisms within their environments. This review will present recent literature which is pertinent to our understanding how adhesion occurs across a broad spectrum of organisms, to our current knowledge of the chemistry of adhesives, and in particular to the significance of adhesion to the fungal infection process. The pathogens *Colletotrichum graminicola* and *Erysiphe graminis* will provide examples of adhesion and its relevance to the infection process as they represent both necrotrophic and biotrophic pathogens with very different requirements for their survival and success as pathogens.

The events that constitute adhesion or attachment can include cell to cell interactions within and between members of the same species, for example during mating [42, 78], the development of successful mycorrhizal relationships [5], and even the successful interactions of lichen symbionts [1]. In addition to organisms within the terrestrial environment, considerable attention has been given to the adhesion of organisms, in particular the algae, within marine environments where attachment to a substratum may be essential for survival [28]. Similarly, the adhesion of aquatic fungi has been investigated extensively [38, 70, 71, 72]. The attachment of phytopathogenic bacteria to plant surfaces has also received increasing attention on the belief that infection is preceded by bacterial recognition of specific host surface structures and is followed by adhesion to the leaf surface, phenomena presumed to be mediated by bacterial fimbriae [73]. Recent investigations suggest that fimbriae in the pathogen *Xanthomonas campestris* pv. *hyacinthi* function as virulence factors and allow the bacterium to associate with, and attach specifically to, stomata of the host [83]. Another important aspect of adhesion is the influence of nutrients on the adhesion of organisms that exist as normal members of fungal populations found in the phylloplane [2].

Recent reviews on fungal adhesion include the specificity of attachment to a host plant [52], adhesion as a prerequisite to pathogenesis [59], the importance of adhesion to pathogenicity of nematophagous fungi [81], and the general concept and

M. Nicole and V. Gianinazzi-Pearson (eds.), Histology, Ultrastructure and Molecular Cytology of Plant-Microorganism Interactions, 117-134.

importance of adhesion to fungal growth, development, and pathogenicity [6, 23, 39]. Jones [39] gives particular attention to the adhesion of aquatic fungi and suggests that adhesion should be considered to occur as either a function of passive or active phases of attachment. In the case of passive attachment or adhesion, many conidia of aquatic marine fungi have appendages or filaments that allow for the entrapment of the propagule with a substratum simply because of its three dimensional configuration. Another form of adhesion that can also be considered as passive is that in which the fungal propagule has associated with it, either with its appendages or with the body of the propagule itself, a material that is "sticky" and acts as an adhesive [38]. The resultant adhesion typically occurs in a non-specific manner to any available substratum. Thus, this form of passive adhesion avoids the involvement of a specific interaction between the fungal propagule and the substratum, or, in the case of pathogens, a host . For plant pathogenic fungi, an often cited example of such passive adhesion is that which occurs with *Magnaporthe grisea* , the causal agent of rice blast, as a result of the release of an adhesive material from the tip of the conidium upon hydration [31]. It should be kept in mind, however, that there are numerous plant pathogenic fungi that exhibit similar mechanisms of passive, non-specific adhesion [59]. A particularly interesting example of passive adhesion is that which occurs with *Basidiobolus ranarum*, an organism often found as a human pathogen in the tropics. In this fungus, a droplet of adhesive material is formed at the tips of conidia known as capilliconidia. This adhesive aids in the dispersal of the organism by a variety of insects [21].

1.1 MECHANISMS OF ADHESION

With few exceptions, notably conidia of the powdery mildew pathogens, most fungi begin the process of adhesion within an aquatic environment. Such an environment could consist of only a film of water on a leaf surface or it could be a truly aquatic environment of either fresh or salt water. This relationship of adhesion within an essentially aquatic medium reflects an important consideration of adhesives in general, namely that adhesives must be water-insoluble! Yet because adhesives must be synthesized within cells, they must also be water-soluble at least until the time of their secretion. Once secreted, one would assume that the chemical and physical properties of the material change to provide it with the properties of an adhesive, that is, it must be sticky and displace water [22, 84]. For example, Dykstra and Bradley-Kerr [21] emphasized that once formed, the adhesive droplet of *B. ranarum* could withstand a variety of harsh treatments and was also undisturbed by exposure to water. Another consideration for adhesion involves the manner in which fungi grow, that is, their common pattern of apical tip growth [3]. Such growth requires that a fungus maintains close contact with a substratum, regardless of whether the substratum is a plant host or an inert material upon which the organism is growing and obtaining nutrients. It is logical that intimate contact requires the adhesion of as much of the fungal hypha as possible to the substratum.

1.1.1 *The fungal sheath and extracellular mucilages*

A common observation is that hyphae, mycelia, and appressoria are typically surrounded by what appears to be an extracellular sheath of material. Numerous examples of such sheaths are available in the literature including observations made in the late 1800s [85]. Visualization of the fungal sheath often requires scanning electron microscopy [39, 59, 72, 74] although significant observations have been made recently using histochemical techniques with both transmission electron microscopy (TEM) and light microscopy [7, 27, 66]. Such sheath material is often referred to as mucilage, is generally assumed to consist of glycoproteins, and usually is considered to have a role in adhesion. Whether sheath materials actually act as adhesives has yet to have been demonstrated, although it seems likely that sheaths are at least involved in adhesion given the fact that they appear to cement the developing germ tube to the substratum and in some cases have been shown to attach the entire germling to the substratum [6, 7, 59].

It is possible that, in addition to adhesion, another function of the hyphal sheath is to protect the developing germ tube. That a sheath or mucilage might function as a protectant is consistent with the extracellular conidial mucilage of *C. graminicola* (described below) for which numerous protective roles have been described [50, 61, 62, 63, 68]. The importance of such sheaths or mucilages has also been pointed out by Hoagland et al. [32] to be critical to the survival of the microalgae known as diatoms. It is interesting to note that one of the proposed roles of the diatom mucilage secretions is to function as antidesiccants, just as the conidial mucilage of *C. graminicola* acts as an antidesiccant [61].

The terms extracellular mucilage or extracellular matrix are also used to refer to materials that are produced in association with the conidia of many fungal species [51, 59, 61]. Unlike sheaths, these extracellular mucilages are generally water-soluble and represent complex mixtures of glycoproteins, enzymes, and low molecular-weight materials. Because of their solubility in water, the extracellular conidial mucilages probably do not function in adhesion. That the conidial mucilage of *C. graminicola* does not have a role in adhesion supports this assumption [55].

Although the conidial mucilage of only one fungus, *C. graminicola*, has been studied in significant detail, it seems reasonable to assume that water-soluble, extracellular mucilages consist of a variety of components each of which has a specific function to ensure the survival of the organism [45, 57, 58]. Significant observations by TEM have been made which show details of the extracellular mucilages associated with conidia of *C. graminicola* and *C. truncatum* [56, 82] and *Discula umbrinella* [79]. These observations reaffirm evidence indicating that the extracellular mucilages are composed primarily of polysaccharide-rich materials.

In contrast to aquatic fungi and terrestrial species whose conidia are produced in acervuli, pycnidia, or similar structures, the asexual conidia of many terrestrial species exist are not surrounded by an obvious sheath or layer of mucilage. Upon germination of these conidia, it is often apparent that a material is released from the conidium which eventually surrounds the developing germ tube. Upon its emergence from the conidium the germ tube is anchored to the underlying substratum by this sheath. Figure 1, observed by interference contrast microscopy, shows that a sheath of material is present from the time the germ tube emerges from a conidium of the maize pathogen *Cochliobolus carbonum* . Similar sheath materials have been demonstrated for the germlings of *Cochliobolus heterostrophus* (*Bipolaris maydis*) as well as numerous other species of fungal pathogens [7, 27, 59]. Of particular importance is the fact that sheath materials are commonly observed to be associated with appressoria, suggesting that the sheath is involved in adhesion [7, 59, 72].

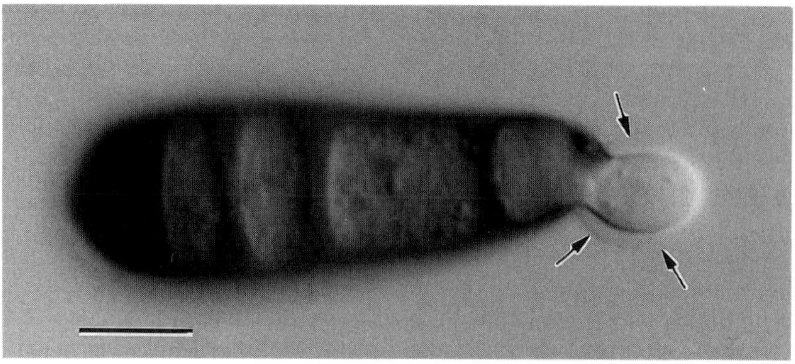

Figure 1. Germ tube emerging from a *Cochliobolus carbonum* conidium. Arrows show that a sheath surrounds the germ tube. Bar equals 20 μm.

It seems clear that the extracellular sheath that is associated with a fungal hypha, germ tube, or appressorium is involved in adhesion. It is also clear that the water-soluble, extracellular mucilages or matrices that are often associated with conidia are not involved in adhesion but may have significant roles in ensuring the survival of the organism [23, 57, 58, 60]. A similar situation exists for the green alga *Closterium, which* also produces an extracellular polysaccharide-rich mucilage. It is unclear whether this mucilage is involved in adhesion, but the forcible extrusion of the mucilage propels the organism through its environment, a phenomenon that enhances the chance that a suitable site for attachment will be encountered [18].

1.1.2 *Adhesion prior to germination*

Adhesion is not associated only with germination or appressorium formation. For example, upon hydration and before germination of *M. grisea* conidia, an adhesive is released from a periplasmic compartment at the tip of the conidium [31]. The adhesive is non-specific in that it adheres to a variety of substrata. Adhesion has also been shown to occur prior to germination for a number of other fungi. Important examples include the adhesion of ungerminated conidia of *C. graminicola* (discussed below) and other *Colletotrichum* species [75]. Of particular importance is the adhesion of ungerminated urediospores of the rust fungus *Uromyces viciae-fabae*. Upon hydration, a material involved in adhesion is released from the wall of the ungerminated urediospore [15]. The material accumulates between urediospores that are in contact with each other and also between the urediospore and the underlying substratum, whether it is a leaf or other substratum. The material forms what the authors referred to as an adhesion pad which sticks to the substratum, a process that resembles a passive mechanism of adhesion. An important observation was that even killed urediospores released materials that formed an adhesion pad, but these pads failed to adhere to the substratum. The authors proposed that the adhesion of the pad, and therefore of the urediospores to a leaf surface, depended on the presence of hydrolytic enzymes released from the urediospore wall along with the materials that comprise the pad. They demonstrated that a cutinase and two non-specific esterases are released from the urediospore wall and presented strong evidence suggesting that these enzymes are required for the adhesion of the spore to the bean leaf. This form of spore adhesion occurs well in advance of germination and therefore in advance of the appearance of sheath materials that can be assumed to anchor the germinated spore to the host surface. Clearly, adhesion prior to germination serves to maintain the propagule on the host surface thereby preventing its displacement prior to germination. A proper evaluation of the importance of adhesion which occurs prior to germination will depend on attempts to determine the strength of such adhesion; to date, few such investigations have been attempted [19, 31, 55, 75].

Adhesion prior to germination also occurs in conidia of nematophagous fungi. For example, the nematophagous fungus *Drechmeria coniospora* presents an excellent example of the complexity of adhesion that occurs prior to germination [16]. An adhesive knob at the distal end of a conidium of this fungus provides a point of attachment for the conidium to the nematode host. Interestingly, adhesion occurs preferentially to the sensory organs of the nematode, the head and posterior region. After contact and adhesion, the spore forms an appressorium which arises from the adhesive knob. Both the adhesive knob and the appressorium are covered with a layer of adhesive material.

Numerous other instances of adhesion prior to propagule germination have been reported and elaborated upon in previous reviews [23, 59]. An example of special interest is that of Schuerger and colleagues [74] who pointed out the importance of agglutination of *Fusarium solani* macroconidia as a phenomenon that occurs prior to germination and ensures the success of the fungal infection process. These authors demonstrated that ungerminated conidia attach to host roots prior to any visual evidence of the presence of a spore mucilage, a phenomenon that suggests that the adhesive is not immediately associated with a mucilage. The ungerminated conidia of this fungus agglutinate in the presence of extracts of host roots. Importantly, agglutination leads to the attachment of conidia to roots and precedes germination and the appearance of a mucilage. Thus, agglutination allows macroconidia to make contact with the host after which germination and the development of the infection process can continue. Similar investigations on the adhesion of macroconidia of *Nectria haematococca* have been outlined in detail previously [23]. Briefly, like *F. solani*, *Nectria* is induced by host nutrients to become adhesion competent, a condition that results in the ability of macroconidia to adhere at their apices to tissue of zucchini fruit prior to their germination [11, 24, 41, 49].

1.1.3. *Adhesion that occurs coincident with germination and adhesion after penetration*

As indicated above, a sheath of material typically found associated with fungal hyphae is generally presumed to be involved in adhesion. Doss et al. [19] demonstrated that the sheath associated with *Botrytis cinerea* germlings remains attached to the substratum even after the removal of the germlings. These authors also demonstrated by atomic force microscopy that the sheath is up to 60 μm thick in the area immediately adjacent to the germ tube. It seems reasonable to assume that hyphal sheaths of other fungi would have similar functions.

Excellent photographic examples of sheath materials can be found in articles by Beckett et al. [4] and Clement et al. [13] for germlings of *U. viciae-fabae*. Similar sheath materials were shown by Chaubal et al. [12] for urediospore germlings of the rust fungus *Puccinia sorghi*. Although there are numerous photographic records of sheath material associated with hyphae, significantly fewer visual records demonstrate the extent to which sheath material can envelop appressoria. An exception to this are the scanning electron micrographs presented by Braun and Howard [6, 7] for sheath materials that surround appressoria of *C. heterostrophus* (Fig. 2). Epstein et al. [24] also demonstrated that a sheath of material covers the convoluted hyphopodia (appressoria) produced by *Gaeumanomyces graminis*.

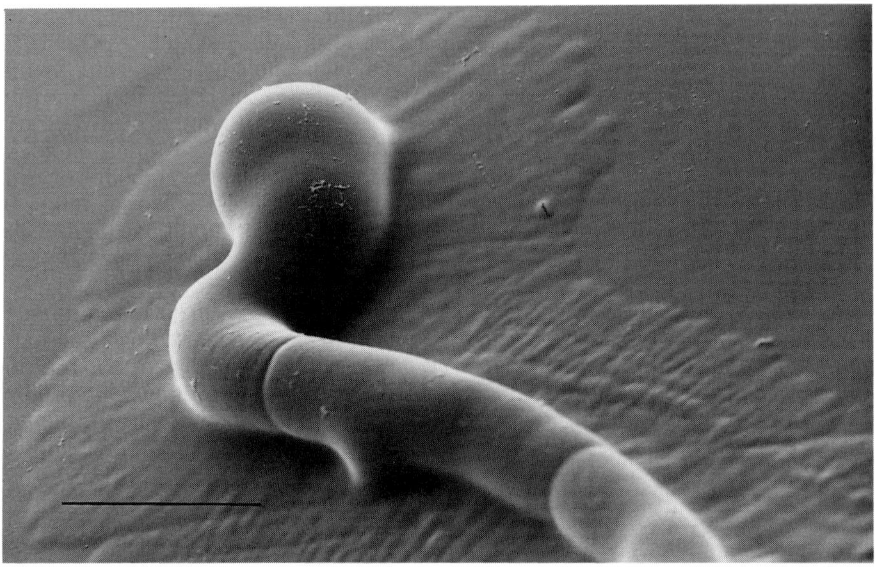

Fig. 2. Scanning electron micrograph of a sheath that surrounds the germ tube and appressorium of the pathogen *Cochliobolus heterostrophus*. Bar represents 10 μm. Photograph by R. Howard [6] (with permission from Academic Press).

Adhesion of the fungus within a host plant has received little attention. A notable exception is the work of Hohl and colleagues who have investigated the adhesion of *Phytophthora megasperma* within its soybean host. Important investigations by this group were initially centered around understanding the surfaces of the fungus and the host soybean cell in order to evaluate the potential for surface recognition and or the presence of receptors that may be involved in the adhesion of the pathogen to the host. It was demonstrated that lectin-hapten interactions accounted for adhesion of the fungus to the host [35, 36, 37]. A subsequent investigation demonstrated conclusively that the lectin is present on the surface of the fungus, and that mannose pretreatment of fungal hyphae inhibited adhesion, suggesting that the lectin on the surface of the fungus had affinity properties similar to that observed for Concanavalin A, [29]. Significant observations were that the adhesive material associated with the fungus is not preformed and that the lining of the intercellular spaces of the host tissue is rich in glucose, mannose and N-acetyl-glucose residues, a factor consistent with the binding pattern of the lectin-hapten relationship between the fungus and the plant.

A recent report by Pain et al. [67] dealt with the identification of a monoclonal antibody that is specific to intracellular hyphae of the biotrophic phase of growth of the bean anthracnose pathogen, *C. lindemuthianum*. Although there is no indication that the material recognized by the antibody is involved in adhesion, it is significant that the antigen is present where an adhesive would be expected to occur, in the interfacial matrix around intercellular hyphae. Moreover, its location was restricted to infection vesicles and primary hyphae that were in contact with host protoplasts, and it was never detected in intracellular hyphae of the fungus.

1.1.4. *Adhesion and hydrophobic interactions*
The attachment of *U. viciae-fabae* urediospores has also been investigated by Beckett and colleagues, who presented evidence that the initial attachment of these spores occurs as a result of hydrophobic interactions of the spore with the substratum in addition to the release of extracellular matrix materials from the spore [4, 13]. Recently, this group reported the importance of hydrophobicity of the spore surface to spore aggregation [14]. They demonstrated that extraction of lipids from the spore wall reduced aggregation and promoted the ability of hydrated spores to attach or adhere to hydrophilic surfaces. Importantly, the authors also suggested that the lipid sheath covering the urediospore spines is hydrophobic and ensures that, upon contact of the spore with the leaf cuticle, the hydrophobic interaction should promote adhesion. At first, these reports would seem to contradict the observations by Deising et al. [15] that hydrolases such as cutinase are involved in the initial stages of passive spore adhesion to the leaf surface; however, they may not be contradictory since neither group addressed the strength of adhesion. Moreover, because Deising et al. [15] used spores that had been washed free of the enzymes present in the spore wall it is to be expected that any hydrophobic coating on the surface of urediospore spines would have been removed.

The importance of hydrophobic interactions to the another rust, *U. appendiculatus*, was also demonstrated by Terhune and Hoch [77]. These authors showed that adhesion of both ungerminated spores and of spore germlings was greatest on relatively hydrophobic surfaces. Of particular significance in this study was the demonstration that the area of contact by a germ tube with the substratum was greater if the substratum surface was hydrophobic. These studies, which were carried out by interference reflection microscopy, provide convincing evidence for the importance of hydrophobic interactions to maximize the surface area of adhesion of the fungal germ tube with the substratum.

The involvement of hydophobic and electrostatic interactions have also been suggested to be of considerable importance to propagule adhesion for some fungi. On the basis of propagule size, Jones [39] questioned the importance of electrostatic interactions to fungal adhesion. However, he does suggest that, since microorganisms have a negative surface potential, as do other surfaces, an electrostatic repulsion exists between the propagules and surfaces, and that factors such as hydrogen bonding and hydrophobic interactions may counteract the forces of electrostatic repulsion. Jones and O'Shea [40] took the position that, although specific ligand-receptor mechanisms have been identified to account for the adhesion of *Candida albicans* to human tissues, it is unreasonable to assume that adhesion depends only on ligand-receptor binding. They investigated the role of electrostatic forces in the adhesion of *C. albicans* and found that the yeast form of the organism behaves as a simple charged colloidal system. In spite of the electronegativity of the fungus, adhesion occurred, leading these authors to the conclusion that adhesion may also occur by nonelectrostatic forces which are attractive, for example nonpolar forces, and overcome electrostatic repulsion.

2. Composition of Fungal Adhesives

In comparison to the "simple" phenomenon of adhesion, very little is known about the chemistry of fungal adhesives except that they are generally presumed to include glycoproteins. Although we know too little at the present time to assume that fungal adhesives are always glycoproteins, the evidence gathered to date suggests that this is the case. For example, the materials described by Pain et al [67] and cited above were found to be glycoproteins, and their possible role in adhesion is not inconsistent with their location around the infection hypha, a structure that would demand close association, and therefore adhesion, with the host. Four high-molecular-weight glycoproteins (mannoproteins) have also been shown to be involved in the adhesion of the human pathogen *C. albicans* to plastics [80]. These adhesives were suggested to be components of the fibrillar surface layer of the germ tube. Mannoglycoproteins (65 kDa) have also been identified as the adhesives from *P. megasperma* that allow the fungus to bind to host cells after penetration [17].

The involvement of high-molecular-weight glycoproteins as adhesives released by encysting zoospores of *Phytophthora cinnamomi* [30] is reviewed in detail elsewhere [59]. An almost identical situation was found to exist for zoospores of *Pythium aphanidermatum* . The adhesive secreted upon zoospore encystment was also found to consist of high-molecular-weight glycoproteins [26]. Similar results have been obtained for the adhesive assumed to be released during secondary zoospore encystment of *Saprolegnia ferax* [20].

The putative adhesive material assumed to be associated with the adhesion of macroconidia of *Nectria haematococca* has been studied in detail by Epstein and colleagues. Initially a 90 kDa glycoprotein was suggested to be the adhesive based on the association of the protein only with conidia that were adhesion competent [49]. Although there is no question that the glycoprotein is involved in adhesion, more recent results suggest that the phenomenon of adhesion is more complicated than previously assumed and that the 90 kDa protein is not itself the only entity involved in adhesion [25]. This is based on the isolation of strains of the fungus that produce macroconidia with a reduced ability to adhere compared to strains that exhibit normal levels of adhesion, Att^- vs. Att^+ strains, respectively. Surprisingly, macroconidia of the Att^- strains produced the 90 kDa protein as well as a conidial tip mucilage that had been assumed to always be associated with adhesion [41].

3. Importance of Adhesion to the Infection Process of *Colletotrichum graminicola* and *Erysiphe graminis*

The importance of adhesion to plant pathogenic fungi is sometimes dismissed as not being significant to disease development. However, results of recent investigations with a variety of phytopathogenic fungi have made it clear that such an assumption overlooks those components of the fungal infection process that occur prior to penetration and may be essential for the initiation of physiological contact. Moreover, they do not take into consideration the importance of continued recognition and physiological contact with the host during pathogen ingress into the tissue.

Important examples of these early relationships that establish the success or failure of the disease interaction can be found within all groups of fungal pathogens. Of particular significance is the work of Hoch et al. [34] that demonstrated that the rust *U. appendiculatus* depends upon signals from the host surface for growth orientation and cell differentiation. Without these signals the fungus fails to form appressoria, to recognize stomates, and to penetrate. We now know that, in addition to the requirement for signal reception, this fungus must also be tightly adhered to the host surface and this adhesion probably depends to some extent on hydrophobic interactions [77].

The infection process includes events that are associated with stages of morphogenesis such as germination and the formation of infection structures such as appressoria. As discussed above, these structures must adhere to the host surface to prevent displacement. Necrotrophic fungi usually penetrate their hosts directly from an appressorium or similar structure with the site of penetration occurring at random, but often with preference for the anticlinal junction of epidermal cells [33]. Biotrophic pathogens such as the rusts do not generally penetrate their hosts randomly. These fungi often form germ tubes that seek out guard cells, form appressoria, and then penetrate through the stomatal opening. Other biotrophs, such as the powdery mildew fungi, penetrate their hosts directly as do the common necrotrophs.

I consider the infection process to include events that occur immediately after the pathogen contacts the host and to include phenomena that occur through the time of penetration. The success of the infection process may first require recognition of the host as a suitable infection court. For fungi such as *U. appendiculatus*, recognition is accomplished after germination by the directed thigmotrophic growth of the germ tube in response to topographic features of the host surface. A recent review by Hoch and Staples [33] discusses the importance of surface recognition and the signals required for infection structure formation. It is now accepted that firm adhesion of the fungal germling to the host surface is required if signals from the host that direct growth and morphogenesis are to be recognized [33, 59]. As discussed above, evidence shows that phenomena essential to the infection process often occur in advance of germination. Moreover, events that are often apparent involve the release of materials from the ungerminated spore upon its contact with the host. That propagules of fungal pathogens release materials in advance of germination suggests that knowledge and understanding of such materials are essential to understanding the success or failure of the infection process. The following portion of the review will center around the infection processes of the necrotrophic pathogen *C. graminicola*, the causal agent of maize anthracnose, and the biotrophic pathogen *Erysiphe graminis* f.sp. *hordei.*, the causal agent of powdery mildew of barley. Examination of these fungi has demonstrated that their infection processes require different physical characteristics of their respective hosts.

3.1 ADHESION OF UNGERMINATED CONIDIA OF *COLLETOTRICHUM GRAMINICOLA*

Conidia of *C. graminicola* require a hydrophobic surface for their establishment on either an artificial surface or on the surface of a maize leaf. Ungerminated conidia begin to adhere within minutes of contact with the substratum. In contrast to this rapid initiation of adhesion, the germination of conidia does not begin until 6 h after contact with the substratum. Our investigations showed that adhesion of ungerminated conidia is essential to the success of disease development as it prevents displacement of conidia from the leaf surface [53].

A goal of our investigations was to understand the basis of adhesion of the ungerminated conidium and its importance to the success of the infection process. Scanning electron microscopy (SEM) and micromanipulation of conidia that had adhered to leaves revealed both the alteration of the leaf cuticle to which conidia had adhered and the presence of possible adhesive materials on the contact surface of ungerminated conidia. The results suggest that this rapid form of adhesion is an active process that depends upon the release of materials by conidia.

The maximum percent adhesion of conidia was consistently found to be 30-40% for conidial populations harvested from 14-day-old cultures. Importantly, this level of adhesion was attained within 30 min of the time of contact. Adhesion was not influenced by conidium concentration but was influenced by conidium age [55]. Adhesion assays were generally performed at 24 °C. Pretreatment of conidia with various metabolic inhibitors had surprisingly different effects on adhesion. For example, neither the respiration inhibitor sodium azide (2 mM) nor the transcription inhibitor actinomycin D (100 μM) had an effect on adhesion. This was in direct contrast to the inhibitors of glycoprotein transport (brefeldin A, 50 μM) and protein synthesis (cycloheximide, 100 μM), which reduced adhesion by as much as 50% . Exposure of conidia to temperatures of 4 or 50 °C during adhesion assays was also found to delay the timing of conidial adhesion, but these temperature extremes did not prevent adhesion. Treatment of conidia with pronase E prevented adhesion completely as did the lectin Concanavalin A. Together, these results suggest that glycoprotein synthesis and secretion is involved in the adhesion of the ungerminated conidium.

We also used a variety of techniques to visualize the materials released from conidia during the time when they become adherent. Conidia released materials that could be visualized by scanning electron microscopy (SEM) in combination with micromanipulation or by light microscopy after staining for proteins or for carbohydrates [54]. SEM of conidia that had adhered to the hydrophobic surfaces of dimethyldichlorosilane-coated or polystyrene-coated glass showed that material had been released onto the substratum and covered the area of the surface which was in direct contact with the surface of a conidium. Gold/silver staining of adhered conidia revealed stain patterns that indicated the release of proteinaceous materials from the tips of conidia as well as around conidia and at the interface of contact between conidia and the substratum (Fig. 3). Staining adhered conidia with fluorescein (FITC)-conjugated lectins revealed that only FITC-Concanavalin A (FITC-ConA) labeled carbohydrates at the contact interface between the conidium and the substratum. Since the ConA lectin completely blocked adhesion, these results indicated that the material released from conidia and presumed to be associated with adhesion contained glucose and/or mannose. A variety of different patterns of labeling by other FITC-conjugated lectins were also observed and indicated that the material released from a conidium has a specific spatial arrangement around the conidium. Together, these results suggest that the material associated with the adhesion of conidia is composed of glycoproteins [54, 55].

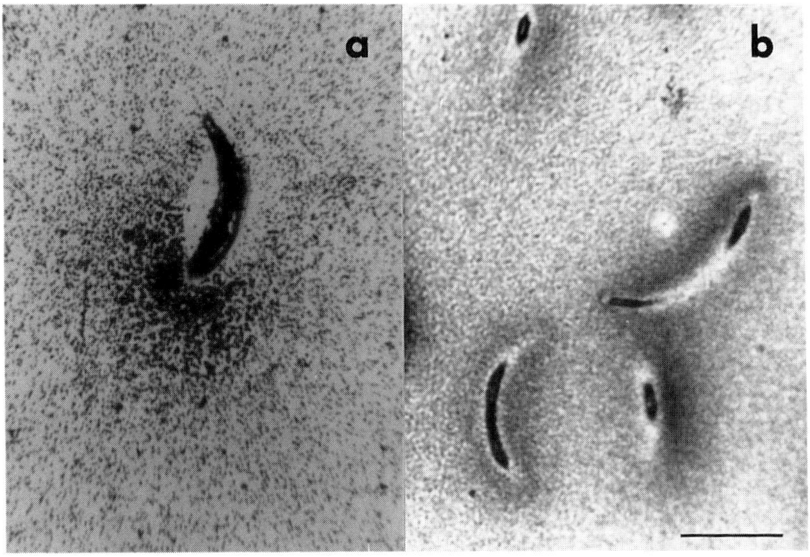

Figure 3. Gold/silver stain for protein released from conidia of *Colletotrichum graminicola*. a) Protein released at the tip of a conidium and b) protein stained at the contact interface of conidia with the substratum (arrows). Bar equals 20 μm.

The function of these putative adhesive materials in the infection process, beyond that of preventing conidial displacement, is unknown. It is important to note that when a conidium germinates, the germ tube also adheres to the substratum surface and that an adhesive is readily evident on the surface of the germ tube [53].

3.2. ADHESION AND THE *ERYSIPHE GRAMINIS* INFECTION PROCESS

In contrast to *C. graminicola*, conidia of *E. graminis* require a hydrophilic substratum for adhesion and subsequently for germination by the formation of an appressorial germ tube. In addition, the *E. graminis* infection process is considerably more complex. Unlike other powdery mildew fungi, *E. graminis* germinates by the formation of two germ tubes. One, the primary germ tube, is short, whereas the other, the appressorial germ tube, is bulbous and elongated [44]. Microscopic observations suggest that the primary germ tube attempts to penetrate the host [44, 47], but whether it actually penetrates through the host cell wall is unknown. An appressorium is formed upon the maturation of the appressorial germ tube and it is from the appressorium that penetration of the host occurs [44].

The barley host responds to the presence of the primary germ tube by the formation of a cytoplasmic aggregate and a papilla, indicating that the primary germ tube actually establishes physiological contact with the leaf [44]. A feature of both the primary germ tube and the appressorium is that they both adhere to the leaf and prevent displacement of the germling from the infection court. Carver et al. [10] have recently demonstrated by SEM the presence of apparent adhesives associated with both the primary germ tube and the appressorium.

The exact role of the primary germ tube is not understood. One suggestion is that it absorbs water and aids germination [8]. It has also been proposed to account for appressorium morphogenesis by recognition of the host surface [9]. However, appressoria can be formed on a biologically inert surface such as cellophane [43] or on the surface of non-host plants [76], neither of which would accurately resemble the surface of a barley leaf. That the primary germ tube functions to take up water and ensure normal appressorium morphogenesis is an attractive hypothesis. However, it does not explain why conidia placed in an arid or humid environment succeed in forming short germ tubes, but fail to form appressorial germ tubes unless they are in full contact with a substratum [10].

3.2.1. *Contact of the Conidium and Preparation of the Infection Court*

When a conidium of *E. graminis* fully contacts a solid surface, it releases a liquid [48]. The release of the liquid was shown to occur on both the hydrophobic barley leaf as well as the hydrophilic surface of a moistened sheet of cellophane. The liquid appears to flow off of the conidium and to bind the conidium to the substratum. The release of the liquid results in a rapid change in the morphology of the conidium surface, a change that begins immediately upon contact of the conidium with the substratum and continues for approximately 30 min when the surface topography of the conidium returns to its native state. Just as with the adhesion of ungerminated *C. graminicola* conidia, these events occur well in advance of the formation of a primary germ tube or the appressorial germ tube. Observations by SEM revealed the liquid exudate as a film that appears to cement the conidium to the surface of the substratum.

The liquid exudate was released in two phases, one within 5 min of contact of the conidium with the substratum and a second between 10 and 15 min of contact [64]. The exudate was shown to contain non-specific esterases which, when incubated on the barley leaf surface, eroded a portion of the leaf cuticle [46]. These results were interpreted as an indication that the material eroded from the leaf cuticle was cutin. Subsequently, one of the esterases in the conidial exudate was confirmed to be a cutinase by assays with ^{3}H-cutin as the substrate [69]. Thus, the *E. graminis* conidial exudate resembles the exudate from hydrated urediospores of *U. viciae-fabae* which also contains cutinase [15].

The importance of enzymatic degradation of cutin in the barley cuticle presented a perplexing problem. One possibility was that the cutinase erodes the leaf cuticle and allows the fungus to penetrate more efficiently. An alternative was that the cutinase is involved in host surface recognition or that it or makes the infection court an acceptable site for germination, appressorium formation, and finally for successful penetration. This too was perplexing since conidia of *E. graminis* can form appressoria on a variety of surfaces provided the surface is hydrophilic [43].

Attempts to discern conditions that allow conidia to form appressoria resulted in observations that appear to explain some of the questions about the importance of the liquid exudate to preparation of the infection court. We found that prior to the release of the liquid exudate the conidium surface is hydrophobic [64]. In addition, conidia with hydrophobic surfaces did not form appressorial germ tubes, although they sometimes formed short germ tubes. When the liquid exudate is released, the surface of the conidium becomes hydrophilic. Importantly, we also found that, for a conidium to form an appressorial germ tube, it must be in contact with a hydrophilic surface. This relationship should present a problem for the fungus since the surface of the barley leaf cuticle is even more hydrophobic than artificial surfaces such as polystyrene. How then, does the *E. graminis* conidium form an appressorium on the hydrophobic surface of the barley leaf? The answer was surprisingly simple! Investigations demonstrated that the liquid exudate released by the conidia makes the leaf surface more hydrophilic [64].

We then suggested that the conidial exudate establishes an area where the conidium can germinate, form an appressorium, and penetrate the barley leaf. Thus, in addition to adhesion of the ungerminated conidium, the exudate appears to prepare an infection court for the fungus. Its release results in adhesion of the conidium and the conversion of the conidium surface from a hydrophobic to a hydrophilic state, a condition which is required if appressorium formation is to occur. Second, the exudate causes the barley leaf surface to become hydrophilic. This is necessary because conidia only form appressoria on a hydrophilic surface. A possible function of the cutinase in the exudate may be to increase the hydrophilicity of the leaf surface by cleaving the ester linkages of the cutin polymer. Such cleavage would greatly reduce the hydrophobicity of the cuticle in an area that immediately surrounds the conidium, the actual site of the infection court.

Acknowledgment: Supported in part by a grant from the National Science Foundation (IBN-9105943). Article 14683 of the Purdue University Agricultural Experiment Station.

References

1. Ahmadjian, V., and Jacobs, J.B. (1983) Algal-fungal relationships in lichens: recognition, synthesis, and development, in L.G. Goff (ed.) *Algal Symbiosis* , Cambridge University Press, London, pp. 147-172.

2. Andrews, J.H., Harris, R.F., Spear, R.N., Lau, G.W., and Nordheim, E.V. (1994) Morphogenesis and adhesion of *Aureobasidium pullulans, Can. J. Microbiol.* **40**, 6-17.

3. Bartnicki-Garcia, S. (1990) Role of vesicles in apical growth and a new mathematical model of hyphal morphogenesis, in I.B. Heath (ed.), *Tip Growth in Plant and Fungal Cells*, Academic Press, New York, pp. 211-232.

4. Beckett, A., Tatnell, J.A., and Taylor, N. (1990) Adhesion and pre-invasion behaviour of urediniospores of *Uromyces viciae-fabae* during germination on host and synthetic surfaces, *Mycol. Res.* **94**, 865-875.

5. Bonfante-Fasolo, P. (1988) The role of the cell wall as a signal in mycorrhizal associations, in S. Scannerini, D. Smith, P. Bonfante-Fasolo, and V. Gianinazzi-Pearson, (eds.), *Cell to Cell Signals in Plant, Animal and Microbial Symbiosis* , Springer-Verlag, Berlin, pp. 219-235.

6. Braun, E., and Howard, R. (1994) Adhesion of fungal spores and germlings to host plant surfaces, *Protoplasma* **181**, 202-212.

7. Braun, E., and Howard, R. (1994) Adhesion of *Cochliobolus heterostrophus* conidia and germlings to leaves and artificial surfaces, *Exper. Mycol.* **18**, 211-220.

8. Carver, T.L.W., and Bushnell, W.R. (1983) The probable role of primary germ tubes in water uptake before infection by *Erysiphe graminis*, *Physiol. Plant Pathol.* **23**, 229-240.

9. Carver, T.L.W., and Ingerson, S.M. (1987) Responses of *Erysiphe graminis* germlings to contact with artificial and host surfaces, *Physiol. Molec. Plant Pathol.* **30**, 359-372.

10. Carver, T.L.W., Ingerson-Morris, S.M., and Thomas, B.J. (1995) Early interactions during powdery mildew infection, *Can. J. Bot.* (in press).

11. Ceasar-TonThat, T-C., and Epstein, L. (1991) Adhesion-reduced mutants and the wild type of *Nectria haematococca*: an ultrastructural comparison of the macroconidial walls, *Exper. Mycol.* **15**, 193-205.

12. Chaubal, R., Wilmot, V.A., and Wynn, W.K. (1991) Visualization, adhesiveness, and cytochemistry of the extracellular matrix produced by urediniospore germ tubes of *Puccinia sorghi*, *Can. J. Bot.* **69**, 2044-2054.

13. Clement, J.A., Martin, S.G., Porter, R., Butt, T.M., and Beckett, A. (1993) Germination and the role of extracellular matrix in adhesion of urediniospores of *Uromyces viciae-fabae* to synthetic surfaces, *Mycol. Res.* **97**, 585-593.

14. Clement, J.A., Porter, R., Butt, T.M., and Beckett, A. (1994) The role of hydrophobicity in attachment of urediniospores and sporelings of *Uromyces viciae-fabae*, *Mycol. Res.* **98**, 1217-1228.

15. Deising, H., Nicholson, R.L., Haug, M., Howard, R.J., and Mendgen, K. (1992) Adhesion pad formation and the involvement of cutinase and esterases in the attachment of uredospores to the host cuticle, *Plant Cell* **4**, 1101-1111.

16. Dijksterhuis, J., Veenhuis, M., and Harder, W. (1990) Ultrastructural study of adhesion and initial stages of infection of nematodes by conidia of *Drechmeria coniospora*, *Mycol. Res.* **94**, 1-8.

17. Ding, H., Balsiger, S., Guggenbühl, C., and Hohl, H.R. (1994) A putative IgG-binding 65 kDa adhesin involved in adhesion and infection of soybeans by *Phytophthora megasperma* f.sp. *glycinea*, *Physiol. Molec. Plant Pathol*, **44**, 363-378.

18. Domozych, C.R., Plante, K., Blais, P., Paliulis, L. and Domozych, D.S. (1993) Mucilage processing and secretion in the green alga *Closterium*, I. Cytology and Biochemistry, *J. Phycol.* **29**, 650-659.

19. Doss, R.P., Potter, S.W., Soeldner, A.H., Christian, J.K., and Fukunaga, L.E. (1995) Adhesion of germlings of *Botrytis cinerea*, *Appl. Environ. Microbiol.* **61**, 260-265.

20. Durso, L., Lehnen, L.P. Jr., and Powell, M.J. (1993) Characteristics of extracellular adhesins produced during *Saprolegnia ferax* secondary zoospore encystment and cystospore germination, *Mycologia* **85**, 744-755.

21. Dykstra, M.J., and Bradley-Kerr, B. (1994) The adhesive droplet of capilliconidia of *Basidiobolus ranarum* exhibits unique ultrastructural features, *Mycologia* **86**, 336-342.

22. Eagland, D. (1988) Adhesion and adhesive performance: The scientific background, *Endeavour, New Series* **12**, 179-184.

23. Epstein L., and Nicholson R.L. (1995) Adhesion of spores and hyphae to plant surfaces, in G. Carroll and P. Tudzynski (eds.), *The Mycota VI: Plant Relationships*, Springer-Verlag, Berlin, (in press).

24. Epstein, L., Kaur, S., Goins, T., Kwon, Y.H., Henson, J.M. (1994) Production of hyphopodia by wild-type and three transformants of *Gaeumannomyces graminis* var. *graminis*, *Mycologia*, **86**, 72-81.

25. Epstein, L., Kwon, Y.H., Almond, D.E., Schached, L.M, and Jones, M.J. (1994) Genetic and biochemical characterization of *Nectria haematococca* strains with adhesive and adhesion-reduced macroconidia, *Appl. Environ. Microbiol.* **60**, 524-530.

26. Estrada-Garcia, M.T., Callow, J.A., and Green, J.R. (1990) Monoclonal antibodies to the adhesive cell coat secreted by *Pythium aphanidermatum* zoospores recognize $200x10^3$ M_r glycoproteins stored within large peripheral vesicles, *J. Cell Sci.* **95**, 199-206.

27. Evans, R.C., Stempen, H., and Frasca, P. (1982) Evidence for a two-layered sheath on germ tubes of three species of *Bipolaris*, *Phytopathology* **72**, 804-807.

28. Fletcher, R.L., and Callow, M.E. (1992) The settlement, attachment and establishment of marine algal spores, *Br. phycol. J.* **27**, 303-329.

29. Guggenbühl, C., and Hohl, H.R. (1992) Mutual adhesion and interactions of isolated soybean mesophyll cells and germ tubes of *Phytophthora megasperma* f.sp. *glycinea*, *Bot. Helv.* **102**, 247-261.

30. Gubler, F., and Hardham, A.R. (1988) Secretion of adhesive material during encystment of *Phytophthora cinammomi* zoospores, characterized by immunogold labeling with monoclonal antibodies to components of peripheral vesicles, *J. Cell Sci.* **90**, 225-235.

31. Hamer, J.E., Howard, R.J., Chumley, F.G., and Valent, B. (1988) A mechanism for surface attachment in spores of a plant pathogenic fungus, *Science* **239**, 288-290.

32. Hoagland, K.D., Rosowski, J.R., Gretz, M.R., and Roemer, S.C. (1993) Diatom extracellular polymeric substance: function, fine structure, chemistry, and physiology, *J. Phycol.* **29**, 537-566.

33. Hoch, H.C. and Staples, R.C. (1991) Signaling for infection structure formation in fungi. in G.T. Cole and H.C. Hoch (eds.), *The fungal spore and disease initiation in plants and animals*, Plenum, New York, pp. 25-46.

34. Hoch, H.C., Staples, R.C., Whitehead, B., Comeau, J., and Wolf, E.D. (1987) Signaling for growth orientation and cell differentiation by surface topography in *Uromyces*, *Science* **235**, 1659-1662.

35. Hohl, H.R., and Balsiger, S. (1986) A model system for the study of fungus - host surface interactions: adhesion of *Phytophthora megasperma* to protoplasts and mesophyll cells of soybean, in B. Lugtenberg, (ed.), Recognition in microbe-plant symbiotic and pathogenic interactions, *Proc. Nato Adv. Res. Workshop. Vol. H4*, pp. 259-273.

36. Hohl, H.R., and Balsiger, S. (1986) Probing the surfaces of soybean protoplasts and of germ tubes of the soybean pathogen *Phytophthora megasperma* f. sp. *glycinea* with lectins, *Bot. Helv.* **96**, 289-297.

37. Hohl, H.R., and Balsiger, S. (1988) Surface glucosyl receptors of *Phytophthora megasperma* f.sp. *glycinea* and its soybean host, *Bot. Helv.* **98**, 271-277.

38. Hyde, K.D., Greenwood, R., and Jones, E.B.G. (1993) Spore attachment in marine fungi. *Mycol. Res.* **97**, 7-14.

39. Jones, E.B.G. (1994) Fungal Adhesion, *Mycol. Res.* **98**, 961-981.

40. Jones, L., and O'Shea, P. (1994) The electrostatic nature of the cell surface of *Candida albicans*: a role in adhesion, *Exp. Mycol.* **18**, 111-120.

41. Jones, M.J., and Epstein, L. (1990) Adhesion of macroconidia to the plant surface and virulence of *Nectria haematococca*, Appl. Environ. Microbiol. **56**, 3772-3778.

42. Knox, R.B., Clarke, A., Harrison, S., Smith, P., and Marchalonis, J.J. (1976) Cell recognition in plants: Determinants of the stigma surface and their pollen interactions, *Proc. Natl. Acad. Sci.* USA **73**, 2788-2792.

43. Kobayashi, I, Tanaka C., Yamaoka, N., and Kunoh, H. (1991) Morphogenesis of *Erysiphe graminis* conidia on artificial membranes, *Trans. Mycolog. Soc. Japan.* **32**, 187-198.

44. Kunoh, H., Ishizaki, H., and Nakaya, K. (1977) Cytological studies of early stages of powdery mildew in barley leaves and wheat leaves: (II) significance of the primary germ tube of *Erysiphe graminis* on barley leaves, *Physiol. Plant Pathol.* **10**, 191-199.

45. Kunoh, H. Nicholson, R.L. and Kobayashi, I. (1991) Extracellular materials of fungal structures: their significance at penetration stages of infection. in K. Mendgen (ed.), *Electron Microscopy of Plant Pathogens*, Springer-Verlag, Berlin, pp. 223-234.

46. Kunoh, H., Nicholson, R.L., Yoshioka, H., Yamaoka, N., and Kobayashi, I. (1990) Preparation of the infection court by *Erysiphe graminis*: Degradation of the host cuticle, *Physiol. Molec. Plant Pathol.* **36**, 397-407.

47. Kunoh, H., Tsuzuki, T., and Ishizaki, H. (1978) Cytological studies of early stages of powdery mildew in barley and wheat. IV. Direct ingress from superficial primary germ tubes and appressoria of *Erysiphe graminis hordei* on barley leaves, *Physiol. Plant Pathol.* **13**, 327-333.

48. Kunoh, H., Yamaoka, N., Yoshioka, H., and Nicholson, R.L. (1988) Preparation of the infection court by *Erysiphe graminis*. I. Contact-mediated changes in morphology of the conidium surface, *Exp. Mycol.* **12**, 325-335.

49. Kwon, Y.H., and Epstein, L. (1993) A 90-kDa glycoprotein associated with adhesion of *Nectria haematococca* macroconidia to substrata, *Molec Plant-Microbe Interact*, **6**, 481-487.

50. Leite, B., and Nicholson, R.L. (1992) Mycosporine-alanine: a self-inhibitor of germination from the conidial mucilage of *Colletotrichum graminicola*. *Exp. Mycol.* **16**, 76-86.

51. Lewis, I., and Cooke, R.C. (1985) Conidial matrix and spore germination in some plant pathogens, *Trans. br. Mycol. Soc.* **84**, 661-667.

52. Manocha, M.S., and Chen, Y. (1990) Specificity of attachment of fungal parasites to their hosts, *Can. J. Microbiol.* **36**, 69-76.

53. Mercure, E.W., Kunoh, H., and Nicholson, R.L. (1994) Adhesion of *Colletotrichum graminicola* to corn leaves: a requirement for disease development, *Physiol. Molec. Plant Pathol.* **45**, 407-420.

54. Mercure, E.W., Kunoh, H., and Nicholson, R.L. (1995) Visualization of materials released from adhered, ungerminated conidia of *Colletotrichum graminicola*, *Physiol. Molec. Plant Pathol.* **46**, 121-135.

55. Mercure, E.W., Leite, B., and Nicholson, R.L. (1994) Adhesion of ungerminated conidia of *Colletotrichum graminicola* to artificial hydrophobic surfaces, *Physiol. Molec. Plant Pathol.* **45**, 421-440.

56. Mims, C.W., Richardson, E.A., Clay, R.P., and Nicholson, R.L. (1995) Ultrastructure of conidia and the conidium aging process in the plant pathogenic fungus *Colletotrichum graminicola*, Int. J. Plant Sci. **156**, 9-18.

57. Nicholson, R.L. (1990) Functional significance of adhesion to the preparation of the infection court by plant pathogenic fungi. in R.E. Hoagland, (ed.), *Biological Weed Control Using Microbes and Microbial Products as Herbicides. American Chemical Society Symposium Series No. 439.* pp. 218-238.

58. Nicholson, R.L. (1992) *Colletotrichum graminicola* and the anthracnose disease of corn and sorghum, in J.A. Bailey and M.J. Jeger (eds.), *Colletotrichum: Biology, Pathology and Control.*, CAB press. pp. 186-202.

59. Nicholson, R.L., and Epstein, L. (1991) Adhesion of fungi to the plant surface: Prerequisite for pathogenesis in G.T. Cole and H.C. Hoch (eds.), *The Fungal Spore and Disease Initiation in Plants and Animals*, Plenum, New York. pp. 3-23.

60. Nicholson, RL., and Kunoh, H. (1995) Early interactions, adhesion and establishment of the infection court by *Erysiphe graminis*, *Can. J. Bot.* (in press).

61. Nicholson, R.L., and Moraes, W.B.C. (1980) Survival of *Colletotrichum graminicola*: importance of the spore matrix, *Phytopathology* **70**, 255-261.

62. Nicholson, R.L., Butler, L.G., and Asquith, T.N. (1986) Glycoproteins from *Colletotrichum graminicola* that bind phenols: Implications for survival and virulence of phytopathogenic fungi, *Phytopathology* **76**, 1315-1318.

63. Nicholson, R.L., Hipskind, J., and Hanau, R.M. (1989) Protection against phenol toxicity by the spore mucilage of *Colletotrichum graminicola, Physiol. Molec. Plant Pathol.* **35**, 243-252.

64. Nicholson, R.L., Kunoh, H., Shiraishi, T., and Yamada, T. (1993) Initiation of the infection process by *Erysiphe graminis*: Conversion of the conidial surface from hydrophobicity to hydrophilicity and influence of the conidial exudate on the hydrophobicity of the barley leaf surface, *Physiol. Molec. Plant Pathol.* **43**, 307-318.

65. Nicholson, R.L., Yoshioka, H., Yamaoka, N., and Kunoh, H. (1988) Preparation of the infection court by *Erysiphe graminis* II. Release of esterase enzyme from conidia in response to a contact stimulus, *Exp. Mycol.* **12**, 336-349.

66. Nicole, M., Chamberland, H., Rioux, D., Lecours, N., Rio, B., Geiger, J.P., and Ouellette, G.B. (1993) A cytochemical study of extracellular sheaths associated with *Rigidoporus lignosus* during wood decay, Appl. Environ. Microbiol. **59**, 2578-2588.

67. Pain, N.A., O'Connell, R.J., Mendgen, K., and Green, J.R. (1994) Identification of glycoproteins specific to biotrophic intracellular hyphae formed in the *Colletotrichum lindemuthianum*-bean interaction, *New Phytol.* **127**, 233-242.

68. Pascholati, SF, Deising, H., Leite, B. Anderson, D., and Nicholson, RL. (1993) Cutinase and non-specific esterase activities in the conidial mucilage of *Colletotrichum graminicola*, *Physiol. Molec. Plant Pathol*, **42**, 37-51

69. Pascholati, S.F., Yoshioka, H., Kunoh, H., and Nicholson, R.L. (1992) Preparation of the infection court by *Erysiphe graminis* f.sp. *hordei*: cutinase is a component of the conidial exudate, *Physiol. Molec. Plant Pathol.* **41**, 53-59.

70. Read, S.J., Moss, S.T., and Jones, E.B.G. (1991) Attachment studies of aquatic hyphomycetes, *Philos. Trans. R. Soc. London*, B. **334**, 449-457.

71. Read, S.J., Moss, S.T., and Jones, E.B.G. (1992) Germination and development of attachment structures by conidia of aquatic hyphomycetes: light microscopic studies, *Can. J. Bot.* **70**, 831-837.

72. Read, S.J., Moss, S.T., and Jones, E.B.G. (1992) Germination and development of attachment structures by conidia of aquatic hyphomycetes: a scanning electron microscope study, *Can. J. Bot.* **70**, 838-845.

73. Romantschuk, M. (1992) Attachment of plant pathogenic bacterial to plant surfaces, *Annu. Rev. Phytopathol.* **30**, 225-243.

74. Schuerger, A.C., and Mitchell, D.J. (1993) Influence of mucilage secreted by macroconidia of *Fusarium solani* f. sp. *phaseoli* on spore attachment to roots of *Vigna radiata* in hydroponic nutrient solution, *Phytopathology* **83**, 1162-1170.

75. Sela-Burlage, M.B., Epstein, L., and Rodriguez, J. (1991) Adhesion of ungerminated *Colletotrichum musae* conidia, *Physiol. Molec. Plant Pathol.* **39**, 345-352.

76. Staub, T., Dahmen, H., and Schwinn, F.J. (1974) Light and scanning electron microscopy of cucumber and barley powdery mildew on host and nonhost plants, *Phytopathology* **64**, 364-372.

77. Terhune, B.T., and Hoch, H.C. (1993) Substrate hydrophobicity and adhesion of *Uromyces* urediospores and germlings, *Exp. Mycol.* **17**, 241-252.

78. Terrance, K., Heller, P., Wu, Y.-S., and Lipke, P.N. (1987) Identification of glycoprotein components of alpha-agglutinin, a cell adhesion protein from *Saccharomyces cerevisiae*, *J. Bacteriol.* **169**, 475-482.

79. Toti, L., Viret, O., Chapela, I.H., and Petrini, O. (1992) Differential attachment by conidia of the endophyte, *Discula umbrinella* (Berk. & Br.) Morelet, to host and non-host surfaces, *New Phytol.* **121**, 469-475.

80. Tronchin, G., Bouchara, J-P., Robert, R., and Senet, J-M. (1988) Adherence of *Candida albicans* germ tubes to plastic: ultrastructural and molecular studies of fibrillar adhesins. *Infect. Immun.* **56**, 1987-1993.

81. Tunlid, A., Jansson, H-B, and Nordbring-Hertz, B. (1992) Fungal attachment to nematodes. *Mycol. Res.* **96,** 401-412.

82. Van Dyke, C.B., and Mims, C.W. (1991) Ultrastructure of conidia, conidium germination, and appressorium development in the plant pathogenic fungus *Colletotrichum truncatum*, *Can. J. Bot.* **69**, 2455-2467.

83. van Doorn, J., Boonekamp, P.M., and Oudega, B. (1994) Partial characterization of fimbriae of *Xanthomonas campestris* pv. *hyacinthi*, *Molec. Piant-Microbe Interac.* **7**, 334-344.

84. Vreeland, V., and Epstein, L. (1995) Analysis of plant-substratum adhesives, in J.F. Jackson and H-F. Linskens (eds.), *Modern Methods of Plant Analysis*, V 17: *Plant cell-wall analysis*, Springer-Verlag, Berlin (in press).

85. Ward, H.M. (1888) A lily-disease, *Annal. Bot.* **2**, 319-382.

CELLULAR ASPECTS OF RUST INFECTION STRUCTURE DIFFERENTIATION.
Spore Adhesion and Fungal Morphogenesis

H. DEISING, S. HEILER, M. RAUSCHER,
H. XU AND K. MENDGEN
Universität Konstanz, Fakultät für Biologie, Lehrstuhl für Phytopathologie, Universitätsstr. 10, D-78434 Konstanz, Germany

1. Introduction

Rust fungi are important plant pathogens, infecting taxonomically diverged plant taxa such as angiospermes, gymnospermes and also some pteridophytes. Though members of all families of spermatophytes are infected by rusts, it is interesting that members of the Poaceae, Leguminosae and Compositae are the dominating hosts. Many of the plants are economically important crops such as maize, wheat, barley, oat, peas and beans. The success of rust fungi is based on their producdtion of spores which spread over great distances. More than 50,000 uredospores can be prodced from a single pustule, and anemochoric transportation over more than 1000 km has been reported [71]. The coffee rust fungus *Hemileia vastatrix* was first discovered in plantations in Ceylon in 1868. Ten years later most coffee plantations had been destroyed, and tea was produced instead of coffee. Likewise, due to destruction of coffee plantations in Java by the rust fungus, caoutchouc rubber production replaced that of coffee. So far, H. vastatrix has destroyed plantations in 25 countries, and since 1970 it is also present in Brazil and Colombia, both of which are among the most important coffee-producing states of the world. In North America wheat stem rust epidemics have been a severe problem from the 18th century on, and regional epidemics occurded in 1904, 1916, 1923, 1925, 1935, 1937, 1953 and 1954 [82]. These epidemics, caused by *Puccinia graminis* f. sp. *tritici*, were severe and resulted in significant crop failures in the U.S.A. These two examples clearly indicate the economic importance of rust fungi and encourage the study of mechanisms that provide the basis for establishment of a successful pathosystem.

In order to produce large spore numbers it is necessary to withdraw nutrients from the host over an extended time, and this requires maintainance of integrity of the inefcted host cells and tissues. This is possible as the fungus causes only very limited destruction of the host cell walls, so that even heavily infected plants stay alive for weeks after the onset of sporulation, and new inoculum is efficiently generated. All fungi that use this strategy are obligate biotrophs. One of the most striking difficulties of working with obligate biotrophs is that they have an absolute requirement for living host cells, and only few exceptions of axenically cultured rusts exist [28, 67].

M. Nicole and V. Gianinazzi-Pearson (eds.), Histology, Ultrastructure and Molecular Cytology of Plant-Microorganism Interactions, 135-156.

The complex and delicate mechanisms active during the establishment of a rust on its host give rise to several questions. How is adhesion of the rust spore to the plant cuticle mediated, and which signals induce differentiation of the first infection structure, the appressorium? After penetration into the intercellular space, what is the molecular basis of avoidance of mycelial degradation by chitinases present in the plant apoplast? During the process of cell wall penetration, how is degradation of the wall restricted to the penetration pore? If enzymes are involved, which substrate specificities do they exhibit, and how is limited degradation coordinated? Are there any clues of a coordination at the molecular or biochemical level of these events by a developmental program in obligate pathogens?

We will in this article review ultrastructural, biochemical and molecular aspects relevant to answer these questions and will compare mechanisms of rust infection with those found in interactions of non-biotrophic plant pathogenic fungi with their hosts.

2. Adhesion of Fungal Spores

The first event in the interaction between a fungal pathogen attacking above-ground plant organs and its host is the contact of spore and cuticle. The spore, or germling, needs to adhere to the cuticle to establish the fungus on the leaf. Thus, adhesion may be regarded as a necessary prerequisite for successful infection [72, 97], Nicholson, this volume). Appressorial adhesion is important for those fungi which differentiate melanized appressoria to build up high turgor pressure and push the penetration hypha through cuticle and epidermal cell wall primarily by mechanical forces. In spite of the importance of adhesion, the mechanisms providing its basis are understood in only a few cases.

Hydrophobic interactions may be initially involved in the adhesion process since both, the fungal spore surface and the plant cuticle are hydrophobic (Nicholson, this volume). By using low angle scanning electron microscopy, Clement et al. [15] provided evidence that this kind of interaction attaches uredospores of the broad bean rust fungus, *Uromyces viciae-fabae*, to the host cuticle immediately after landing. In other fungi, hydrophobic rodlet-shaped proteins have been detected on the spore surface. The rodlets consist of moderately hydrophobic proteins that were first discovered as gene products abundantly expressed during emergence of aerial hyphae and fruiting bodies of the basidiomycete *Schizophyllum commune* [101]. These proteins, due to their hydrophobic nature, have been termed hydrophobins. Rodlet formation results from the assembly of hydrophobins at the interface of the hyphal wall and air [104]. Genes encoding hydrophobins have been cloned from a number of fungi such as *S. commune, Neurospora crassa, Aspergillus nidulans, Metarhizium anisopliae,* and plant pathogenic fungi such as *Ophiostoma ulmi, Cryphonectria parasitica,* and *Magnaporthe grisea* [100]. From *M. grisea* the MPG1 gene homologous to other hydrophobins was cloned as a gene abundantly expressed during infection [94]. By gene replacement, a hydrophobin-deficient mutant was constructed which was neither able to form mature appressoria nor to infect its host, rice, efficiently. This mutant would offer the opportunity to see, by low angle scanning electron microscopy, whether hydrophobins act by mediating adhesion of the appressorium to the cuticle.

A non-specific mutagenesis approach was taken to address the question of which mechanisms are active in adhesion of conidia of *Nectria haematococca* to the host surface. Treatment of microconidia of this pathogen with *N* -methyl-*N*' -nitro-*N* -nitrosoguanidine gave rise to two mutants showing reduced adhesion to zucchini fruits and polystyrene. While both, mutant like wild type macroconidia caused disease when inoculated into wounded zucchini fruits, virulence was reduced on fruits with an intact surface, indicating that adhesion is important in disease development [49]. Differential screening of a DNA library of wild type *N. haematococca* with DNA probes complementary to mRNA from wild type and from adhesion-reduced mutants would offer the opportunity to clone and thus identify genes encoding products which play an important role in adhesion [1, 95].

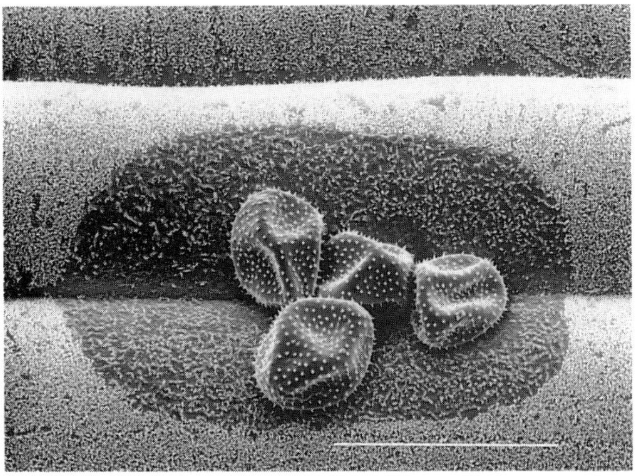

Figure 1. Uredospores of *Puccinia hordei* on barley cuticle. Note alteration of wax platelets surrounding ungerminated spores, indicating enzymatic degradation. Photograph by N.D. Read [79] (with permission from Springer-Verlag). Bar = 40 mm.

In a number of biochemical studies adhesive substances in fungal spores and germ tubes have been identified as polysaccharides and proteins or glycoproteins (for review, see [72]. Interestingly, spores are often imbedded in mucilagenous material or, if they are dispersed as dry spores by wind, they release a liquid upon contacting a hard surface like the cuticle (powdery mildews) or after incubation in high humidity (rusts). The composition of liquids released by fungal spores has been a matter of research for several years. Nicholson and co-workers have shown that the mucilagenous matrix embedding conidia of the maize pathogen *C. graminicola* contains several enzymes such as invertase, b-glucosidase, peroxidase, DNase and esterase [72, 74]. Importantly, several of the *Colletotrichum* esterases found in the mucilage have cutinase activity [74]. These enzymes could alter the structure of epicuticular wax platelets or cutin underneath the spore. In this context, Nicholson and Epstein [72] emphasize that "cutinase activity could theoretically produce a surface with different properties for attachment". Also in other fungi cutinase activity was found in extracellular spore material. For example, Pascholati et al. [75] demonstrated cutinase activity in the liquid released from conidia of the powdery mildew fungus, *Erysiphe graminis* f. sp. *hordei*. Scanning electron microscopy investigations have shown surface erosions of leaf areas to which partially purified esterases had been applied. The erosion was similar to those seen on barley cuticle in contact with powdery mildew conidia [61].

Also, around uredospores of *Puccinia hordei*, the structure of the epicuticular wax was shown to be altered (Figure 1), and these alterations have been attributed to the action of enzymes [79]. Since surface erosion was only seen around ungerminated, but never around germinated uredospores, it can be speculated that intracellular esterases leached out of the spores, due to desintegrity of membranes (N.D. Read, personal communication). The scanning electron micrograph shows, however, that enzymes capable of altering the structure of epicuticular wax (and possibly cutin) are associated with uredospores. It can thus be speculated that in general esterases and/or cutinases are factors important in the establishment of spore adhesion to the plant cuticle.

The first experimental proof that esterases and cutinase are important in establishment of adhesion comes from work with the broad bean rust fungus *U. viciae-fabae* [21]. After incubation of uredospores in high humidity, the spore surface changes, viscous material forms and assembles at sites where the spore surface contacts the plant cuticle. After removal of the spore, the material remains on the cuticle as an adhesive pad. This adhesion pad is not only formed by living spores, but also after autoclaving uredospores, indicating that pad formation is a passive process. The pads of autoclaved spores, however, show only reduced levels of adhesion to the broad bean cuticle. Three esterases, one of them exhibiting cutinase activity, are released from the broad bean rust uredospore surface upon contact with aqueous solutions. The enzymes were partially purified by non-denaturing polyacrylamide gel electrophoresis and eluted from the gel. Adhesion of autoclaved spores was restored not only by adding the mixture of enzymes released by uredospores, but also by adding esterases and the cutinase individually, which had previously been separated by non-denaturing polyacrylamide gel electrophoresis. Addition of diisopropyl fluorophosphate, an inhibitor of serin esterases, inhibited complementation of adhesion, which demonstrated the specificity of the assay.

3. Differentiation of Fungal Infection Structures

After successful establishment of the spore on the host surface, germination takes place under appropriate environmental conditions. Many plant pathogenic fungi differentiate more or less complex infection structures to invade the host tissue. Direct penetration through the host cuticle and cell wall, as in the case of *Colletotrichum* and *Magnaporthe* species, or penetration through the stomatal pore as in the case of most rusts is mediated by such infection structures [69]. Germ tubes growing on the plant surface apparently show directional growth [80] and appressorium formation is often initiated predominantly at favourable sites. Therefore, an interesting question is which signals are required for temporal and spatial orientation of pathogenic fungi, leading to infection structure differentiation and successful infection.

3.1. SIGNALLING IN INFECTION STRUCTURE DIFFERENTIATION

Read et al. [80, 1992] described that rust uredospores germinate from pores on the side of the spore facing the substratum, indicating that even spores perceive signals from their environment and especially their substratum. As early as 1934, Johnson [47] described the directional growth of germ tubes of *Puccinia graminis* f. sp. *tritici* as a response to the parallel arrangement of anticlinal epidermal walls of wheat leaf cells, and similar responses have since been discussed for several different rust species pathogenic to members of the Poaceae [2]. In rusts of dicotyledonous plants, orientation based on anticlinal epidermal cell walls is not possible as their host plants do not exhibit an arrangement of anticlinal walls comparable to that of grasses. However, Lewis and Day [65, 1972] hypothesize that rust germ tubes use the grid structure of wax platelets on the cuticle for growth orientation, and Edwards and Bowling [26] have demonstrated that pH gradients may direct germ tube growth of *U. viciae-fabae*.

The external signals which trigger infection structure differentiation in different fungi are quite diverse. Colletotrichum and Magnaporthe conidia as well as rust basidiospores form appressoria on solid surfaces such as glass slides, wax paper, polystyrene with different degrees of efficiency [63] or even on high-percent agar [30] and apparently do not require special structural surface features as do rust uredo- and aeciosporelings. Factors such as surface hydrophobicity or hardness, light and possibly additional external signals have been discussed as important for *Magnaporthe* appressorium differentiation [46, 105]. In *C. gloeosporioides,* a pathogen of papaya, tomato and other fruits, and in C. musae, a banana pathogen, a pre-penetration phase of latency is overcome by ethylene, the host's ripening hormone, which provides an external signal [29]. Concentrations of ≤1mg ethylene/liter, which are easily reached in ripening fruits, cause conidial germination, branching of the germ tubes and formation of up to six appressoria from a single spore. In the same fungus, wax isolated from the avocado fruit surface induced appressorium formation and gene expression [45, 76]. Likewise, Jelitto et al. [46] take the fact that *M. grisea* forms appressoria more efficiently on the rice cuticle than on artificial hydrohobic substrata as evidence for the importance of external infection structure-inducing factors associated with the host plant. Perception of these external signals may initiate an internal signal transduction that leads to infection structure differentiation. Based on appressorium formation on non-inductive surfaces in the presence of cAMP, its analogues 8-bromo-cAMP and *N*-6-monobutyryl-cAMP, and 3-iso-butyl-1-methylxanthine, an inhibitor of phosphodiesterase, Lee and Dean [63] suggested that cAMP acts as an internal signal of appressorium induction in *Magnaporthe*. cAMP has also been reported to regulate morphogenesis, which determines pathogenicity in the corn smut fungus *Ustilago maydis* [32].

In contrast to *Colletotrichum* and *Magnaporthe,* most rust uredosporelings position the appressorium above the stomata [27, 37], and it is known that a surface stimulus is necessary and sufficient to induce appressorium formation as well as differentiation of subsequent infection structures. Dickinson [24] and somewhat later Maheshwari et al. [68] have shown that certain *Puccinia*-species such as *P. graminis* f. sp. *tritici*, *P. helianthi, P. sorghi, P. antirrhini* and also the bean rust fungus *Uromyces phaseoli* differentiate appressoria in the absence of their hosts on oil-collodion membranes providing a surface stimulus. Later other materials such as polystyrol replicas [40] or scratched polyethylen membranes [20, 90] have been used to induce infection structure differentiation. A comprehensive review on different methods used for *in vitro*-differentiation of rust infection structures has been published by Read et al. [80].

Using polystyrene replicas of microfabricated silicon wafers of defined surface structures, Hoch et al. [40] have shown that the dimensions of the surface structures causing maximal rates of appressorium differentiation correspond closely to the dimension of the lips of stomatal guard cells. It has now been demonstrated with many different rusts that sensing the stimulus provided by this lip induced appressorium formation [2, 40, 93].

In the appressorium a first round of mitotic nuclear division takes place, giving rise to four nuclei in this structure. From the base of the appressorium a penetration hypha grows through the stomatal pore and differentiates into a substomatal vesicle in the substomatal chamber of the leaf [69]. In the vesicle, the four nuclei undergo a second round of mitosis, resulting in eight nuclei in the vesicle. The nuclei migrate into the elongating vesicle, now called the infection hypha. Upon contact with a mesophyll cell a haustorial mother cell containing two of the nuclei is differentiated (Figure 2; [20, 93]. The haustorial mother cell penetrates the wall of the plant mesophyll cell to differentiate the primary haustorium which serves to take up nutrients [35], and successful establishment of a haustorium terminates the metabolically independent phase of rust infection structure differentiation. The haustorial mother cell has, as compared with substomatal vesicle or infection hypha, a multilayered and thus stronger cell wall, reminiscent of the appressorium [69]. The reinforcement of the haustorial mother cell wall may have led to the discussion whether rust fungi produce cell wall-degrading enzymes at all [55] (see below).

3.2. BIOCHEMICAL ANALYSES OF FUNGAL INFECTION STRUCTURES

In order to understand infection structure differentiation and the infection process of rust fungi, biochemical and molecular investigations had to be carried out. While a significant body of literature is available on various aspects of appressorium differentiation in the wheat stem rust fungus *Puccinia graminis* f. sp. *tritici* (for review, see [37, 38], stages of infection structures differentiated subsequently have received only limited attention, probably due to difficulties with producing in vitro infection structures in quantities allowing biochemical analyses. The use of scratched polyethylene sheets allowed the production of *Uromyces* infection structures sufficient to analyze changes in the fungus associated with infection structure differentiation on the biochemical and the molecular level.

3.2.1. In vitro Systems Used to Study Infection Structures in the Absence of the Host

If changes in metabolism or composition of rust infection structures are to be analyzed by biochemical or molecular techniques, an *in vitro*-system is needed which allows temporary cultivation of the fungus and infection structure differentiation in the absence of the host. Media have been described for the axenic culture of different rusts [28], but in such cultures infection structures are not formed so that morphogenetic studies can not be performed. Differentiation of infection structures can, apart from using thigmo-inductive substrata as described above (thigmo-differentiation), be induced by application of chemicals such as potassium- or calcium-ions or sucrose [39, 52, 53] (chemo-differentiation). In rusts belonging to the genera *Puccinia* (e.g. *P. graminis* f. sp. *tritici* or *P. coronata* [56, 70] or *Uromyces* (*U. appendiculatus* [91] differentiation can also be induced by application of a heat shock (thermo-differentiation).

In this context, it is important to address which mode of *in vitro* induction resembles that occurring in planta most closely. So far, the only publication allowing this comparison is by Stark-Urnau and Mendgen [93] who, using light microscopy, found no significant differences between *in planta*- and thigmo- (*in vitro*-) induced aecidiospore- and uredospore-derived infection structures of the cowpea rust fungus *U. vignae*.

Since comparisons of *in planta*- with chemo-, thermo- or thigmo-differentiated infection structures are not yet available on the biochemical or molecular level it appears reasonable to use thigmo-induced structures in such experiments, as thigmo-induction resembles the situation *in planta* most closely. For our biochemical and molecular studies we have chosen the broad bean rust fungus *U. viciae-fabae* which differentiates all infection structures - appressoria, substomatal vesicles, infection hyphae and haustorial mother cells - on thigmo-inductive polyethylene membranes within 24 h. This system can thus be used to analyze basically all events occurring prior to penetration of the host cell wall and establishment of a successful pathosystem. Figure 2 shows infection structures differentiated *in vitro* on thigmo-inductive polyethylene membranes. The fact that infection structure differentiation occurs in a synchronized fashion [20] makes the system suitable to relate alterations detected by biochemical or molecular methods to developmental changes.

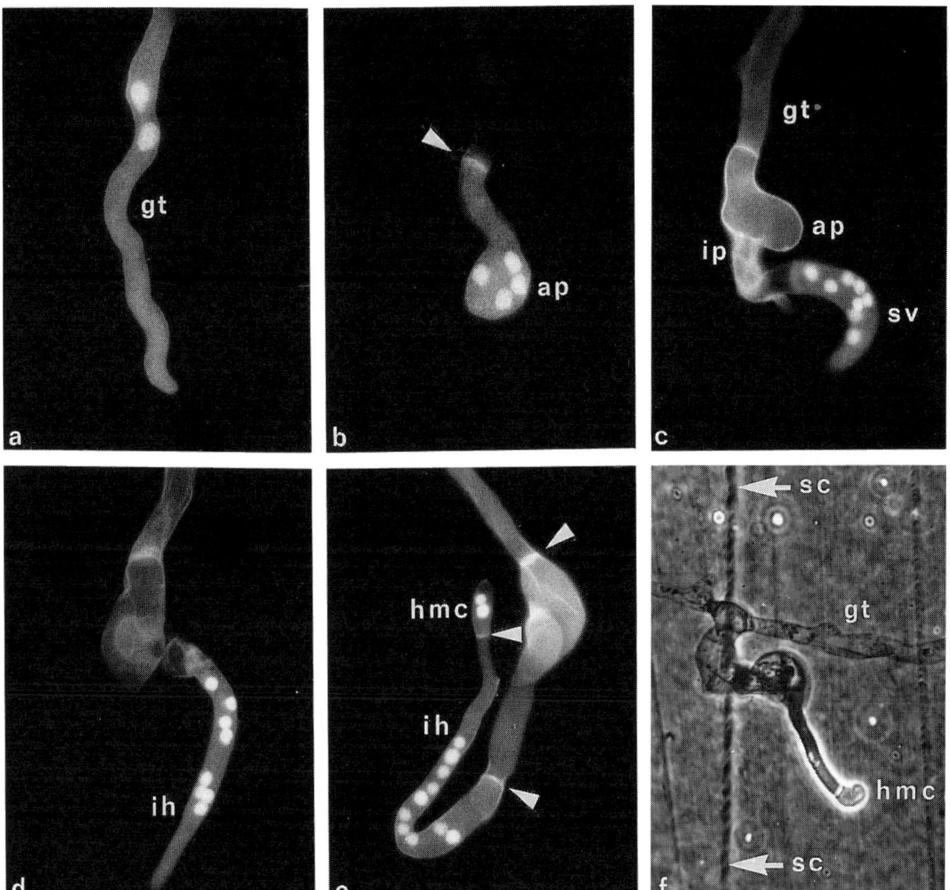

Figure 2. In vitro differentiation of infection structures of *Uromyces viciae-fabae* on thigmo-inductive polyethylene membranes. Germ tube (a), appressorium (b), substomatal vesicle (c), infection hypha (d), and haustorial mother cell (e) are sequentially differentiated. Nuclear conditions of infection structures are demonstrated by fluorescence microscopy after DAPI-Calcofluor-staining (a-e). Staining with trypan blue in lactophenol-glycerol shows a scratch on the polyethylene membrane inducing infection structure differentiation. ap, appressorium; gt, germ tube; hmc, haustorial mother cell; ih, infection hypha; ip, infection peg; sc, scratch on polyethylene membrane; sv, substomatal vesicle. From: Deising et al. [20] (with permission from Springer-Verlag).

3.2.2. Changes in Protein Pattern and Gene Expression during Infection Structure Differentiation

In different phytopathogenic fungi, proteins have been described which are qualitatively or quantitatively altered during morphogenesis. Examples are different *Colletotrichum* species, e.g. *C. magna* (teleomorph: *Glomerella magna*) [6], and rust fungi. In rusts, differentiation of infection structures requires both, RNA and protein biosynthesis, as has been demonstrated for *P. graminis* f. sp. *tritici* with different inhibitors [25]. Huang and Staples [44] and Staples et al. [92] looked for qualitative and quantitative changes in the protein pattern of the bean rust fungus *U. appendiculatus* during formation of the germ tube, appressorium and substomatal vesicle. In these studies, however, protein pattern changes in infection structures relevant for penetration of the plant cell wall and establishment of the fungus on its host, i.e. in infection hypha and haustorial mother cell, have not been analyzed. Using one-dimensional SDS-PAGE, Huang and Staples [44] showed that two proteins of 18.5 and 24 kDa are synthesized during appressorium differentiation, and another protein of 23 kDa is synthesized when vesicles are formed. In another study, two-dimensional PAGE revealed that 15 proteins are altered during formation of these developmental stages of the bean rust fungus [92].

In the wheat stem rust fungus *P. graminis* f. sp. *tritici*, application of a heat shock can be used to induce infection structure differentiation [25]. Kim et al. [56] reported on two newly formed proteins of 20 und 30 kDa after two-dimensional PAGE, and Wanner et al. [98] demonstrated synthesis of two proteins of 21 and 35 kDa by one-dimensional SDS-PAGE after 35S-methionine-pulse-labelling in differentiated infection structures of this fungus. In the flax rust *Melampsora lini*, increased rates of incorporation of 35S-methionine into 15 proteins has been found after application of a heat shock [86]. In contrast to *P. graminis*, however, heat shock does not induce infection structure differentiation in this fungus, indicating that heat shock-proteins may contribute to alterations of the protein pattern in thermo-induced rusts.

TABLE 1: Analyses of protein alterations during infection structure differentiation of *Uromyces viciae-fabae*. Proteins were isolated from infection structures of different morphogenic stages and subjected to high-resolution two-dimensional polyacrylamide gel electrophoresis. Newly formed proteins are marked +, protens which disappear are marked -. Proteins which are increased are marked↑, those which are decreased in intensity are marked↓.

Structure formed	+	-	↑	↓	**Number of alterations**
germ tube	4	2	13	12	**31**
appressorium	1	-	8	-	**9**
substomatal vesicle	-	-	3	1	**4**
infection hypha	9	-	15	12	**36**
haustorial mother cell	-	1	6	6	**13**

In *U. appendiculatus* infection structure differentiation can be induced by both, thigmo-inductive membranes and by application of a heat shock. Staples et al. [91] showed that none of the 15 proteins synthesized in the bean rust fungus after thigmo-induction was detectable in the rust after thermo-induction. Deuterium oxide, which blocks thermo-induced differentiation, also blocks synthesis of six thermo-induction-specific proteins. None of these proteins were found in thigmo-induced structures [91].

In contrast to the rust fungi mentioned above, *U. viciae-fabae* differentiates all infection structures including haustorial mother cells on thigmo-inductive polyethylene membranes [20]. High-resolution two-dimensional PAGE and silver staining of proteins revealed that 55 out of 733 proteins resolved were altered during infection structure differentiation. Interestingly, the study showed that two phases of pronounced alterations exist (table 1). While 31 alterations were seen during uredospore germination, only nine and four changes were detected when appressoria and substomatal vesicles were formed. Pronounced changes occurred again when the fungus differentiated infection hyphae and haustorial mother cells, which are *in planta* formed immediately prior to penetration of the host leaf cell wall. Thirty-six proteins were altered when infection hyphae were formed and thirteen when haustorial mother cells developed [20].

It is possible that proteins formed at late stages of infection structure differentiation may be important for expression of fungal virulence. Thus, it would be important to characterize and identify such proteins, e.g. by micro-sequence analyses. A major difficulty is the provision of sufficient amounts of proteins for such analyses.

The analysis of alterations in the mRNA population during fungal morphogenesis and preparation of infection is a promising approach to identify molecules important in host-pathogen interactions. By differential screening of a genomic DNA library of the rice blast fungus *M. grisea*, two infection structure-specific genes were cloned [64]. In the obligate biotrophs *Bremia lactucae* and *P. graminis*, highly abundant and developmentally regulated genes were cloned and analyzed [50, 66]. Genes preferentially expressed when appressoria or substomatal vesicles are formed were obtained by screening a genomic DNA library of *U. appendiculatus* using cDNA probes enriched in differentiation-specific sequences by cascade hybridization [7]. So far, two of the infection structure-specific genes of the bean rust fungus, INF24 and INF56, have been characterized [8, 106].

Studies on the protein level had indicated that major changes in the protein pattern of *U. viciae-fabae* occur at spore germination and when penetration of the host cell wall is prepared, i.e. when infection hyphae and haustorial mother cells are differentiated [20]. In order to clone cDNAs of transcripts formed during late stages of infection structure differentiation a cDNA library was constructed in λZAPII [87] using mRNA isolated from 20 h old structures which had differentiated haustorial mother cells. Probes used for differential screening were cDNAs from 7h and 20h old infection structures. At these stages, appressoria and haustorial mother cells have been formed. Phage clones representing mRNAs preferentially synthesized during late infection structure differentiation were obtained and subjected to *in vivo* excision, and differential hybridization using the cDNA probes was veryfied on the colony level. pBluescript phagemids harboring differentially hybridizing cDNAs were used as probes in Northern blots. Respective *Uromyces* genes were designated *rif32, rif 21* and *rif 16* (rust infection). *rif21* and *rif 16* transcripts were significantly increased in 24h old induced structures (haustorial mother cells) but neither detected in 7h old induced (appressoria) and 24h non-induced structures (germ tubes)[22]. Detailed analyses have shown that *rif16* formation is first detectable in 18h old induced infection structures, i.e. immediately prior to penetration of the host cell wall. In contrast, *rif32* transcripts appear to increase with age of structures, and to be independent of fungal morphogenesis.

The *rif21* cDNA sequence shows significant homology with *con6*, a

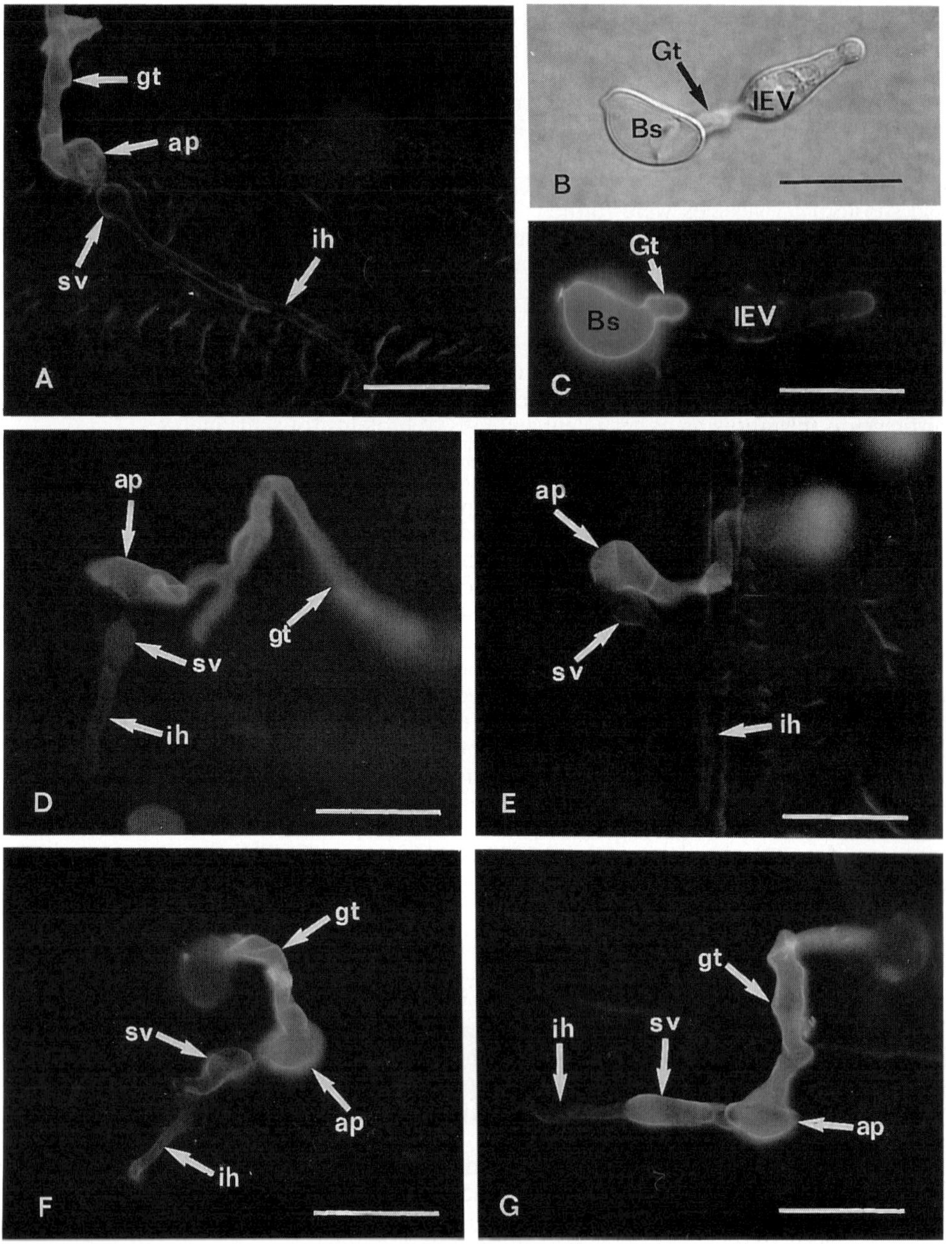

Figure 3. FITC-wheat germ agglutinin-staining of different uredospore- and basidiospore-derived *in vitro* differentiated rust infection structures.
A-C, *Uromyves vignae*; D, *U. viciae-fabae*; E, *U. appendiculatus*; F, *Phragmidium mucronatum*; G, *Puccinia carduorum*. Infection structures were stained with trypan blue in lactophenol (B) or with FITC-WGA (A, C-G)[20]. Abbreviations in A and D-G are as in Figure 2. Bs, basidiospore; Gt, germ tube; IEV, intraepidermal vesicle. Bars are 20 μm (B, C) or 50 μm (A, D-G)

conidiation-specific gene of the ascomycete *Neurospora crassa*. In this fungus, *con-6* transcripts are found at high concentration when conidia are formed; upon germination transcript levels decrease and in vegetative non-sporulating mycelia *con-6* mRNA is not detectable [102]. Since the function of *con-6* is unknown, sequence homology does not allow one to predict the function of the *rif2* gene products. Thus, the function of the differentiation-specific genes of the broad bean rust, *rif 16* and *rif 21*, remains to be determined. However, transcript formation at late stages of infection structure differentiation suggests that the gene products may be important in host-pathogen interactions.

3.2.3. Are Changes in Infection Structure Surface Carbohydrate a Prerequisite for Successful Infection?

Penetration into the intercellular space protects the rust fungus from factors present on the cuticle, e.g. anatgonigstic microorganisms or unfavourable climatic conditions, but at the same time it exposes its cell surface to antifungal enzymes such as chitinases or ß-1,3-glucanases which are present in the plant apoplast. Rust fungi thus need mechanisms that allow for exposure to these plant defense systems.

Sentandreu et al. [85] hypothesize that fungal "cell-wall growth is governed by a morphogenetic pattern which responds to a preestablished program. This programme would control not only the biosynthesis of wall polymers, but also their interactions and assembly, both of which occur external to the plasmalemma". Such a developmental program is clearly responsible for changes in the surface carbohydrates observed during rust infection structure differentiation [31]. The most prominent carbohydrate polymer on structures formed by the fungus on the plant cuticle, i.e. germ tube and appressorium, is chitin. However, on substomatal vesicles, infection hyphae and haustorial mother cells, all of which are formed in the intercellular space of the leaf, chitin is difficult to detect by FITC-labelled wheat germ agglutinin (WGA). Restriction of chitin exposition to those structures which are not in direct contact with intercellular or intracellular host defense enzymes seems to be a general phenomenon in rust fungi. Figure 3 demonstrates that in both, uredospore- and basidiospore-derived infection structures of different rusts chitin is labeled much less intensively on structures formed in the plant cell or the intercellular space. Labeling with antibodies raised to chitosan, the deacetylation product of chitin, indicated that rust infection structures formed in the intercellular space, but not germ tubes and appressoria expose chitosan on their surfaces [34].

Conversion of chitin to chitosan is catalyzed by the enzyme chitin deacetylase. This enzyme has been characterized from zygomycetes, since chitosan represents a major cell wall polymer in these fungi [51]. In plant pathogenic fungi, however, this enzyme has received attention only in *Colletotrichum lindemuthianum* and *C. lagenarium* [54, 89]. Comparison of the kinetics of enzyme and infection structure formation in C. lagenarium suggests that chitin deacetylase in this fungus is expressed after appressorium differentiation. By electron microscopy chitin was not detectable on infection pegs and young intracellular vesicles of *C. lindemuthianum* by gold-labelled WGA during the initial interaction with French bean [73]. Fully expanded vesicles and appressorial germ tubes (resembling infection pegs) formed on Formvar membranes in the absence of the host plant exhibit chitin on their surfaces, indicating the necessity of exclusion of chitin from surfaces intimately contacting the host.

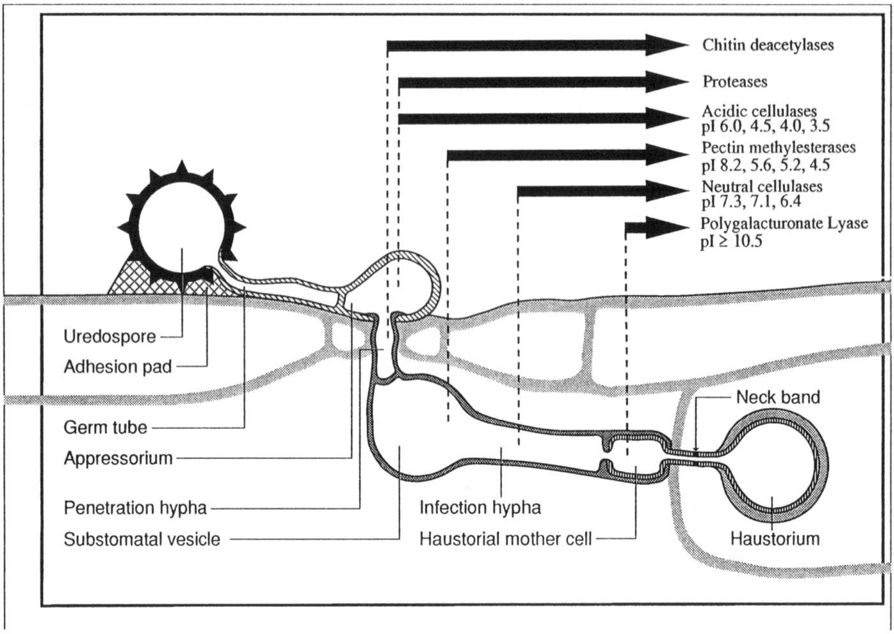

Figure 4. Infection structures of the broad bean rust fungus *Uromyces viciae-fabae.* After Mendgen and Deising [69], with modifications.

In the rust *U. viciae-fabae*, chitin deacetylase activity became detectable by a radiometric assay when appressoria had matured and substomatal vesicles began to form [23](Figure 4). Using substrate-including SDS gels, five extracellular chitin deacetylase forms of different apparent molecular masses (48.1, 30.7, 25.2, 15.2 and 12.7 kDa)were separated electrophoretically. The fact that no chitin deacetylase activity was found in 24h old germ tubes indicates that formation of the enzyme is morphogenetically controlled in the broad bean rust fungus.

An indication of the importance of chitin modification by deacetylases in plant pathogenic fungi comes from the work of Ride and Barber [81]. These authors demonstrated that wheat *endo* -chitinases preferentially hydrolyze highly acetylated chitin, and chitinase activities decrease with decreasing degree of substrate acetylation. The results suggest that chitin modification is critical to the success of the pathogen, and thus the developmental program of chitin deacetylase formation may be critical to the establishment of the fungus-host interaction.

3.2.4. The Role of Turgor Pressure and Enzymes in the Penetration Process

Cell wall-degrading enzymes are thought to be of special importance in fungi that do not differentiate melanized appressoria and are thus not able to produce significant turgor pressure to support penetration. As early as 1886, De Bary [18] postulated that extracellular fungal enzymes may aid the infection process.

Mutants produced by conventional mutagenesis protocols or targeted gene inactivation were used to address the question of the importance of cell wall-degrading enzymes. In accordance with the structure of the plant cell wall, cellulases and xylanases, pectin-degrading enzymes and, in recent years, proteases have received attention.

However, fungi that are able to build up appressorial pressure may be able to penetrate through cuticle and cell wall without support of cell wall-degrading enzymes. A series of experiments with *Colletotrichum* species and with *Magnaporthe grisea* indicate that in these fungi melanin biosynthesis and incorporation into the appressorium is essential for infection of the host.

3.2.4.1. The Importance of Turgor Pressure. Electron micrographs of appressoria of *Colletotrichum* species such as *C. lagenarium* and *C. lindemuthianum*, and of the rice blast fungus *Magnaporthe grisea* show an electron-dense melanin layer external to the plasmalemma [42, 57]. While this layer is absent in the melanin-deficient *C. lagenarium* mutant (albino mutant 79215) it can be seen in the mutant treated with scylatone [57]. This intermediate of the pentaketide melanin biosynthetic pathway is thus able to restore melanization of the albino mutant. During differentiation of infection structures, melanin synthesis is confined to appressoria; since spores, germ tubes, and penetration hyphae are never melanized it can be assumed that melanin biosynthetic enzymes are subject to morphogenetic control [59]. The thin and smooth melanin layer found in mature appressoria is different from granular or amorphous melanins of spores or hyphae where it is thought to protect cells from unfavourable environmental conditions [58]. By contrast, in fungi forming melanized appressoria, melanin is a factor essential for penetration, as indicated by several lines of evidence as described below.

Rigidification of appressorial walls in many fungi by incorporation of melanin suggests that turgor pressure may be important for penetration in these fungi. *M. grisea* is able to penetrate hard synthetic non-biodegradable surfaces, showing that pressure may be sufficient for penetration [43]. Howard and co-workers [43] demonstrated that in melanized appressoria of the rice blast fungus a turgor pressure of more than 8.0 MPa can be generated, and penetration of the rice leaf was completely inhibited by application of an extracellular osmotic pressure of approximately 6 MPa. Indirectly, these results indicate that melanin is important for generation of the turgor pressure needed for penetration of the host cuticle and cell wall. By cloning a melanin biosynthetic gene essential for appressorial penetration by complementation, a direct molecular approach was taken by Kubo et al. [60]. These authors transformed the *C. lagenarium* UV-induced albino mutant 79215, which is deficient in pentaketide cyclization to yield 1,3,6,8-tetrahydroxynaphthalene, with wild type DNA. Seven out of approximately 10,000 transformants restored the ability to synthesize melanin. While the melanin-deficient mutant was not able to cause disease on cucumber leaves, virulence of wild type and transformants with restored ability to synthesize melanin were not distinguishable. These results clearly show that melanin biosynthesis resulting in build up of substantial turgor pressure is essential for penetration by melanized appressoria. However, the results do not exclude that enzymes play a role during penetration.

3.2.4.2. Regulation of Cell Wall-Degrading Enzymes. Different nutritional groups of plant pathogenic fungi, i.e. obligate biotrophs, hemi-biotrophs, and necrotrophs obviously differ considerably in the mechanism of regulation of cell wall-degrading enzymes and, as a consequence, in the degree of tissue destruction which they cause [16]. While most obligate biotrophs differentiate relatively complex infection structures, many necrotrophs do not form such structures but rather penetrate directly with the germ tube developing into a functional infection hypha which is morphologically barely distinguishable from the germ tube itself. Only few obligate biotrophs have been studied so far, but it became evident from the work with rust fungi that morphogenetic control of cell wall-degrading enzymes and the characteristics of the enzymes formed are important features in restricting the degree of cell wall degradation to the minimum needed for penetration [19, 69]. While in these fungi morphogenesis and enzyme synthesis follow a pre-established program, growth and enzyme formation is highly adaptive in necrotrophs. Cell wall-degrading enzymes are synthesized in large amounts in response to the wall polymer present, resulting in extended tissue destruction [16]. The presence of a single polymer, e.g. polygalacturonic acid, may be sufficient to induce a variety of different enzymes, e.g. polygalacturonases, pectin- or polygalacturonate lyases and pectin methylesterases, and one gene may code for several different forms of an enzyme [13]. Generally, in necrotrophs several pectic enzymes can be synthesized, depending on the culture conditions [9, 48].

Necrotrophs are able to cope with new or changing environmental (nutritional) conditions very efficiently, with the molecular basis for such adaptation being unknown. In different plants challenged by fungal pathogens or pectin fragments as elicitors, several defense-related genes such as polygalacturonase inhibiting protein, proteinase inhibitor IIK, phenylalanin ammonium lyase, chalkone synthase and 4-coumarate:CoA ligase are activated, suggesting a common regulatory mechanism. In fact, almost identical elements in the promoter regions of genes encoding these enzymes have been found [14]. It therefore seems possible that in plant pathogenic fungi, a fragment of a plant cell wall polymer could result in the formation of a trans-acting element which may activate a number of genes encoding wall-degrading enzymes simultaneously. This strategy may allow more flexibility of growth, and perhaps also in the host range.

The hemibiotrophic *C. lindemuthianum* differentiates appressoria to invade the plant directly through cuticle and cell wall [4]. However, after an initial biotrophic phase of development the fungus causes extended tissue maceration and thus becomes a necrotroph in late stages of disease development. The bean anthracnose fungus has been demonstrated to form pectic and other cell wall-degrading enzymes in response to the availability of its substrate [103], and this regulation is typical of necrotrophs. During *in planta* -growth, only low pectin lyase activity was measured when primary hyphae had invaded few host cells. Activity increased drastically when secondary hyphae developed and the fungus entered the necrotrophic stage of development. It would be interesting to analyze immuno-electronmicroscopically whether or not these or others enzymes are formed during the initial biotrophic growth phase and whether they are restricted to the vicinity of penetration hyphae.

In the obligately biotrophic broad bean rust fungus, developmental control of a number of cell wall-degrading enzymes has been demonstrated (see Figure 4). While acidic cellulases and proteases are present on uredospores, their activities are increased when the fungus differentiates appressoria and penetrates into the substomatal cavity. Later, during differentiation of infection hyphae, neutral *endo* -cleaving cellulases are synthesized [36]. Substrate-inclusion SDS-PAGE showed that the rust synthesizes and secretes a complex pattern of proteases from substomatal vesicles and further developed structures [78]. Interestingly, the proteases show specificity for fibrous hydroxyproline-rich structural proteins similar to those formed by plants in response to fungal pathogens or elicitors [88]. Formation of pectin methylesterases begins with differentiation of substomatal vesicles, and polygalacturonate lyase is synthesized when haustorial mother cells develop. Importantly, formation of all carbohydrate polymer-degrading enzymes of the rust fungus is developmentally controlled. For induction of polygalacturonate lyase both the developmental stage of the fungus and the presence of substrate are necessary [19]. Since the lyase can only utilize largely deesterified substrate, the different pectin methylesterases may play an important role in preparation of the infection court, and in restriction of cell wall destruction.

In contrast to penetration sites of haustorial mother cells which are hidden in the mesophyll of the host, basidiospores form appressoria on the leaf and directly penetrate the cuticle and cell wall. Therefore basidiospores are more suitable to investigate changes occurring in the plant cell wall in the vicinity of the penetrating hypha. Figure 5 shows a penetration site of a basidiospore of the cowpea rust *U. vignae*, labelled with a gold-conjugated monoclonal antibody designated JIM5 which is directed against polygalacturonic acid. The plant cell wall clearly shows two layers. The side facing the cuticle is strongly labelled indicating the presence of polygalacturonic acid, while the side facing the plasmalemma shows only few gold particles. After treatment with commercially available pectin methylesterase, the inner layer is also strongly labelled by the antibody, suggesting that the inner layer of the cell wall contains highly methylated polygalacturonic acid, i.e. pectin (Figure 5, insert). Interestingly, in the vicinity of the penetration hypha, labelling is uniform across the plant cell wall, suggesting that during penetration pectin methylesterase is secreted and converts pectin to polygalacturonic acid. This non-methylated polymer represents a substrate suitable for polygalacturonate lyase known from uredospore infection structures of the broad bean rust fungus [19].

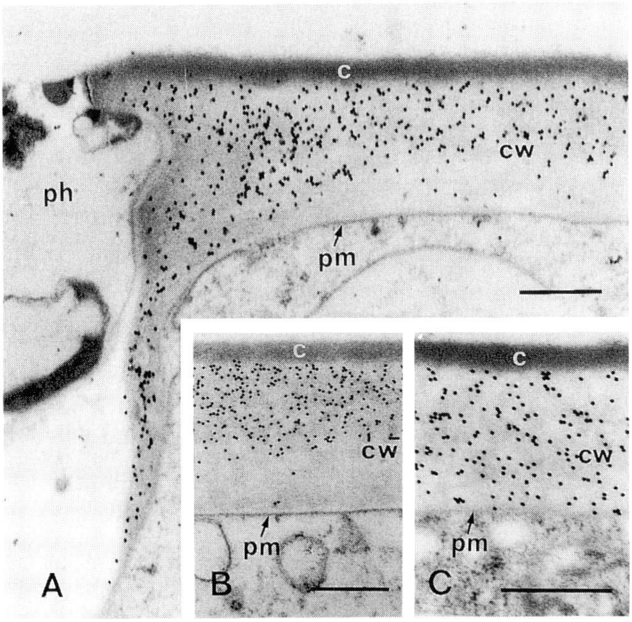

Figure 5. Labelling of basidiospore penetration sites with JIM5, a monoclonal antibody directed against polygalacturonic acid (A). Untreated control sections (B) show two cell wall layers after JIM5 labelling. In sections treated with pectin methylesterase (C) uniform labeling across the cell wall can be observed. c, cuticle; cw, cell wall; ph, penetration hypha; pm, plasmalemma. Bars are 0.5 μm .

The physico-chemical properties of the enzymes suggest that the potentially destructive enzymes such as polygalacturonate lyase, but also certain cellulases and pectin methyl esterases are tightly bound to the plant cell wall exchangers [19]. This may be important to restrict cell wall degradation to a minimum needed to allow growth of the penetration hypha through the cell wall. In contrast to findings with necrotrophic fungi, catabolite repression does not seem to be involved in regulation of cell wall-degrading enzymes of this biotrophic rust fungus [19, 36].

3.2.4.3. How Important are Cell Wall-Degrading Enzymes? Different experimental approaches have been used to challenge the importance of cell wall-degrading enzymes in the penetration process, with most attention paid to pectin-degrading enzymes.

Kolattukudy and co-workers[17, 33] raised anitisera against two polygalacturonases and one polygalacturonate lyase formed by the pea pathogen *Fusarium solani* f. sp. *pisi*. The antisera were included in the infection droplet, and from inhibition of infection by antibodies to polygalacturonate lyase it was concluded that the enzyme was important for penetration and virulence. The antibodies against the hydrolases did not have any effect.

In addition, protection of plants by naturally occuring inhibitors of pectic enzymes was taken as an indication of the importance of pectic enzymes. The presence of polygalacturonase-inhibiting proteins have been demonstrated in a number of plants [12, 14] and are thought to be potentially important to prevent infection. In *Phaseolus*, polygalacturonase activity was not detectable after infection with *C. lindemuthianum* [103] and polygalacturonase inhibitors may be a factor which determines host specificity of certain pathogens [11, 41]. However, since these inhibitors are abundantly synthesized by plants after pathogen attack [96] it can be assumed that polygalacturonases are important to the invading pathogen. Likewise, epicatechin, a phenolic present in high concentration in unripe avocado fruits, was found to completely inhibit both polygalacturonase and polygalacturonate lyase activity of the pathogen *C. gloeosporioides* [77, 99]. Only when the inhibitor concentration was reduced during fruit ripening and sank below a certain threshold, the fungus was able to macerate the tissue. An advantage of the relative broad effectiveness of the inhibitor is that all pectic enzymes have been inactivated simultaneously, which has not yet been achieved by classical or targeted mutagenesis approaches. Scott-Craig et al. [84] used targeted insertional mutagenesis to inactivate a gene encoding *endo* -polygalacturonase of the maize pathogen *Cochliobolus heterostrophus*. Virulence of the resulting mutant, probably due to *exo*-polygalacturonase activity was not affected by mutagenesis. Likewise, Kusserow [62] inactivated a *Penicillium olsonii* gene encoding a basic (pI 8.9) polygalacturonase I, creating mutants which only expressed a highly active acidic (pI 4.6) polygalacturonase II. After inoculation of *Arabidopsis thaliana* with *P. olsonii* wild type and PG I-deficient mutant, no difference in virulence was detected. Bowen et al. [10] inactivated the *pnlA* (pectin lyase) gene of the apple rot fungus *Glomerella cingulata* by targeted mutagenesis. Again, the mutant, probably due to other *pnl* -genes, showed no reduction in virulence.

Using targeted mutagenesis, other carbohydrate depolymerase genes were inactivated. Walton and co-workers created *C. heterostrophus* mutants deficient in xylanase I and II activity [3]. In spite of 90% reduction in xylanase activity, the mutant was not reduced in virulence, probably because of xylanase III activity that remained. Also, targeted gene disruption of *EXG1*, encoding *exo*-ß-1,3-glucanase of this fungus did not reduce virulence [83].

Due to high specificity, targeted mutagenesis has not resulted in mutants deficient in a whole group of enzymes, e.g. all pectic enzymes or all pectin or polygalacturonate lyases or hydrolases. Therefore it is difficult to evaluate the importance of these enzymes.

Ball et al. [5] UV-mutagenized the hemibiotrophic oilseed rape pathogen *Pyrenopeziza brassicae* and isolated mutants which were apathogenic and deficient in protease activity. Backcrosses of a mutant and the wild type strain indicated that the protease-positive phenotype co-segregated with pathogenicity. Furthermore, after transformation of the mutant with wild-type genomic DNA, a transformant was found which was both protease-positve and pathogenic. While these data suggest that protease is a pathogenicity factor in *P. brassicae*, targeted mutagenesis has not been performed. Such an approach could prove the significant role played by protease in *Pyrenopeziza*.

Taken together, by inactivating genes by molecular techniques it seems possible to directly analyze the role of different cell wall-degrading enzymes. Results obtained so far clearly show that the deletion of single genes will only have limited effects, and understanding the role of pectin-, cellulose- or xylose-degrading enzymes will be possible only after inactivating entire groups of enzymes and thus creating true pectic enzyme-, cellulase- or xylanase-deficient strians.

4. Conclusions

In this paper we have described events occuring in plant pathogenic fungi during infection structure differentiation, and we compared mechanisms of initial infection of fungi of different nutritional groups. In obligate biotrophs such as rust fungi many events which are thought of as essential for the establishment of a pathosystem are regulated differentiation-specifically, i.e. they follow a strict morphogenetic program (Figure 4). Thus, various differentiation-specific proteins/peptides have been detected after high-resolution two-dimensional polyacrylamide gel electrophoresis, and differential screening of a cDNA library has allowed the isolation of cDNA clones complementary to stage-specific mRNA species. Also formation of enzymes are under morphogenetic control. Chitin deacetylase, an enzyme which converts chitin to chitosan, is synthesized when the fungus penetrates through the stomatal pore. The modified fungal cell wall polymer is less accessable to degradation by plant chitinases and chitin deacetylase may thus play an important role in the protection of the fungus. Cell wall-degrading enzymes of the broad bean rust fungus are not substrate-inducible and catabolite-repressable as known for many necrotrophs, but strictly developmentally controlled. Proteases, cellulases, pectin methyl esterases and a polygalacturonate lyase are formed sequentially after the fungus has invaded the leaf. All of these enzymes are synthesized during defined stages of infection structure differentiation, with polygalacturonate lyase being the only enzyme requiring substrate in addition to morphogenesis (Figure 4).

The early stages of infection by obligate biotrophs are thus governed by a developmental program, and with respect to biochemical adaptation to their environment, these fungi appear to be less flexible as compared to most necrotrophs. The tight morphogenetic control may also determine the narrow range of hosts that rusts are able to parasitize.

Acknowledgement: We thank the Deutsche Forschungsgemeinschaft (DFG) for financial support (Me 523/13-1, Me 523-14-1, and Me 523/14-2), and our colleagues, present and past, for critical and stimulating discussions and for skilful help in the laboratory. We are indebted to Dr. P. Knox, University of Leeds, UK, for the donation of the monoclonal antibody JIM5. Especially we thank Dr. R.L. Nicholson, Purdue University, West Lafayette, IN, U.S.A. for critically reading the manuscript and for stimulating discussions throughout the work.

5. References

1. Ace, C.I., Balsamo, M., Le, L.T. and Okulicz, W.C. (1994) Isolation of progesterone-dependent complementary desoxyribonucleic acid fragments from rhesus monkey endometrium by sequential subtractive hybridization and polymerase chain reaction, *Endocrinology* **134**, 1305-1309.
2. Allen, E.A., Hazen, B.E., Hoch, H.C., Kwon, Y., Leinhos, G.M.E., Staples, R.C., Stumpf, M.A. and Terhune, B.T. (1991) Appressorium formation in response to topographical signals by 27 rust species, *Phytopathology* **81**, 323-331.
3. Apel, P.C., Panaccione, D.G., Holden, F.R. and Walton, J.D. (1993) Cloning and targeted gene disruption of XYL1, a ß-1,4-xylanase gene from the maize pathogen *Cochliobolus carbonum, Molecular Plant-Microbe Interactions* **6**, 467-473.
4. Bailey, J.A., O'Connell, R.J., Pring, R.J. and Nash, C. (1992) Infection strategies of *Colletotrichum* species, in J.A. Bailey and M.J. Jeger (eds.), *Colletotrichum* : Biology, Pathology and Control, Commonwealth Agricultural Bureau International, Wallingford, Oxon, U.K., pp. 88-120.
5. Ball, A.M., Ashby, A.M., Daniels, M.J., Ingram, D.S. and Johnstone, K. (1991) Evidence for the requirement of extracellular protease in the pathogenic interaction of *Pyrenopeziza brassicae* with oilseed rape, *Physiological and Molecular Plant Pathology* **38**, 147-161.
6. Bhairi, S., Buckley, E.H. and Staples, R.C. (1990) Protein synthesis and gene expression during appressorium formation in *Glomerella magna, Experimental Mycology* **14**, 207-217.
7. Bhairi, S., Staples, R.C., Freve, P. and Yoder, O.C. (1988) Analysis of a developmental gene expressed during differentiation of germlings of the bean rust fungus, in R. Palacios and D.P.S. Verma (eds.), *Molecular Genetics of Plant-Microbe Interactions*, APS Press, St. Paul, Minnesota, pp. 271-272.
8. Bhairi, S.M., Staples, R., C., Freve, P. and Yoder, O.C. (1989) Characterization of an infection structure-specific gene from the rust fungus *Uromyces appendiculatus, Gene* **81**, 237-243.
9. Blais, P., Rogers, P.A. and Charest, P.M. (1992) Kinetic of the production of polygalacturonase and pectin lyase by two closely related formae speciales of *Fusarium oxysporum, Experimental Mycology* **16**, 1-7.
10. Bowen, J.K., Templeton, M.D., Sharrock, K.R., Crowhurst, R.N. and Rikkerink, E.H.A. (1995) Gene inactivation in the plant pathogen *Glomerella cingulata* : Three strategies for the disruption of the pectin lyase gene pnlA, *Molecular and General Genetics* **246**, 196-205.
11. Brown, A.E. and Adikaram, N.K.B. (1982) The differential inhibition of pectic enzymes from *Glomerella cingualta* and *Botrytis cinerea* by a cell wall protein from *Capsicum annuum* fruit, *Phytopathologische Zeitschrift* **105**, 27-38.
12. Bugbee, W.M. (1993) A pectin lyase inhibitor protein from cell walls of sugar beet, *Phytopathology* **83**, 63-68.
13. Caprari, C., Bergmann, C., Migheli, Q., Salvi, G., Albersheim, P., Darville, A., Cervone, F. and De Lorenzo, G. (1993) *Fusarium moniliforme* secretes four *endo*-polygalacturonases derived from a single gene product, *Physiological and Molecular Plant Pathology* **43**, 453-462.
14. Cervone, F., De Lorenzo, G., Caprari, C., Clark, A.J., Desiderio, A., Devoto, A., Leckie, F., Nuss, L., Salvi, G. and Toubart, P. (1993) The interaction between fungal endopolygalacturonase and plant cell wall PGIP (polygalacturonase-inhibiting protein), in B. Fritig and M. Legrand (eds.), *Mechanisms of Plant Defense Responses*, Kluwer Academic Publishers, Dordrecht, The Netherlands, pp. 64-67.
15. Clement, J.A., Porter, R., Butt, T.M. and Beckett, A. (1994) The role of hydrophobicity in attachment of urediniospores and sporelings of *Uromyces viciae-fabae, Mycological Research* **98**, 1217-1228.
16. Cooper, R.M. (1984) The role of cell wall-degrading enzymes in infection and damage, in R.K.S. Wood and G.J. Jellis (eds.), *Plant Diseases: Infection, Damage and Loss*, Blackwell Scientific Publications, Oxford, pp. 13-27.
17. Crawford, M.S. and Kolattukudy, P.E. (1987) Pectate lyase from *Fusarium solani* f.sp. *pisi* : purification, characterization, *in vitro* translation of the mRNA, and involvement in pathogenicity,

Archives of Biochemistry and Biophysics **258**, 196-205.
18. De Bary, A. (1886) Über einige Sclerotinien und Sclerotien-Krankheiten, *Botanische Zeitung* **44**, 376-480.
19. Deising, H., Frittrang, A.K., Kunz, S. and Mendgen, K. (1995) Regulation of pectin methylesterase and polygalacturonate lyase activity during differentiation of infection structures in *Uromyces viciae-fabae*, *Microbiology* **141**, 561-571.
20. Deising, H., Jungblut, P.R. and Mendgen, K. (1991) Differentiation-related proteins of the broad bean rust fungus *Uromyces viciae-fabae*, as revealed by high resolution two-dimensional polyacrylamide gel electrophoresis, *Archives of Microbiology* **155**, 191-198.
21. Deising, H., Nicholson, R.L., Haug, M., Howard, R.J. and Mendgen, K. (1992) Adhesion pad formation and the involvement of cutinase and esterases in the attachment of uredospores to the host cuticle, *The Plant Cell* **4**, 1101-1111.
22. Deising, H., Rauscher, M., Haug, M. and Heiler, S. (1995) Differentiation and cell wall-degrading enzymes in the obligately biotrophic rust fungus *Uromyces viciae-fabae*, *Canadian Journal of Botany* in press,
23. Deising, H. and Siegrist, J. (1995) Chitin deacetylase activity of the rust *Uromyces viciae-fabae* is controlled by fungal morphogenesis, *FEMS Microbiology Letters* **127**, 207-212.
24. Dickinson, S. (1949) Studies in the physiology of obligate parasitism. II. The behaviour of the germ tubes of certain rusts in contact with various membranes, *Annals of Botany* **13**, 219-236.
25. Dunkle, L.D., Maheshwari, R. and Allen, P.J. (1969) Infection structures from rust urediospores: Effect of RNA and protein synthesis inhibitors, *Science* **163**, 481-482.
26. Edwards, M.C. and Bowling, D.J.F. (1986) The growth of rust germ tubes towards stomata in relation to pH gradients, *Physiological and Molecular Plant Pathology* **29**, 185-196.
27. Emmett, R.W. and Parbery, D.G. (1975) Appressoria, *Annual Review of Phytopathology* **13**, 147-167.
28. Fasters, M.K., Daniels, U. and Moerschbacher, B.M. (1993) A simple and reliable method for growing the wheat stem rust fungus, *Puccinia graminis* f. sp. *tritici*, in liquid culture, *Physiological and Molecular Plant Pathology* **42**, 259-265.
29. Flaishman, M.A. and Kolattukudy, P.E. (1994) Timing of fungal invasion using host's ripening hormone as a signal, *Proceedings of the National Academy of Sciences of the U.S.A* **91**, 6579-6583.
30. Freytag, S., Bruscaglioni, L., Gold, R.E. and Mendgen, K. (1988) Basidiospores of rust fungi (*Uromyces* species) differentiate infection structures *in vitro*, *Experimental Mycology* 12, 275-283.
31. Freytag, S. and Mendgen, K. (1991) Carbohydrates on the surface of urediniospore- and basidiospore-derived infection structures of heteroecious and autoecious rust fungi, *The New Phytologist* **119**, 527-534.
32. Gold, S., Duncan, G., Barrett, K. and Kronstad, J. (1994) cAMP regulates morphogenesis in the fungal pathogen *Ustilago maydis*, *Genes & Development* **8**, 2805-2816.
33. González-Candelas, L. and Kolattukudy, P.E. (1992) Isolation and analysis of a novel inducible pectate lyase gene from the phytopathogenic fungus *Fusarium solani* f. sp. *pisi* (*Nectria haematococca*, mating population VI), *Journal of Bacteriology* **174**, 6343-6349.
34. Hadwiger, L.A. and Line, R.F. (1981) Hexosamine accumulations are associated with the terminated growth of *Puccinia striiformis* on wheat isolines, *Physiological Plant Pathology* **19**, 249-255.
35. Harder, D.E. and Chong, J. (1991) Rust haustoria, in K. Mendgen and D.-E. Lesemann (eds.), *Electron Microscopy of Plant Pathogens*, Springer-Verlag, Berlin, Heidelberg, pp. 235-250.
36. Heiler, S., Mendgen, K. and Deising, H. (1993) Cellulolytic enzymes of the obligately biotrophic rust fungus *Uromyces viciae-fabae* are regulated differentiation-specifically, *Mycological Research* **97**, 77-85.
37. Hoch, H.C. and Staples, R.C. (1987) Structural and chemical changes among the rust fungi during appressorium development, *Annual Review of Phytopathology* **25**, 231-247.
38. Hoch, H.C. and Staples, R.C. (1991) Signaling for infection structure formation in fungi, in G.T. Cole and H.C. Hoch (eds.), *The Fungal Spore and Disease Initiation in Plants and Animals*, Plenum Press, New York, London, pp. 25-46.
39. Hoch, H.C., Staples, R.C. and Bourett, T. (1987) Chemically induced appressoria in *Uromyces appendiculatus* are formed aerially, apart from the substrate, *Mycologia* **79**, 418-424.
40. Hoch, H.C., Staples, R.C., Whitehead, B., Comeau, J. and Wolf, E.D. (1987) Signaling for growth orientation and cell differentiation by surface topography in *Uromyces*, *Science* **235**, 1659-1662.
41. Hoffman, R.M. and Turner, J.G. (1984) Occurrence and specificity of an endopolygalacturonase inhibitor in *Pisum sativum*, *Physiological Plant Pathology* **24**, 49-59.
42. Howard, R.J. and Ferrari, M.A. (1989) Role of melanin in appressorium function, *Experimental Mycology* **13**, 403-418.

43. Howard, R.J., Ferrari, M.A., Roach, D.H. and Money, N.P. (1991) Penetration of hard substances by a fungus employing enormous turgor pressures, *Proceedings of the National Academy of Sciences of the U.S.A.* **88**, 11281-11284.
44. Huang, B.-F. and Staples, R.C. (1982) Synthesis of proteins during differentiation of the bean rust fungus, *Experimental Mycology* **6**, 7-14.
45. Hwang, C.-S. and Kolattukudy, P.E. (1995) Isolation and characterization of genes expressed uniquely during appressorium formation by *Colletotrichum gloeosporioides* conidia induced by the host surface wax, *Molecular and General Genetics* **247**, 282-294.
46. Jelitto, T.C., Page, H.A. and Read, N.D. (1994) Role of external signals in regulating the pre-penetration phase of infection by the rice blast fungus, *Magnaporthe grisea*, *Planta* **194**, 471-477.
47. Johnson, T. (1934) A tropic response in germ tubes of uredospores of *Puccinia graminis tritici*, *Phytopathology* **24**, 80-82.
48. Johnston, D.J. and Williamson, B. (1992) Purification and characterization of four polygalacturonases from *Botrytis cinerea*, *Mycological Research* **96**, 343-349.
49. Jones, M.J. and Epstein, L. (1990) Adhesion of macroconidia to the plant surface and virulence of *Nectria haematococca*, *Applied and Environmental Microbiology* **56**, 3772-3778.
50. Judelson, H.S. and Michelmore, R.W. (1990) Highly abundant and stage-specific mRNAs in the obligate pathogen *Bremia lactucae*, *Molecular Plant-Microbe Interactions* **3**, 225-232.
51. Kafetzopoulos, D., Martinou, A. and Bouriotis, V. (1993) Bioconversion of chitin to chitosan: Purification and characterization of chitin deacetylase from *Mucor rouxii*, *Proceedings of the National Academy of Sciences of the U.S.A.* **90**, 2564-2568.
52. Kaminskyj, S.G. and Day, A.W. (1984) Chemical induction of infection structures in rust fungi. I. Sugars and complex media, *Experimental Mycology* **8**, 63-72.
53. Kaminskyj, S.G.W. and Day, A.W. (1984) Chemical induction of infection structures in rust fungi. II. Inorganic ions, *Experimental Mycology* **8**, 193-201.
54. Kauss, H., Jeblick, W. and Young, D.H. (1982/1983) Chitin deacetylase from the plant pathogen *Colletotrichum lagenarium*, *Plant Science Letters* **28**, 231-236.
55. Keon, J.P.R., Byrde, R.J.W. and Cooper, R.M. (1987) Some aspects of fungal enzymes that degrade plant cell walls, in G.F. Pegg and P.G. Ayres (eds.), *Fungal Infection of Plants*. Symposium of the British Mycological Society, Cambridge University Press, Cambridge, New York, New Rochelle, Melbourne, Sydney, pp. 133-157.
56. Kim, W.K., Howes, N.K. and Rohringer, R. (1982) Detergent-soluble polypeptides in germinated uredospores and differentiated uredosporelings of wheat stem rust, *Canadian Journal of Plant Pathology* **4**, 328-333.
57. Kubo, Y. and Furusawa, I. (1986) Localization of melanin in appressoria of *Colletotrichum lagenarium*, *Canadian Journal of Microbiology* **32**, 280-282.
58. Kubo, Y. and Furusawa, I. (1991) Melanin biosynthesis. Prerequisite for successful invasion of the host by appressoria of *Colletotrichum* and *Pyricularia*, in G.T. Cole and H.C. Hoch (eds.), *The Fungal Spore and Disease Initiation in Plants and Animals*, Plenum Publishing Corporation, New York, pp. 205-218.
59. Kubo, Y., Furusawa, I. and Yamamoto, M. (1984) Regulation of melanin biosynthesis during appressorium formation in *Colletotrichum lagenarium*, *Experimental Mycology* **8**, 364-369.
60. Kubo, Y., Nakamura, H., Kobayashi, K., Okuno, T. and Furusawa, I. (1991) Cloning of a melanin biosynthetic gene essential for appressorial penetration of *Colletotrichum lagenarium*, *Molecular Plant-Microbe Interactions* **4**, 440-445.
61. Kunoh, H., Nicholson, R.L., Yoshioka, H., Yamaoka, N. and Kobayashi, I. (1990) Preparation of the infection court by *Erysiphe graminis* : Degradation of the host cuticle, *Physiological and Molecular Plant Pathology* **36**, 397-407.
62. Kusserow, H., Untersuchungen zur Bedeutung von Polygalakturonasen für das Infektions- und Kolonisationsverhalten von *Penicillium olsonii* auf *Arabidopsis thaliana*. (1994) Doctoral thesis, Freie Universität Berlin.
63. Lee, Y.-H. and Dean, R.A. (1993) cAMP regulates infection structure formation in the plant pathogenic fungus *Magnaporthe grisea*, *The Plant Cell* **5**, 693-700.
64. Lee, Y.-H. and Dean, R.A. (1993) Stage-specific gene expression during appressorium formation of *Magnaporthe grisea*, *Experimental Mycology* **17**, 215-222.
65. Lewis, B.G. and Day, J.R. (1972) Behaviour of uredospore germ-tubes of *Puccinia graminis tritici* in relation to the fine structure of wheat leaf surfaces, *Transactions of the British Mycological Society* **58**, 139-145.
66. Liu, Z., Szabo, L.J. and Bushnell, W.R. (1993) Molecular cloning and analysis of abundant and stage-specific mRNAs from *Puccinia graminis*, *Molecular Plant-Microbe Interactions* **6**, 84-91.

67. Maclean, D.J. (1982) Axenic culture and metabolism of rust fungi, in K.J. Scott and A.K. Chakravorty (eds.), *The Rust Fungi*, Academic Press, London, New York, pp. 37-120.
68. Maheshwari, R., Allen, P.J. and Hildebrandt, A.C. (1967) Physical and chemical factors controlling the development of infection structures from urediospore germ tubes of rust fungi, *Phytopathology* **57**, 855-862.
69. Mendgen, K. and Deising, H. (1993) Infection structures of fungal plant pathogens - a cytological and physiological evaluation, *The New Phytologist* **124**, 193-213.
70. Mendgen, K. and Dressler, E. (1983) Culturing *Puccinia coronata* on a cell monolayer of the *Avena sativa* coleoptile, *Phytopathologische Zeitschrift* **108**, 226-234.
71. Nagarajan, S. and Singh, D.V. (1990) Long-distance dispersion of rust pathogens, *Annual Review of Phytopathology* **28**, 139-153.
72. Nicholson, R.L. and Epstein, L. (1991) Adhesion of fungi to the plant surface, in G.T. Cole and H.C. Hoch (eds.), *The Fungal Spore and Disease Initiation In Plants And Animals*, Plenum Press, New York, London, pp. 3-23.
73. O'Connell, R.J. and Ride, J.P. (1990) Chemical detection and ultrastructural localization of chitin in cell walls of *Colletotrichum lindemuthianum, Physiological and Molecular Plant Pathology* **37**, 39-53.
74. Pascholati, S.F., Deising, H., Leite, B., Anderson, D. and Nicholson, R.L. (1993) Cutinase and non-specific esterase activities in the conidial mucilage of *Colletotrichum graminicola, Physiological and Molecular Plant Pathology* **42**, 37-51.
75. Pascholati, S.F., Yoshioka, H., Kunoh, H. and Nicholson, R.L. (1992) Preparation of the infection court by *Erysiphe graminis* f.sp. *hordei* : Cutinase is a component of the conidial exudate, *Physiological and Molecular Plant Pathology* **41**, 53-59.
76. Podila, G.K., Rogers, L.M. and Kolattukudy, P.E. (1993) Chemical signals from avocado surface wax trigger germination and appressorium formation in *C. gloeosporioides, Plant Physiology* **103**, 267-272.
77. Prusky, D., Gold, S. and Keen, N.T. (1989) Purification and characterization of an endopolygalacturonase produced by *Colletotrichum gloeosporioides, Physiological and Molecular Plant Pathology* **35**, 121-133.
78. Rauscher, M., Mendgen, K. and Deising, H. (1995) Extracellular proteases of the rust fungus *Uromyces viciae-fabae, Experimental Mycology* **19**, 26-34.
79. Read, N.D. (1991) Low temperature scanning electron microscopy of fungi and fungus-plant interactions, in K. Mendgen and D.-E. Lesemann (eds.), *Electron Microscopy of Plant Pathogens*, Springer-Verlag, Berlin, Heidelberg, New York, pp. 17-29.
80. Read, N.D., Kellock, L.J., Knight, H. and Trewavas, A.J. (1992) Contact sensing during infection by fungal pathogens, in J.A. Callow and J.R. Green (eds.), *Perspectives in Plant Cell Recognition*, Cambridge University Press, Cambridge, pp. 137-172.
81. Ride, J.P. and Barber, M.S. (1990) Purification and characterization of multiple forms of endochitinase from wheat leaves, *Plant Science* **71**, 185-197.
82. Roelfs, A.P. (1985) Epidemiology in North America, in A.P. Roelfs and W.R. Bushnell (eds.), *The Cereal Rusts*, Academic Press, Orlando, San Diego, New York, pp. 403-434.
83. Schaeffer, H.J., Leykam, J. and Walton, J.D. (1994) Cloning and targeted gene disruption of EXG1, encoding exo-ß-1,3-glucanase, in the phytopathogenic fungus *Cochliobolus carbonum, Applied and Environmental Microbiology* **60**, 594-598.
84. Scott-Craig, J.S., Panaccione, D.G., Cervone, F. and Walton, J.D. (1990) Endopolygalacturonase is not required for pathogenicity of *Cochliobolus carbonum* on maize, *The Plant Cell* **2**, 1191-1200.
85. Sentandreu, R., Mormeneo, S. and Ruiz-Herrera, J. (1994) Biogenesis of the fungal cell wall, in J.G.H. Wessels and F. Meinhardt (eds.), *The Mycota. I. Growth, Differentiation and Sexuality*, Springer-Verlag, Berlin, Heidelberg, pp. 111-124.
86. Shaw, M., Boasson, R. and Scrubb, L. (1985) Effect of heat shock on protein synthesis in flax rust uredosporelings, Canadian Journal of Botany 63, 2069-2076.
87. Short, J.M., Fernandez, J.M., Sorge, J.A. and Huse, W.D. (1988) λZAP: A bacteriophage λ expression vector with *in vivo* excision properties, *Nucleic Acids Research* **16**, 7583-7600.
88. Showalter, A.M. (1993) Structure and function of plant cell wall proteins, *The Plant Cell* **5**, 9-23.
89. Siegrist, J. and Kauss, H. (1990) Chitin deacetylase in cucumber leaves infected by *Colletotrichum lagenarium, Physiological and Molecular Plant Pathology* **36**, 267-275.
90. Staples, R.C., Grambow, H.J., Hoch, H.C. and Wynn, W.K. (1983) Contact with membrane grooves induces wheat stem rust uredospore germlings to differentiate appressoria but not vesicles, *Phytopathology* **73**, 1436-1439.
91. Staples, R.C., Hoch, H.C., Freve, P. and Bourette, T.M. (1989) Heat shock-induced development of

infection structures by bean rust uredospore germlings, *Experimental Mycology* **13**, 149-157.
92. Staples, R.C., Yoder, O.C., Hoch, H.C., Epstein, L. and Bhairi, S. (1986) Gene expression during infection structure development by germlings of the rust fungi, in J.A. Bailey (ed.), *Biology and Molecular Biology of Plant-Pathogen Interactions*, Springer-Verlag, Berlin, Heidelberg, pp. 331-341.
93. Stark-Urnau, M. and Mendgen, K. (1993) Differentiation of aecidiospore- and uredospore-derived infection structures on cowpea leaves and on artificial surfaces by *Uromyces vignae*, *Canadian Journal of Botany* **71**, 1236-1242.
94. Talbot, N.J., Ebbole, D.J. and Hamer, J.E. (1994) Identification and characterization of MPG 1, a gene involved in pathogenicity from the rice blast fungus Magnaporthe grisea, *The Plant Cell* **5**, 1575-1590.
95. Timberlake, W.E. (1986) Isolation of stage- and cell-specific genes from fungi, in J. Bailey (ed.), *Biology and Molecular Biology of Plant-Pathogen Interactions*, Springer-Verlag, Berlin, Heidelberg, pp. 343-357.
96. Toubart, P., Desiderio, A., Salvi, G., Cervone, F., Daroda, L. and De Lorenzo, G. (1992) Cloning and characterization of the gene encoding the endopolygacturonase-inhibiting protein (PGIP) of *Phaseolus vulgaris* L., *The Plant Journal* **2**, 367-373.
97. Walton, J.D., Bronson, C.R., Panaccione, D.G., Braun, E.J. and Akimitsu, K. (1995) *Cochliobolus*, in K. Kohmoto, U.S. Singh and R.P. Singh (eds.), *Pathogenesis and Host Specificity in Plant Diseases. Histopathological, Biochemical, Genetic and Molecular Bases*, Elsevier Science Ltd., Oxford, New York, Tokyo, pp. 65-81.
98. Wanner, R., Förster, H., Mendgen, K. and Staples, R.C. (1985) Synthesis of differentiation-specific proteins in germlings of the wheat stem rust fungus after heat shock, *Experimental Mycology* **9**, 279-283.
99. Wattad, C., Dinoor, A. and Prusky, D. (1994) Purification of pectate lyase produced by *Colletotrichum gloeosporioides* and its inhibition by epicatechin: A possible factor involvrd in the resistance of unripe avocado fruits to anthracnose, *Molecular Plant-Microbe Interactions* **7**, 293-297.
100. Wessels, J.G.H. (1994) Developmental regulation of fungal cell wall formation, *Annual Review of Phytopathology* **32**, 413-437.
101. Wessels, J.G.H., de Vries, O.M.H., Asgeirsdóttir, S.A. and Schuren, F.H.J. (1991) Hydrophobin genes involved in formation of aerial hyphae and fruit bodies in *Schizophyllum commune*, *The Plant Cell* **3**, 793-799.
102. White, B.T. and Yanofsky, C. (1993) Structural characterization and expression analysis of the *Neurospora* conidiation gene con-6, *Developmental Biology* **160**, 254-264.
103. Wijesundera, R.L.C., Bailey, J.A., Byrde, R.J.W. and Fielding, A.H. (1989) Cell wall degrading enzymes of *Colletotrichum lindemuthianum* : their role in the development of bean anthracnose, *Physiological and Molecular Plant Pathology* **34**, 403-413.
104. Wösten, H.A.B., de Vries, O.M.H. and Wessels, J.G.H. (1993) Interfacial self-assembly of a fungal hydrophobin into a hydrophobic rodlet layer, *The Plant Cell* **5**, 1567-1574.
105. Xiao, J.Z., Ohshima, A., Watanabe, T., Kamakura, T. and Yamaguchi, I. (1994) Studies on cellular differentiation of Magnaporthe grisea. Physico-chemical aspects of substratum surfaces in relation to appressorium formation, *Physiological and Molecular Plant Pathology* **44**, 227-236.
106. Xuei, X., Bhairi, S., Staples, R.C. and Yoder, O.C. (1992) Characterization of INF56, a gene expressed during infection structure development of *Uromyces appendiculatus*, *Gene* **110**, 49-55.

STRUCTURAL AND FUNCTIONAL ASPECTS OF MYCOBIONT - PHOTOBIONT RELATIONSHIPS IN LICHENS COMPARED WITH MYCORRHIZAE AND PLANT PATHOGENIC INTERACTIONS

ROSMARIE HONEGGER
University of Zürich
Institute of Plant Biology
Zollikerstrasse 107
CH - 8008 Zürich, Switzerland

1. Peculiarities of the lichen symbiosis

1.1. LICHEN MYCOBIONTS

Lichen-forming fungi are, like plant pathogens or mycorrhizal fungi, a taxonomically heterogeneous, polyphyletic group of nutritional specialists which acquire fixed carbon from a living photoautotrophic partner (Tab. 1; for molecular systematics see [12]). In the lichen symbiosis the photobiont is not a multicellular structure as in fungal symbioses with plants, but a population of minute green algal and/or cyanobacterial cells. These are housed and maintained within the thallus of the mycobiont which, in most cases, is the quantitatively predominant exhabitant. According to Article 13.1d of the International Code of Botanical Nomenclature [16] the species names of lichens refer to the fungal partner. However, the taxonomic description of lichen-forming fungi is based on sexual reproductive structures which are exclusively formed in the symbiotic state, and on characteristics of the symbiotic phenotype (morphotype, chemotype etc.), *i.e.* the form in which we find and collect lichens in nature. Axenically cultured, aposymbiotic phenotypes differ very significantly from symbiotic ones. Therefore it is customary among lichenologists to refer to the mycobiont of a particular lichen species even when the aposymbiotic, cultured phenotype is considered.

Lichenization is a successful nutritional strategy, one out of five fungal species being lichenized (Tab. 1). For comparison: approximately one out of five fungal species is a plant pathogen, but only one out of 12 fungi forms mycorrhizae. There are no fundamental differences between lichenized and non-lichenized fungi except for the manifold adaptations of the former to the cohabitation with a population of minute photobiont cells. Many lichen-forming fungi are physiologically facultative biotrophs, *i.e.* they can be cultured under sterile conditions (review: [1]). Axenically cultured lichen

M. Nicole and V. Gianinazzi-Pearson (eds.), Histology, Ultrastructure and Molecular Cytology of Plant-Microorganism Interactions, 157-176.

Table 1. Orders of Ascomycotina and Basidiomycotina, and classes of Deuteromycotina, which include lichenized taxa [18], [30].

Order Class	Nutritional strategies	Thallus anatomy in lichenized taxa	Abbreviations
ASCOMYCOTINA[1, 2, 3]			
Arthoniales	l, nl (*lp, sap*)	ns	f — fungicolous
Caliciales	l, nl (*f, lp, sap*)	ns, s	l — lichenized
Dothideales	nl (*sap, pp, lp*), l	ns	lp — lichenicolous (lichen parasites)
Graphidales	l, nl (*lp*)	ns	
Gyalectales	l	ns	myc — mycorrhiza
Lecanorales	l, nl (*sap, lp*)	ns, s	nl — non - lichenized
Leotiales	nl (*sap, pp, lp*), l	ns, s	ns — non - stratified
Lichinales	l	ns, s	pp — plant pathogens
Opegraphales	l, nl (*lp, sap*)	ns, s	s — internally stratified
Ostropales	nl (*sap, pp, lp*), l	ns	sap — saprotrophic
Patellariales	nl (*sap, lp*), l	ns	
Peltigerales	l	s, ns	1) approx. 98% of lichen-forming fungi are Ascomycetes
Pertusariales	l	ns	
Pyrenulales	l, nl (*sap*)	ns	2) 16 out of 46 orders of Ascomycotina include lichenized taxa
Teloschistales	l, nl (lp)	ns, s	
Verrucariales	l, nl (*sap, lp*)	ns, s	
			3) ~46% of Ascomycotina are lichenized (approx. 13,250 spp.)
BASIDIOMYCOTINA[4, 5]			
Aphyllophorales	nl (*sap, myc, pp, f, lp*), l	ns, s	4) approx. 0.4% of lichen-forming fungi are Basidiomycetes
Agaricales	nl (*sap, myc, pp, lp*), l	ns, s	
			5) only ~0.3% of Basidiomycotina are lichenized (approx. 50 spp.)
DEUTEROMYCOTINA[6, 7]			
Coelomycetes	nl (*sap, pp, f, lp*), l	ns	6) approx. 1.6% of lichen-forming fungi are Deuteromycetes
Hyphomycetes	nl (*sap, pp, f, lp*), l	ns	
			7) ~0.3% of Deuteromycotina are lichenized (~200 spp.)
STERILE TAXA (with no known reproductive structures)[8]		ns, s	8) approx. 75 spp. (11 genera)

Approx. 55% of lichen-forming fungi form non-stratified (crustose, microfilamentous etc.) thalli
20% form either squamulose or placodioid thalli, and
22% form either foliose, or fruticose, internally stratified thalli

Señor Luis -

Aqui está, el diseño del ensayo.

- John

far, the most important concept I gained
ound myself continuously searching for
ndscape evolution throughout the regions
o 407 this past semester, I knew that
hology. Familiarity and a deeply rooted r
graphy has brought this application to gr
ord since I returned to the university in fa
anding and have achieved academic exce
nese past two years is 3.65.

Geography, I am most interested in pur
ng stratigraphic evidence in river valle
ition within the Central Plains or northern
ronments in which landscape evoluti
rch is important in order to discover p
d how this has affected the land surface
ow us to proceed toward better climatic

mycobionts are morphologically completely different from the lichen thallus, *i.e.* the symbiotic phenotype. Unfortunately it is not possible to resynthesize morphologically complex lichens routinely by combining axenically cultured isolates under sterile conditions (reviews: [13], [31]). The factors which trigger the expression of the symbiotic phenotype in compatible combinations are unknown (reviews: [28], [29]). This is a serious handicap for a wide range of experimental studies and one among other reasons why many aspects of lichen biology are so very poorly understood.

1.2. LICHEN PHOTOBIONTS

An estimated 85% of lichen-forming fungi are symbiotic with selected green algae, approximately 10% with cyanobacteria, and 3-4%, the so-called cephalodiate species, associate simultaneously with green algae and nitrogen-fixing cyanobacteria (reviews: [49], [27]). Approximately 100 spp. of lichen photobionts are so far known to science. In less than 2% of lichen species has the photobiont ever been identified at the species level. Taxonomic identification necessitates the isolation and culturing under defined conditions since many lichen photobionts change their morphology in the symbiotic state (review: [49]).

Many lichen photobiont taxa, especially representatives of the Trentepohliales (genera *Cephaleuros, Phycopeltis, Trentepohlia*), are common and widespread in nature. Some of them are even considered as pests in subtropic and tropic agriculture since they settle on long-living leaves not only of wild species, but also of a wide range of economically important plants such as coffee, tea, cocoa, rubber etc. where they have a considerable shading effect [17]. The growth of these epiphyllous algae is slowed down when they are getting colonized by foliicolous lichen mycobionts (approximately 480 spp. have so far been described [10]). Needless to say that growers don't like these biologically very interesting, often quite bizarre foliicolous lichens either! However, the most common green algal lichen photobionts in arctic/alpine to temperate and even desert ecosystems are representatives of the small genus *Trebouxia* de Puymaly (Microthamniales *fide* Friedl and Zeltner [11]). *Trebouxia* spp. are not abundant in aerophilic algal communities, but they are the photobionts of more than 50% of all lichen species.

Considering, on one hand, the enormous biomass of *Trebouxia* cells which are bound, for example, in the reindeer lichens which cover the ground of subarctic to arctic tundras, or in the various crustose, placodioid and foliose lichen taxa on alpine rock surfaces and, on the other hand, the paucity of the same algae in the free-living state it is obvious that *Trebouxia* spp. gain a distinct ecological advantage from the symbiotic way of life, their ecological fitness being increased when living within the partly very elaborate thalli of lichen-forming fungi. This is the reason why most lichenologists consider morphologically advanced lichens as mutualistic symbioses irrespective of the substantial gain of algal photosynthates by the heterotrophic partner. However, among more primitive microlichens there are numerous forms of interactions which might be best regarded as mild forms of fungal parasitism on algal or cyanobacterial cells. So far it

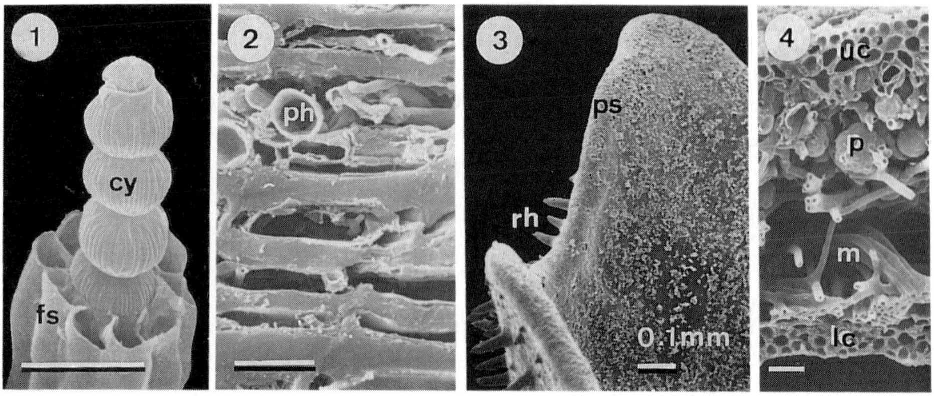

Figures 1 - 4. Examples of micro- and macrolichens. *Fig. 1.* Microfilamentous thallus of the lichenized basidiomycete *Dictyonema sericeum* (Telephoraceae, Aphyllophorales). *fs*: fungal sheath (dissected); *cy*: cyanobacterial filament (*Scytonema* sp.). For further details see [42]. *Fig. 2.* Crustose thallus of *Thelotrema lepadinum* (Graphidales, Ascomycotina) within the bark of *Ilex europaeus*. *ph*: photobiont: the coccoid phenotype of *Trentepohlia* sp. *Fig. 3.* Detail of a marginal lobe of the foliose macrolichen *Parmelia sulcata* (Lecanorales, Ascomycotina). The marginal, pseudomeristematic rim (*ps*) is slightly bulging. Rhizinae (*rh*) are seen on the strongly melanized lower thalline surface (see Fig. 6a). *Fig. 4.* Vertical cross-section of the foliose *Xanthoria parietina* (Teloschistales, Ascomycotina), wlth conglutinate upper (*uc*) and lower cortex (*lc*) around the gas-filled algal (*ph*) and medullary (*m*) layers.

Magnification bars equal **10 µm** unless otherwise stated.

unknown why so many lichen-forming fungi from diverse orders associate with these particular unicellular green algae which are obviously rather poor competitors outside the thalli of their fungal partner.

A somewhat similar situation is found among green algal endosymbionts of freshwater invertebrates (e.g., *Hydra viridis*) and protists (e.g., *Paramecium bursaria*). These *Chlorella* spp. are rare outside their heterotrophic exhabitants. Since they are continuously releasing fixed carbon into the medium they are very poor competitors. This peculiarity has a strong signalling effect during the establishment of the symbiosis (review: [44]). Within the lichen thallus *Trebouxia* cells do release photosynthates (mainly the acyclic polyol ribitol = adonitol) as has been shown in a series of experiments using the so-called "inhibition" or "isotope trapping technique" (reviews: [47], [25], [9]), and with ^{13}C NMR [39]. However, axenically cultured *Trebouxia* cells do not normally release significant amounts of soluble carbohydrates into the medium.

1.3. SPECIFICITY AND SELECTIVITY

As it is not possible to resynthesize lichens routinely by combining isolates of the mycobiont and photobiont under axenic conditions our present, rather limited knowledge on the range of acceptable photobiont species per fungal taxon originates mainly from isolation studies. Lichen thalli were collected in different regions and their photobiont was isolated and cultured. Presently available data indicate that lichen mycobionts are moderately to highly specific with regard to their photobiont, specificity being defined by the taxonomic relatedness of acceptable partners (moderate: several species from one genus are accepted; high: only one species is accepted; reviews: [24], [31]). Mycobionts which are symbiotic with *Trebouxia* spp. are highly selective with regard to their photobiont since they do not accept the most common free-living algae; instead they associate with species which are not abundant outside lichen thalli [14].

Resynthesis experiments [2] and field observations [43] indicate that germ tubes or other free hyphae of lichen mycobionts can secure their survival by forming inconspicuous crusts, so-called pre-thalli, with ultimately incompatible algal cells. However, the symbiotic phenotype is expressed in compatible combinations only (review: [28]). A wide range of lichen-forming fungi overcome this dilemma by producing vegetative symbiotic propagules such as soredia or isidia, small fungal packages containing juvenile cells of the photoautotrophic partner. Thallus fragmentation is likely to be a common and widespread mode of dispersal of the symbiotic state.

1.4. SYMBIOTIC PHENOTYPES

Best known to biologists are the so-called macrolichens, either leaf-like (foliose; Figs.3, 5, 6), band-shaped (examples: "mousse de chêne" [*Evernia prunastri*] and "mousse d' arbre" [*Pseudevernia furfuracea*], famous ingredients in the perfume industry), or shrubby thalli (fruticose; examples: "reindeer lichens" or "beardlichens") with an interesting internal stratification (Figs. 4, 5, 14-15). Main building blocks of macrolichens are hydrophilic, conglutinate pseudoparenchyma, usually forming peripheral cortical layers, and a system of loosely interwoven aerial hyphae with hydrophobic wall surfaces some of which are in close contact with the photobiont cell population (see 2.1.). These aerial hyphae create a gas-filled thalline area, usually as an algal and medullary layer. Different growth patterns can be distinguished in macrolichens (review: [28]). Most species have a marginal/apical pseudomeristem with high cell turnover rates in the mycobiont and photobiont. These newly formed cells gain their mature dimensions in the adjacent elongation zone. In non-growing, basal thalline areas there is very limited cell turnover in either partner. Upon careful analysis of cell turnover rates and cell dimensions in different thalline areas it becomes evident that the photobiont is under strict positional control by the fungal partner, the regulatory mechanisms being unknown.

Morphologically advanced macrolichens are the most complex vegetative structures in the fungal kingdom. They are the product of a long cohabitation of the partners and a very sophisticated adaptation of the heterotrophic exhabitant to the special requirements of its photoautotrophic symbiont. In contrast to parasitic and mutualistic fungal symbioses with plants, *i.e.* with multicellular, quantitatively predominant photo-autotrophic hosts, it is the **fungal** partner of these macrolichens which competes for space above ground, mimics plant-like structures and secures the gas exchange and adequate illumination of the photobiont cell population which is kept in the gas-filled

The symbiotic phenotype of a *Parmelia* sp. (Lecanorales)

Morphology

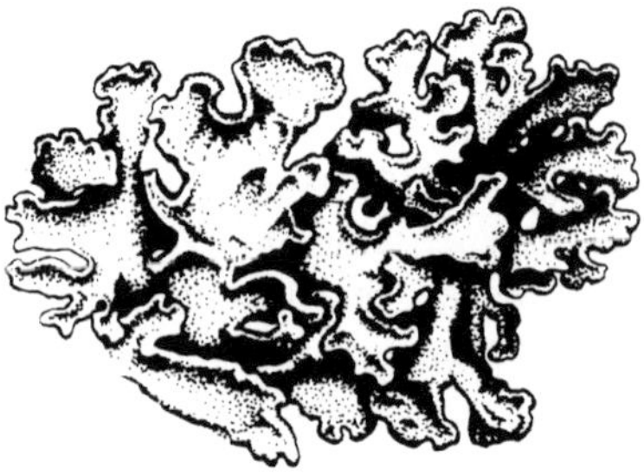

A foliose, dorsiventrally organised fungal thallus which harbours, in its interior, a population of *Trebouxia* cells (unicellular green algae)

Anatomy

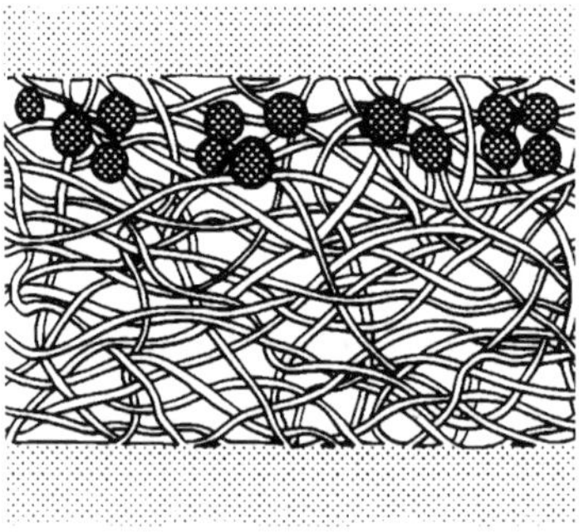

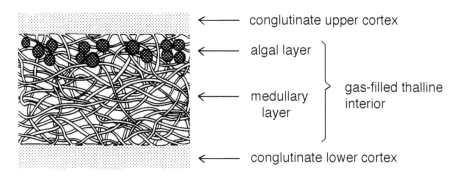

- conglutinate upper cortex
- algal layer } gas-filled thalline interior
- medullary layer } gas-filled thalline interior
- conglutinate lower cortex

Mycobiont - photobiont interface: an intraparietal haustorium

(intra: within; paries: wall)

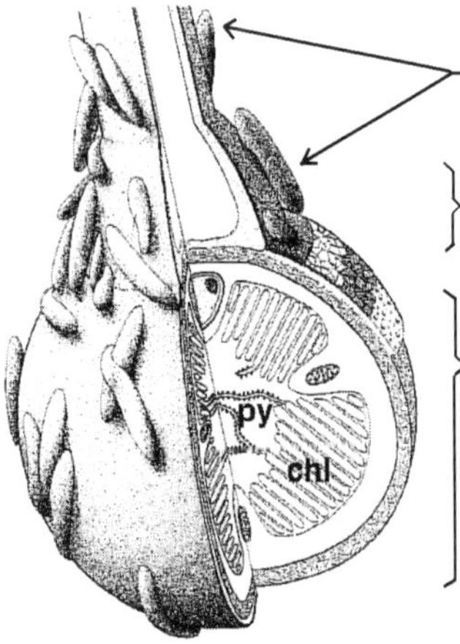

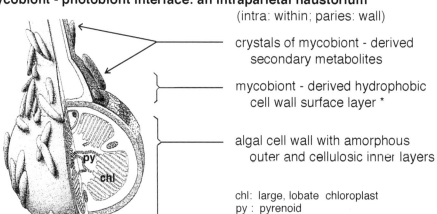

- crystals of mycobiont - derived secondary metabolites
- mycobiont - derived hydrophobic cell wall surface layer *
- algal cell wall with amorphous outer and cellulosic inner layers

chl: large, lobate chloroplast
py : pyrenoid

* for further details see Honegger (1991a, b)

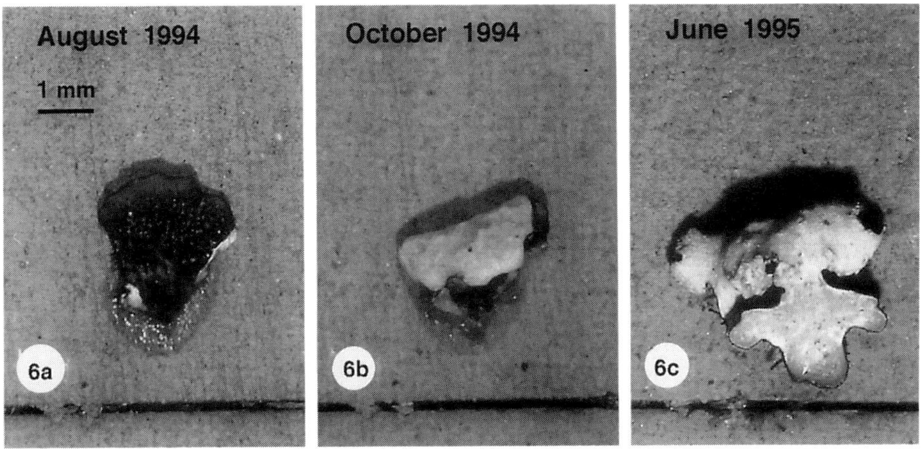

Figures 6a - c. Correction of spatial disturbances by means of growth processes within the fungal partner. Marginal thallus lobes of the foliose macrolichen *Parmelia sulcata* were fixed to a ceramic substratum in an upside-down position *(6a)*. The lobe is actively bending backwards by means of growth processes within the pseudomeristematic marginal rim and adjacent elongation zone *(6b)* and is growing arch-like until its marginal rim is touching the substratum *(6c)*. For further details see [29].

thalline interior in a similar position as the palisade parenchyma in plants. The photobiont cell population of macrolichens has no access to water and dissolved mineral nutrients except via the fungal partner (Figs.4, 11, 14, 16 ; reviews: [25], [26]). Mycobionts of dorsiventrally organized macrolichens are capable of sensing, in a yet unknown manner, spatial disturbances and to correct them by means of growth processes (Figs. 6a-c), thus securing the adequate illumination of the photobiont cell population [29]. However, it is important to note that only one out of four lichen mycobionts has this amazing morphogenetic capacity (see Tab. 1). The majority of lichen species are so-called microlichens, often quite inconspicuous, crustose (Fig. 2), microglobose or microfilamentous structures (Fig. 1) on or within the substratum.

Figure 5. (opposite page) Diagram illustrating structural aspects of the symbiotic relationship in representatives of the Parmeliaceae (Lecanorales).

2. Mycobiont - photobiont interactions

2.1. MORPHOLOGY OF DIFFERENT TYPES OF MYCOBIONT - PHOTOBIONT INTERFACES

Different types of mycobiont-photobiont interactions have been recorded in lichens, morphological complexity of the symbiotic phenotype, and on the cell wall structure and composition of the photobiont (reviews: [25], [27]). There are no free photobiont cells in naturally grown lichen thalli; each cell is connected to a specialized fungal element (see below) which has a triple function: 1) it is the site where soluble carbohydrates of photobiont origin are mobilized; 2) it provides water and dissolved mineral nutrients via passive, apoplastic transport from the thalline surface to the photobiont cells, and 3) short distance shifting of photobiont cells within the thalline interior and thus optimal positioning with regard to gas exchange and illumination is performed by means of intercalary growth processes within the contacting fungal structure (reviews: [20], [25], [26]).

Not found in lichens are extensive fungal haustorial structures with a large exchange surface in close contact with the plasma membrane of the photoautotroph partner, as seen in numerous biotrophic plant pathogenic interactions (e.g., Erysiphales, Uredinales; Fig. 7) and mycorrhizal symbioses (arbuscular [Fig. 8], ericoid, arbutoid and orchid). The majority of lichen mycobionts do not pierce the cell wall of their photobiont. The most common and widespread mode of interaction are intraparietal haustoria within cellulosic cell walls of *Trebouxia* spp. (Figs. 5, 11). These are more than appressoria since the fungal partner forms a small infection peg into the algal cell wall [19], [22], [25]. Intragelatinous fungal protrusions within the gelatinous sheath of cyanobacterial colonies are commonly found in cyanobacterial lichens (Fig. 20; reviews: [24], [25], [27]). Wall-to-wall apposition is typically found in micro- and macrolichens with green algal photobionts of the genera *Coccomyxa* or *Elliptochloris* whose trilaminar outermost cell wall layer contains enzymatically non-degradable, sporopollenin-like biopolymers (Fig. 13) [4], [25]. Wall-to wall apposition is also the most common mode of interaction in micro- and macrolichens with green algal photobionts of the genera *Myrmecia* or *Dictyochloropsis* (Figs. 16-19) whose cell walls are structurally very similar to those of *Coccomyxa* and *Elliptochloris* spp, but their trilaminar outermost layer contains no hydrolysis-resistant polymer ("intermediate type" *sensu* Brunner and Honegger [4]; reviews: [24], [25], [27]). Only in relatively primitive crustose lichens are finger-like, transparietal fungal protrusions (Fig. 10) regularly found in contact with the plasma membrane of the photobiont cell [19], [22], 27].

At first sight it is difficult to understand why lichen-forming fungi do not try to form large, branched or lobate transparietal haustorial structures as seen in powdery mildews or rusts (Fig. 7) or in arbuscular (Fig. 8) and a range of other mycorrhizal fungi. The photobiont cell population of macrolichens amounts for 20% or less of thalline biomass. Therefore it would seem advantageous for the fungal partner to create an

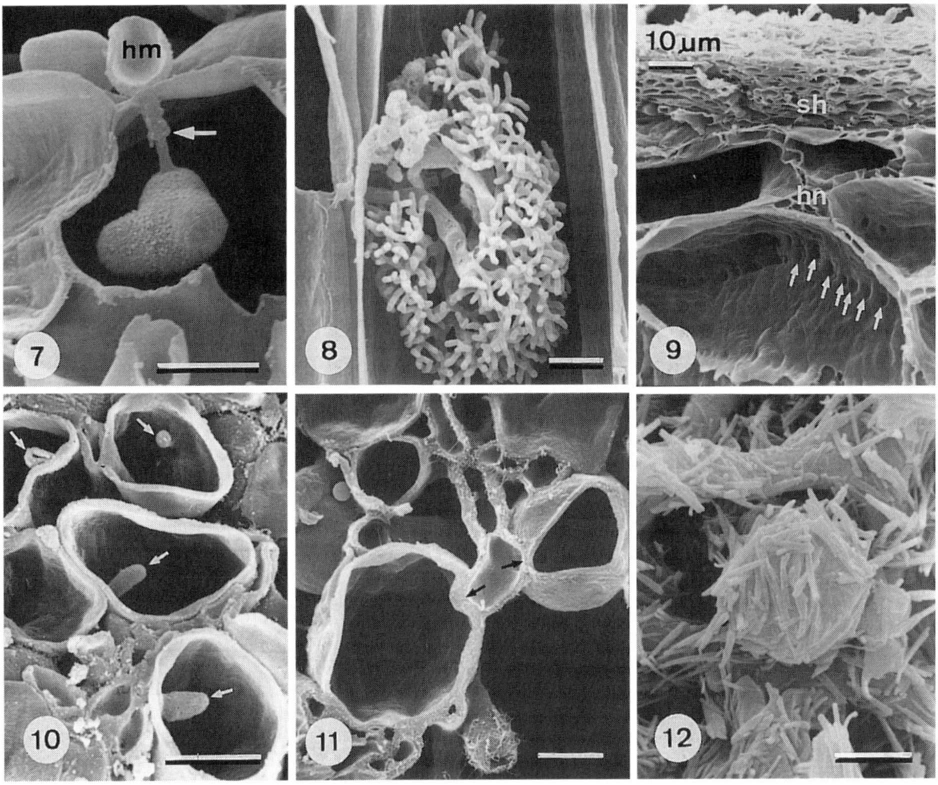

Figures 7 - 12. Examples of mycobiont-photobiont interfaces in diverse biotrophic fungal symbioses with photoautotrophic partners (for preparative techniques see [21]). *Fig. 7.* Lobate haustorium with narrow neck (arrow points to neck-band) of coffee rust, *Hemileia vastatrix,* in a spongy mesophyll cell of *Coffea arabica. hm*, haustorial mother cell. *Fig. 8.* The dichotomously branched arbuscule of *Glomus mossae* in a cortical cell of the root of *Zea mays*. *Fig. 9.* An ectomycorrhiza of *Pinus sylvestris*, formed by an unidentified basidiomycete, with conglutinate peripheral sheath (*sh*) around the root and the Hartig net (*hn*) around rhizodermal and cortical cells. Arrows point to plasmodesmata which are interlinking adjacent cortical cells. *Fig. 10.* Finger-like transparietal fungal protrusions (arrows) in cells of the *Trebouxia* photobiont of the crustose microlichen *Lecanora chlarotera* (Lecanorales, Ascomycotina). *Fig. 11.* Intraparietal haustoria (arrows) in the foliose macrolichen *Xanthoria parietina* (photobiont: *Trebouxia arboricola). Fig. 12.* Crystals of olivetoric acid, a carboxylic acid derivative, are covering the cell wall surfaces of the algal and medullary layers of *Cetrelia olivetorum* (Parmeliaceae, Lecanorales), thus enhancing its hydrophobicity (freeze-dried specimen).

Magnification bars equal **5 µm** unless otherwise stated.

exchange surface as large as possible in close contact with the plasma membrane of these photobiont cells. However, the main reason for this difference is likely to be found in the completely different water relations. Lichen mycobionts and photobionts are poikilohydrous organisms (see 3.1). The main fluxes of solutes between the partners occur passively during the often quite dramatic wetting and drying cycles, not between adequately hydrated, metabolically active partners as in biotrophic fungal symbioses with higher plants. During severe drought stress events the plasma membranes of mycobiont and photobiont cells seem to be leaky. About 10% of total soluble compounds were recovered from washing fluids during the rehydration phase. These soluble compounds of fungal and algal or cyanobacterial origin were lost into the apoplast whence they could be recovered for analysis [40], [7].

2.2. MYCOBIONT-DERIVED, HYDROPHOBIC CELL WALL SURFACE PROTEINS

Key feature for the functioning of the different types of fungal interactions with photobiont cells in taxonomically diverse lichens is a rather inconspicuous element: the thin, proteinaceous, hydrophobic fungal cell wall surface layer which is passively spreading from the growing hypha over the wall surface of the algal or cyanobacterial cell at the very first contact, thus sealing the photobiont with a mycobiont-derived, water-repellent coat [19], [20], [23], [25], [26]. In freeze-fracture preparations this hydrophobic cell wall surface layer appears as a fine pellicule composed of bundles of semicrystalline rodlets (Fig. 13). These may, in many taxa, be obscured by mycobiont-derived, phenolic secondary metabolites (review: [8]) which crystallize on and partly within this water-repellent coat, thus enhancing its hydrophobicity (Figs. 5, 12). Such water-repellent wall surface layers are typically found in the gas-filled thalline interior which is built up by aerial hyphae. The proteinaceous nature of the rodlet layer was identified with histochemical techniques [23]. A proteinaceous, hydrophobic cell wall surface layer was so far found in the thalline interior of macrolichens originating from different orders of Ascomycotina, but crystalline, phenolic secondary metabolites are not ubiquitous in these zone.

Why is this hydrophobic lining of the mycobiont and photobiont wall surfaces in the thalline interior so important for the functioning of the symbiotic relationship in lichens? It forces water and dissolved nutrients to flow within the apoplastic continuum between the partners during the regularly occurring wetting and drying cycles. Even in fully water saturated thalli is free water confined to the symplast and the apoplast of mycobiont and photobiont cells, as can be directly visualized by means of cryotechniques (Figs. 14, 16, 20; see 3.2.2).

Hydrophobic, proteinaceous wall surface layers are not a peculiarity of aerial hyphae of lichen mycobionts; they have been found on a wide range of aerial fungal structures (hyphae, conidiophores, conidia etc.) of non-lichenized fungi and are assumed to be an ubiquitous feature of aerial fungal cells [6] (reviews: [50], [48]). In recent years such hydrophobic fungal wall surface layers have attracted considerable interest among cell and molecular biologists and plant pathologists since they reveal quite unique properties (e.g., self-assembly at the liquid/air interface) and are likely to play important roles in adhesion, solute translocation and in symbiotic (parasitic, mutualistic) interactions. So-called hydrophobins [51], a group of small, cysteine-rich proteins (90 -

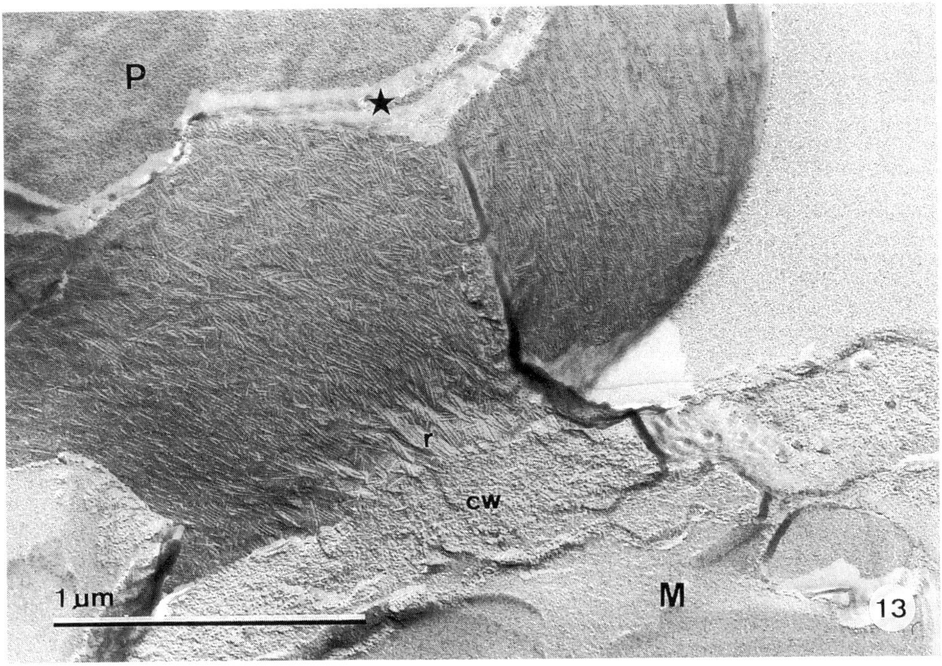

Figure 13. TEM of a freeze-etch preparation of the mycobiont-photobiont interface in the foliose macrolichen *Solorina saccata* (Peltigerales, Ascomycotina). The mycobiont (*M*) contacts its green algal photobiont (*Coccomyxa glaronensis) (PH)* by means of simple wall-to-wall appositions. The semi-crystalline rodlet layer (*r*) of the fungal cell wall (*cw*) surface is passively spreading over the algal cell wall surface. The asterisk refers to the fractured trilaminar outermost algal wall layer which contains a hydrolysis-resistant, sporopollenin-like biopolymer (for further details see [4], [25]).

150 amino acids in length, with 8 cysteine residues, the 2nd and 3rd forming a doublet) have been characterized in diverse asco- and basidiomycetes. Low amino acid homology was found in hydrophobins from different fungal taxa. Hydrophobins form complexes at the wall surface of aerial fungal cells which are insoluble in hot SDS-solutions. Best investigated are developmentally regulated hydrophobins of the basidiomycete *Schizophyllum commune* whose SC3 protein polymerizes at the surface of aerial hyphae, thus forming the semicrystalline, hydrophobic wall surface layer with distinct rodlet pattern (review: [50]). Although the structural similarity of the rodlet layer of lichen-forming fungi with the hydrophobin layer of non-lichenized taxa is obvious these particular proteins remain to be analyzed in detail in the lichen system.

From a functional point of view the mycobiont-photobiont interface in macrolichens bears some similarity with ectomycorrhizae. In both systems is the surface of the photoautotrophic partner sealed by fungal elements, and solute exchange occurs via the apoplastic continuum between the partners. In ectomycorrhizae this sealing of the donor of carbohydrates is performed at the organ level by means of the peripheral sheath and at the cellular level by means of the Hartig net around rhizodermal and cortical cells (Fig. 9). In the lichen symbiosis the fungal partner is sealing the surface of cells (as in green algal lichens) or colonies (as in cyanobacterial lichens) of the photoautotrophic partner with a proteinaceous hydrophobic coat. However, the main difference between macrolichens and all other fungal symbioses with photoautotrophs is the very peculiar symbiotic phenotype of the mycobiont. In no other fungal symbiosis does the heterotrophic partner create such an amazing "culturing chamber" for its minute photoautotrophic partner, and no other mycobiont secures the photosynthetic activity of its symbiont by positioning each individual photobiont cell at an optimal place with regard to gas exchange and illumination.

3. Water relations

3.1. POIKILOHYDRIC *vs.* HOMOIOHYDRIC WATER RELATIONS

One of the main differences between lichens and other biotrophic fungus-plant interactions are the poikilohydric water relations in both partners. Fungi, of course, have no means of controlling their water relations, but the majority of mycorrhizal and plant pathogenic species associate with homoiohydric hosts and nutrients are exchanged between metabolically active cells of fully or at least adequately hydrated partners. In the

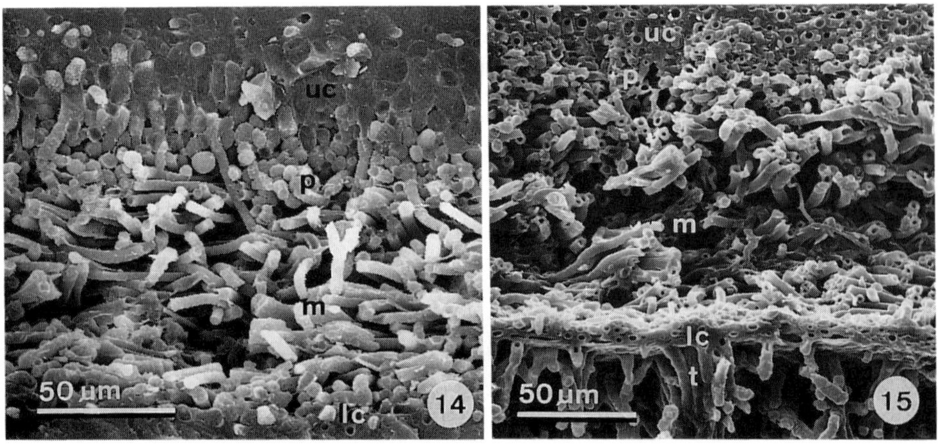

Figures 14 - 15. LTSEM of freeze-fractured *Lobaria virens* (Peltigerales, Ascomycotina) (photobiont: *Dictyochloropsis reticulata*) in the fully hydrated (approx. 220% water · dw^{-1}) and in the desiccated state (approx. 19% water · dw^{-1}). The conglutinate pseudoparenchyma of the upper (*uc*) and lower cortical layers (*lc*) surround the gas-filled thalline interior with the algal (*p*) and medullary layers (*m*).). *Fig. 14.* Vertical cross-fraction of a fully hydrated sample **Note**: there is no free water on the cell wall surfaces of either symbiont. The intercellular space is gas-filled. *Fig. 15.* Drought stress-induced shrinkage in the vertical thalline axis is obvious at the organismic level.

lichen symbiosis both partners are continuously subjected to dramatic fluctuations of cellular water contents between saturation and desiccation (<20% water dw^{-1}). The desiccation and rehydration processes may occur within minutes, and the recovery of the metabolic activities after drought stress events proceeds within seconds to minutes upon rehydration as concluded from measurements of gas exchange and chlorophyll fluorescence (reviews: [36], [37]). In the dry state many lichen mycobionts and photobionts tolerate temperature extremes. Desiccated thallus fragments of the foliose *Xanthoria parietina*, a macrolichen of temperate climates, grew normally after having been shock-frozen in subcooled LN_2, sputter-coated with an alloy of gold and palladium, and examined in the frozen-hydrated state under high vacuum conditions at 20 kV on the cold stage of a LTSEM [29]. These physiological properties have fascinated generations of ecophysiologists who were, and still are, carrying out their field work in climatically most extreme habitats such as deserts or the Antarctis, in places where various lichen species proliferate but where higher plants are at their physiological limits [36], [37]. It is noteworth that an estimated 8% of terrestrial ecosystems, *i.e.* the arctic/alpine and antarctic zones, are lichen-dominated [38].

3.2. DROUHGT STRESS - INDUCED STRUCTURAL ALTERATIONS AT THE ORGANISMIC LEVEL

3.2.1. *Cryotechniques* vs. *conventional preparative procedures*

For obvious reasons it is not possible to explore thalline water relations and the fate of cells and their membrane systems under drought stress by means of conventional preparative techniques since desiccated cells rehydrate in the first preparative step, *i.e.* in aqueous chemical fixatives. Cryotechniques such as low temperature scanning electron microscopy (LTSEM) of frozen-hydrated specimens and TEM of cryofixed, freeze-substituted samples are excellent tools for studying structural aspects of thalline water relations and of drought stress-induced alterations at the cellular level since specimens can be physically immobilized at any level of hydration by means of cryofixation (review: [33]).

3.2.2. *The location of free water within lichen thalli*

LTSEM techniques allow the location and direct visualization of free water within lichen thalli, a matter of considerable debate among ecophysiologists and cell biologists. For quite a while free water was assumed to accumulate within the thalline interior of lichens and to fill the spaces between the cells (*e.g.*, [35]), a serious handicap for photosynthetic gas exchange. Based on the discovery of a hydrophobic wall surface layer on fungal and algal cells of macrolichens it was postulated that the thalline interior remains gas-filled even in water-saturated lichens [20]. LTSEM preparations of fully hydrated lichens [3], [32], [45] unequivocally supported the view of permanently gas-filled medullary and algal layers since there is no free water on the wall surfaces of fungal and algal cells in the thalline interior (Figs. 14, 16, 20).

3.2.3. *Drought stress-induced alterations in thalline morphology*
In contrast to other drought-tolerant organisms such as bryophytes or "resurrection plants" among pteridophytes and angiosperms, all of which shrivel dramatically during desiccation, most lichen thalli do not change their external morphology during drought stress events; exceptions are found among foliose species which are not tightly attached to the substratum and roll their marginal lobes, as seen in *Peltigera* spp. or in vagant and other spp. among the Parmeliaceae and Cladoniaceae. However, thalline shrinkage in the vertical axis (Figs. 15, 18) is obvious, and the fungal and algal cells of lichen thalli undergo very significant structural and physiological changes during desiccation (see 3.3). Conglutinate pseudoparenchyma of the cortical layers and/or of internal strands, which are translucent, elastic and flexible in the fully hydrated state, become opaque, hard and often very brittle during drought stress events.

3.3. DROUHGT STRESS - INDUCED STRUCTURAL ALTERATIONS AT THE CELLULAR LEVEL

Protoplasts of mycobiont and photobiont cells change their volume quite dramatically when the thalline water contents drop to approximately 20% dw^{-1}. The organelles of drought stressed cells shrivel but their membrane systems are structurally well preserved as concluded from TEM studies of freeze-substituted specimens (Fig. 19) [32], [29], [33]. Preservation and protection of cellular membrane and enzyme systems during desiccation and rapid repair of drought stress-induced membrane damage during rehydration are pre-requisites for the very rapid recovery of metabolic activities upon rehydration as is typically observed in lichens (reviews: [36], [37]).The biochemical and genetic bases of these biologically remarkable features are unknown in lichens.

3.4. CYTOPLASMIC GAS BUBBLES IN DROUGHT-STRESSED FUNGAL CELLS

Fungal and algal cells of lichen thalli are facing spatial problems during desiccation. Neither partner plasmolyses when the protoplast volume is strongly reduced at very low water contents. Viable fungal and algal or cyanobacterial cells keep their plasma membrane in close contact with the wall or the murein sacculus, respectively, during drought stress events. The different types of photobiont cell walls are dramatically deformed (Figs. 18, 19, 21), but the relatively rigid fungal walls cannot keep peace with the changes in protoplast volume. A rapidly expanding, but reversible cytoplasmic gas bubble of unknown origin and contents allows drought stressed fungal protoplasts to shrink and, at the same time, keep contact with the cell wall (Fig. 19) [29], [46]. Such drought stress-induced intracellular "air bags" are not restricted to lichenized fungi; they have been observed in LM preparations of a range of ascospores of saprotrophic species [5], [34], [41], in the conidiophore of *Venturia inaequalis* and in freeze-fractured LTSEM preparations of basidiomata of *Schizophyllum commune* (Honegger, unpubl. results). In ultrathin sections of freeze-substituted specimens no biomembrane could be resolved around cytoplasmatic gas bubbles whose smooth lining deserves

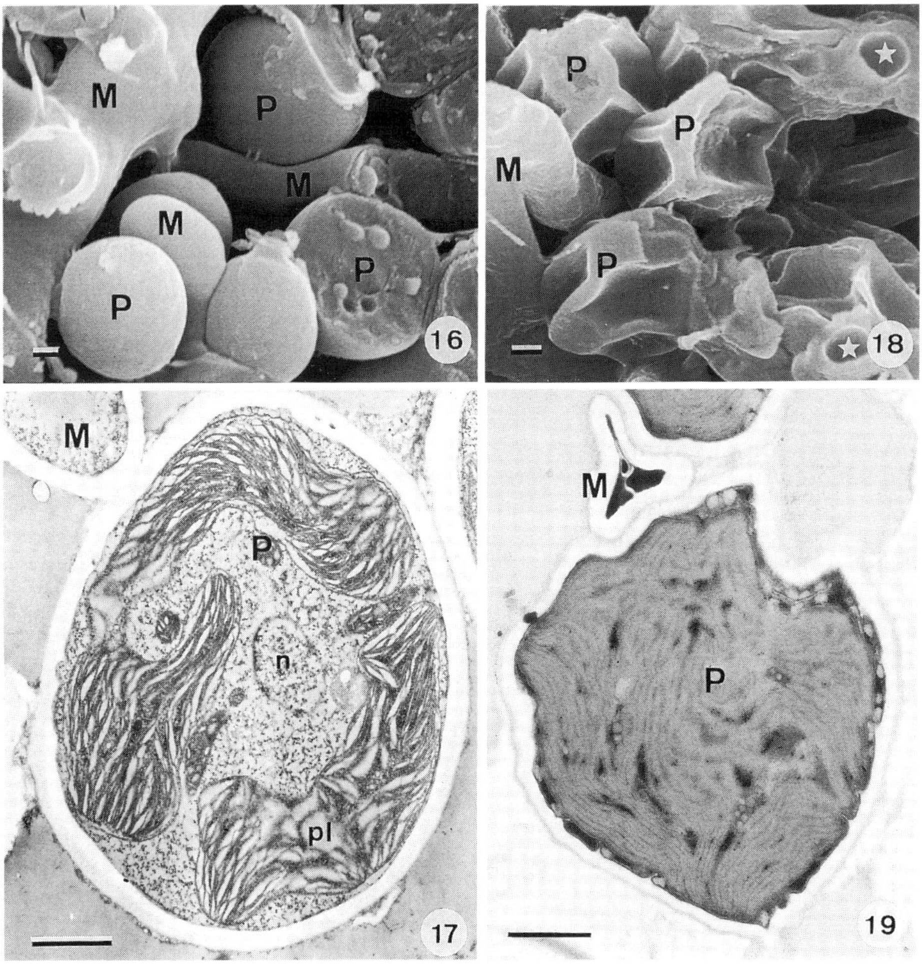

Figures 16 - 19. LTSEM of freeze-fractured and TEM of freeze-substituted *Lobaria virens* (photobiont: *Dictyochloropsis reticulata*) in the fully hydrated and the desiccated state. *M*: mycobiont; *P*: photobiont. *Fig. 16.* Detail of the mycobiont-photobiont interface in a fully hydrated sample (approx. 220% water · dw^{-1}). The fungal partner contacts the *Dictyochloropsis* cells by means of wall-to-wall appositions. *Fig. 17.* TEM of the mycobiont-photobiont interface in a fully hydrated sample. *pl*: profiles of the lobate-reticulate, peripheral chloroplast. *n*: nucleus. *Fig. 18.* LTSEM of the mycobiont-photobiont interface in a desiccated specimen (approx. 19% water · dw^{-1}). The photobiont cells are strongly deformed. Asterisks point to drought stress-induced cytoplasmic gas bubbles in the fungal partner. *Fig. 19.* TEM of the mycobiont-photobiont interface in a desiccated sample. The plasma membrane of the photobiont (*pm*) is in close contact with the strongly deformed cell wall (*cw*).

Magnification bars equal **1µm** unless otherwise stated.

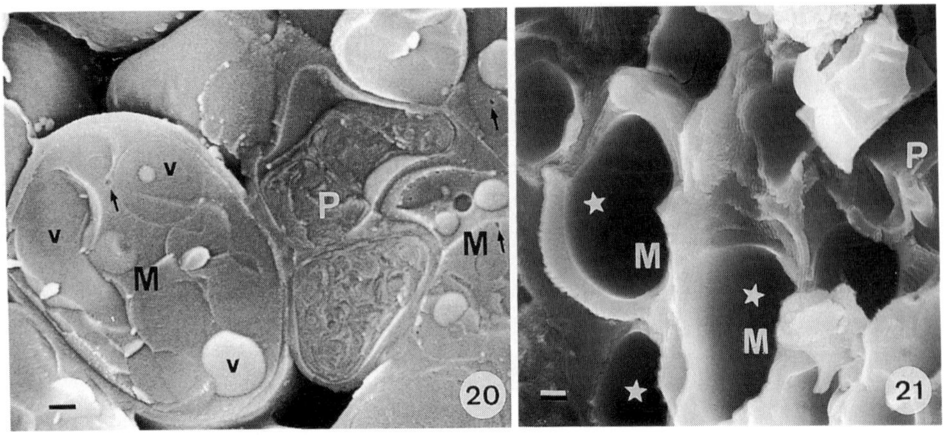

Figures 20 - 21: LTSEM of freeze-fractured *Peltigera canina* (photobiont: *Nostoc punctiforme*) in the fully hydrated and the desiccated state. *M*: mycobiont; *P*: photobiont. *Fig. 20*. LTSEM of the mycobiont-photobiont interface and adjacent cortical cells in a water-saturated sample (approx. 350% water · dw^{-1}). A finger-like intragelatinous protrusion is seen in close contact with the *Nostoc* cells. Numerous vacuoles (v) are seen in the fractured fungal cells. Arrows point to minute cavities which are likely to correspond to concentric bodies (see Fig. 22). **Note**: there is no free water on the cell wall surfaces of either symbiont. The intercellular space is gas-filled. *Fig. 21*. LTSEM of a similar thalline area as in Fig. 20, but in a desiccated specimen (approx. 21% water · dw^{-1}). Colonies of the photobiont are shrivelled and strongly deformed. Asterisks refer to large, drought stress-induced cytoplasmic gas bubbles in freeze-fractured fungal cells.

Magnification bars equal **1µm.**

further investigation [29]. A positive correlation between cell wall rigidity and the formation of cytoplasmic gas bubbles was already observed by Ingold [34]. It remains to be seen whether such gas bubbles develop anywhere in the cytoplasm or at preformed sites.

3.5. CONCENTRIC BODIES

Almost all lichen-forming and a range of desiccation tolerant, relatively long-living plant pathogenic and saprotrophic ascomycetes contain concentric bodies, cytoplasmic organelles of unknown ontogeny and function (review: [28]). These globose structures of 300-400 nm in diameter consist of a proteinaceous, electron dense shell with peripheral rays around an electron-transparent center (Figs. 22-23). It is known from freeze-etch preparations that the electron-transparent center of concentric bodies is gas-filled (Fig. 22) [29]. A whole series of questions awaits scientific exploration:

1) are concentric bodies and drought stress-induced cytoplasmic gas bubbles linked to each others? It is at least imaginable that concentric bodies might be the compressed

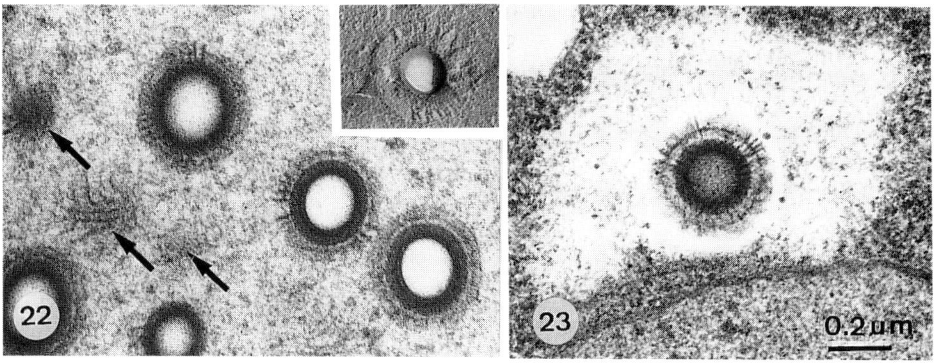

Figures 22 - 23. Concentric bodies, as seen in ultrathin sections of chemically fixed specimens. *Fig. 22.* Cluster of concentric bodies within a cortical cell of the lichenized ascomycete *Peltigera canina* (arrows point to oblique sections). *Insert*: a concentric body of the same cell type and species in a freeze-etch preparation. The hole in the replica, as usually seen when concentric bodies have been fractured in the supra-equatorial plane, refers to a gas-filled center of this organelle. *Fig. 23.* Slightly oblique section of a concentric body within the conidiophore of the plant pathogenic ascomycete *Venturia inaequalis* (collected on *Malus baccata*). Concentric bodies were first shown to occur in *V. inaequalis* by Granett [15].

Same magnification in *Figs. 22* and *23*.

remains of former bubbles. If so, where did the proteinaceous lining of the bubble come from?

2) are the proteins of the semicrystalline shell of concentric bodies related to hydrophobins (see 2.2) some of which have been shown to self-assemble at the liquid/air interface [52]?

3) are concentric bodies persisting or is there a turnover within the cell?

4) are concentric bodies restricted to ascomycetes, or do they also occur in basidiomycetes and zygomycetes?

4. Outlook

It will be a most fascinating task in future research to characterize the hydrophobic cell wall surface proteins of lichen-forming fungi and to compare them with hydrophobins of non-lichenized fungal taxa. Interactions of hydrophobic cell wall surface proteins with cell surfaces of photoautotrophic (as seen in the lichen symbiosis) or heterotrophic partners are likely to play important functional roles in a wide range of fungal symbioses with plants and animals. The structural, biochemical and genetic bases of the remarkable desiccation tolerance of lichen mycobionts and photobionts and the comparison of these parameters with data from bryophytes and "resurrection plants" is another area of biological interest. In this context the enigmatic concentric bodies and the drought stress-induced cytoplasmic gas bubbles of lichenized and non-lichenized fungi await further exploration.

5. Acknowledgements

My sincere thanks are due to the organizers of the 1994 meeting of the *Société Française de Phytopathologie* for their interest in the lichen symbiosis and for providing space for a lichen chapter in this volume; to Sibylle Erni for skilfully preparing the line drawings and to Jean-Jacques Pittet for composing Fig. 5 on the computer; to Marcel Peter for technical assistance; to Verena Kutasi who did all the darkroom work; and to my husband, Dr. Thomas Honegger, for his patient help with the formating of this manuscript.

6. References

1. Ahmadjian, V.: *The Lichen Symbiosis*, John Wiley, New York, 1993.
2. Ahmadjian, V and Jacobs, J.B.: Relationship between fungus and alga in the lichen *Cladonia cristatella* Tuck., *Nature* **389** (1981), 169-172.
3. Brown, D.D., Rapsch, S., Beckett, A. and Ascaso, C.: The effect of desiccation on the cell shape in the lichen *Parmelia sulcata* Taylor, New Phytol. **105** (1987), 295-299.
4. Brunner, U. and Honegger. R.: Chemical and ultrastructural studies on the distribution of sporopollenin-like biopolymers in 6 genera of lichen phycobionts, *Can. J. Bot.* **63** (1985), 2221-2230.
5. deBary, A.: *Morphologie und Physiologie der Pilze, Flechten und Myxomyceten*, W. Engelmann, Leipzig, 1866.
6. deVries, O.M.H., Fekkes, M.P., Wösten, H..B., and Wessels, J.G.H.: Insoluble hydrophobin complexes in the walls of *Schizophyllum-commune* and other filamentous fungi, *Arch. Microbiol.* **159** (1993), 330-335.
7. Dudley, S.A. and Lechowicz, M.J.: Losses of polyol through leaching in subarctic lichens, *Plant Physiol.* **83** (1987), 813-815.
8. Fahselt, D.: Secondary biochemistry of lichens, *Symbiosis* **16** (1994), 117-165.
9. Fahselt, D.: Carbon metabolism in lichens, *Symbiosis* **17** (1994), 127-182.
10. Farkas, E.E. and Sipman, H.J.M.: Bibliography and checklist of foliicolous lichenized fungi up to 1992, *Tropical Bryology* **7** (1993), 93-148.
11. Friedl, T. and Zeltner, C.: Assessing the relationships of lichen algae and the Microthamniales (Chlorophyta) with 18S rRNA gene sequence comparisons, *J. Phycol.* **30** (1994), 500-506.
12. Gargas, A., DePriest. P.T., Grube, M. and Theler, A.: Multiple origins of lichen symbioses in fungi suggested by SSU rDNA phylogeny, *Science* **268** (1995), 1492-1495.
13. Galun, M.:Lichenization, in M. Galun (ed.), *CRC Handbook of Lichenology*, vol. 3,: CRC Press, Boca Raton, (1988), 153-169.

14. Galun, M and Bubrick, P.: Physiological interactions between the partners of the lichen symbiosis, in H.F. Linskens and J. Heslop-Harrison (eds.), *Cellular Interactions. Encyclopedia of Plant Physiology*, Springer, Berlin, (1984), 362-401.
15. Granett, A.L. Ultrastructural studies on concentric bodies in the ascomycetous *fungus Venturia inaequalis*, *Can. J. Bot.* **52** (1974), 2137-2139.
16. Greuter, W. (ed.) *International Code of Botanical Nomenclature*. Koeltz, Königstein, 1988.
17. Hawksworth, D.L.: Effects of algae and lichen-forming fungi on tropical crops, in V.P. Agnihotry, K.A. Sarbhoy and D. Kumar (eds), *Perspectives of Mycopathology*, Malhotra Publishing House, New Delhi, (1988), 76-83.
18. Hawksworth, D.L. and Honegger, R.: The lichen thallus: a symbiotic phenotype of nutritionally specialized fungi and its response to gall producers, in M.A.J. Williams (ed.), *Plant Galls*, Clarendon Press, Oxford, (1994), 77-98.
19. Honegger, R.: Cytological aspects of the mycobiont-phycobiont relationship in lichens. Haustorial types, phycobiont cell wall types, and the ultrastructure of the cell wall surface layers in some cultured and symbiotic myco- and phycobionts, *Lichenologist* **16** (1984), 111-127.
20. Honegger, R.: Fine structure of different types of symbiotic relationships in lichens, in D.H. Brown (ed.), *Lichen Physiology and Cell Biology*, Plenum Press, New York, (1985), 287-302.
21. Honegger, R.: Scanning electron microscopy of the fungus-plant cell interface: a simple preparative technique, *Trans. Br. mycol. Soc.* **84** (1985), 530-533.
22. Honegger, R.: Ultrastructural studies in lichens. I. Haustorial types and their frequencies in a range of lichens with trebouxioid phycobionts, *New Phytol.* **103** (1986), 785-795.
23. Honegger, R.: Ultrastructural studies in lichens. II. Mycobiont and photobiont cell wall surface layers and adhering crystalline lichen products in four Parmeliaceae, *New Phytol.* **103** (1986), 797-808.
24. Honegger, R.:Surface interactions in lichens, in W. Wiessner, D.G. Robinson and R.C. Starr (eds), *Experimental Phycology 1. Cell walls and Surfaces, Reproduction, Photosynthesis*, Springer, Berlin, (1990), 40-54.
25. Honegger, R.: Functional aspects of the lichen symbiosis, *Ann. Rev. Plant Physiol. Plant Mol. Biol.* **42** (1991), 553-578.
26. Honegger, R.: Haustoria-like structures and hydrophobic cell wall surface layers in lichens, in K. Mendgen and D.-E. Lesemann (eds.), *Electron Microscopy of Plant Pathogens*, Springer, Berlin (1991), 277-290.
27. Honegger R.: Lichens: mycobiont-photobiont relationship, in W. Reisser (ed.), *Algae and Symbioses. Plants, animals, fungi, viruses, interactions explored*, Biopress, Bristol, (1992), 255-275.
28. Honegger R.: Developmental Biology of Lichens, *New Phytol.* **125** (1993), 659-677.
29. Honegger, R.:Experimental studies with foliose macrolichens: fungal responses to spatial disturbance at the organismic level and to spatial problems at the cellular level during drought stress events, *Can. J. Bot.* **73**(Suppl.1) (1995), in press
30. Honegger, R.: Mycobiont, in T.H. Nash (ed.), *Lichen Biology*, Cambridge University Press, Cambridge (1995), 24-36.
31. Honegger, R.: Morphogenesis, in T.H. Nash (ed.), *Lichen Biology*, Cambridge University Press, Cambridge (1995), 65-87.
32. Honegger R. and Peter, M.: Routes of solute translocation and the location of water in heteromerous lichens visualized with cryotechniques in light and electron microscopy, *Symbiosis* **16** (1994), 167-186.
33. Honegger, R., Peter, M. and Scherrer, S.: Drought stress-induced structural alterations at the mycobiont-photobiont interface in a range of foliose macrolichens, *Protoplasma*, submitted.
34. Ingold, C.T:. A gas phase in viable fungal spores, *Nature* **177** (1956), 1242-1243.

35. Jahns, H.M.: Morphology, reproduction and water relations - a system of morphogenetic interactions in *Parmelia saxatilis*, *Nova Hedwigia Beih.* **79** (1984), 715-737.
36. Kappen, L.: Ecophysiological Relationships in different climatic regions, in M. Galun (ed.), *CRC Handbook of Lichenology*, vol. 2. CRC Press, Boca Raton, (1988), 37-100.
37. Kappen, L.: Lichens in the antarctic region, in E.I. Friedmann (ed.), *Antarctic Microbiology*, Wiley-Liss, New York, (1993), 433-490.
38. Larson, D.W.: The absorption and release of water by lichens, *Bibl. Lichenol.* **25** (1987), 351-360.
39. Lines, C.E.M., Ratcliffe, R.G., Rees, T.A.V. and Southon, T.E.: A ^{13}C NMR study of photosynthate transport and metabolism in the lichen *Xanthoria calcicola* Oxner, *New Phytol.* **111** (1989), 447-456.
40. MacFarlane, J.D. and Kershaw, K.A.: Some aspects of carbohydrate metabolism in lichens, in D.H. Brown (ed.), *Lichen Physiology and Cell Biology*, Plenum Press, New York, (1985), 1-8.
41. Milburn, J.A.: Cavitation and osmotic potentials of *Sordaria* ascospores, *New Phytol.* **69** (1970), 133-141.
42. Oberwinkler, F.: Fungus-alga interactions in basidiolichens, *Nova Hedwigia Beih.* **79** (1984), 739-774.
43. Ott, S.: Reproductive strategies in lichens, *Bibl. Lichenol.* **25** (1987), 81-93.
44. Reisser, W.: Endosymbiotic associations of algae with freshwater protozoa and invertebrates, in W. Reisser (ed.), *Algae and Symbioses: Plants, Animals, Fungi, Viruses,Interactions explored*, Biopress, Bristol, (1992), 1-19.
45. Scheidegger, C.: Low-temperature scanning electron microscopy: the localization of free and perturbed water and its role in the morphology of the lichen symbionts, *Crypt. Bot.* **4** (1994), 290-299.
46. Scheidegger, C, Schroeter, B. and Frey, B.: Structural and functional processes during water vapour uptake and desiccation in selected lichens with green algal photobionts, *Planta* **197** (1995), in press.
47. Smith, D.C. and Douglas, A.E.: The Biology of Symbiosis, Edward Arnold, London, 1987.
48. Templeton, M.D., Rikkerink, E.H.A. and Beever, R.E.: Small, cysteine-rich proteins and recognition in fungal-plant interactions, *Mol. Plant-Microbe Interact.* **3** (1994), 320-325.
49. Tschermak-Woess, E.: The algal partner, in M. Galun (ed.), *Handbook of Lichenology*, vol. 1, CRC Press, Boca Raton, (1988), 39-92.
50. Wessels, J.G.H.: Wall growth, protein excretion, and morphogenesis in fungi, *New Phytol.* **123** (1993), 397-413.
51. Wessels, J.G.H., deVries, O.M.H., Asgeirsdottir, S.A. and Springer, J.: The *thn* mutation of *Schizophyllum commune*, which suppresses formation of aerial hyphae, affects expression of the SC3 hydrophobin gene, *J. Gen. Microbiol.* **137** (1991), 2439-2445.
52. Wösten, H.A.B, Asgeirsdottir, S.A., Krook, J.H., Drenth, J.H.H. and Wessels, J.G.H.: The fungal hydrophobin *Sc3P* self-assembles at the surface of aerial hyphae as a protein membrane constituting the hydrophobic rodlet layer, *Eur. J. Cell Biol.* **63** (1994), 122-129.

ROOT DEFENCE RESPONSES IN RELATION TO CELL AND TISSUE INVASION BY SYMBIOTIC MICROORGANISMS : CYTOLOGICAL INVESTIGATIONS

V. GIANINAZZI-PEARSON, A. GOLLOTTE, C. CORDIER and S. GIANINAZZI
Laboratoire de Phytoparasitologie INRA/CNRS, SGAP, INRA, BV 1540, 21034 Dijon Cédex, France

1. Introduction

Although in nature plants are constantly submitted to attack by microorganisms, only a very limited number of these succeed in colonising plant tissues and an even smaller proportion establish biotrophic relations with their hosts. In the case of mutualistic symbioses, where both organisms derive benefits from each other, interactions are characterised by an advanced cellular adaptation which must somehow be based on a highly evolved physiological and genetical co-ordination between the partners. With the exception of stem nodules, mutualistic plant symbioses with microorganisms concern principally root tissues. Endocellular root symbioses fall into two main categories : nodules and mycorrhizas. The former furnish fixed nitrogen to the plant, and the latter ensure the host's mineral nutrition as well as protection against certain soil-borne pathogens [for references see 1,2]. Both symbioses result from the proliferation of microorganisms within well-defined root domains, bacteria in the first case and fungi in the second. When the microbial symbiont grows within host tissue, it does so in a controlled fashion and its development is accompanied by extensive morphogenetic, and in the case of nodules also organogenetic, events at both tissue and cellular levels. In spite of extensive research on endocellular root symbioses, the mechanisms underlying plant control over infection by symbiotic microorganisms, or those by which roots distinguish these from pathogens, still have to be fully elucidated. It is becoming evident from quantitative molecular and biochemical analyses that defence-related molecules are elicited at a much lower level and/or only during the early phases of establishment of root symbioses as compared to when plants exert control over the development of pathogens [for references see 3,4]. Such low or transient elicitation may be due to overall low activation of defence responses, to their suppression or to very localised activation of defence-related events, the extent of which may still be sufficient to contribute to plant control over the microbial symbiont. Direct observation of cellular events *in planta* is an essential approach for distinguishing between the relative validity of these different hypotheses.

M. Nicole and V. Gianinazzi-Pearson (eds.), Histology, Ultrastructure and Molecular Cytology of Plant-Microorganism Interactions, 177-191.

In this chapter, we review results from cytological investigations of plant defence reactions in endocellular root symbioses and discuss possible means of their regulation. We have paid particular attention to defence-like reactions in arbuscular mycorrhizas and nodules, which share a number of common features in that 1) the target cells for microbial proliferation derive from the inner cortex, 2) the invading microorganisms form a specialised apoplastic interface for exchange of nutrients with the host protoplast, and 3) microbial development is restricted to an extracytoplasmic compartment delimited by a host membrane possessing similar molecular components [5,6,7,8].

2. Host Cell Reactions Associated With Reciprocal Compatibility in Arbuscular Mycorrhiza

Arbuscular mycorrhizas are a ubiquitous symbiotic association between soil-borne fungi of the Glomales and species belonging to over 80% of plant genera [9]. Arbuscular mycorrhizal fungi have been colonising plant tissues since some of the first terrestrial plant species appeared [10,11]. These microorganisms must therefore have dealt very early on with problems of non-elicitation or avoidance of plant-defence responses, a tactic which has enabled them to spread to an extensive number of plant taxa and form the most widespread symbiosis to have evolved in the plant kingdom.

Colonisation of host root tissues by the symbiotic fungi follows a complex, morphologically well-defined scenario, beginning with the formation of an appressorium on the root surface, followed by intercellular or intracellular penetration of epidermal and hypodermal tissues, and terminating in the development of highly branched intracellular haustoria (arbuscules) specifically within parenchyma cells of the root cortex (Figure 1).

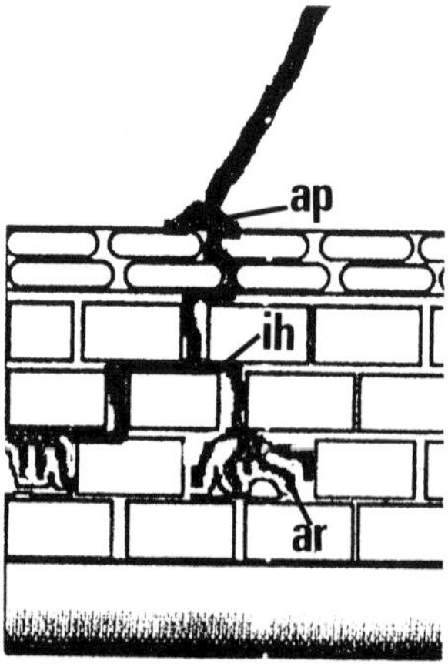

Figure 1. Diagrammatic representation of root colonisation by an arbuscular mycorrhizal fungus forming an appressorium (ap) at the root surface, intercellular hyphae (ih) and intracellular arbuscules (ar).

At the cytological level, most host plants show remarkably little reaction to the first steps of colonisation, with no significant modifications in either the contents or wall structure of epidermal or hypodermal cells in contact with invading hyphae (Figures 2 and 3). If the cells of the outer root tissue are colonised, hyphae remain unbranched and the only sign of host reaction is the continuous deposition of wall material by the surrounding host membrane as the fungus crosses the cell. This contrasts with the extensive morphological modifications induced by arbuscule formation in parenchyma cortical cells. Here, host cytoplasm and organelles proliferate around the developing hyphae and again material continuous with the inner host cell wall is deposited by the surrounding host membrane in response to the invading fungus. Cytochemical analyses together with affinity probing using antibodies, lectins or enzymes have shown that, as in the outer root cells, this material is essentially primary plant wall, being made up of β (1-4) glucans, pectins (Figure 4), hemicelluloses and glycoproteins [12,13]. In addition, β (1-3) glucans which are components of callose, a typical defence-related molecule, can also be immunolocalised in this wall material in parenchyma cells (Figure 5). However, the occurrence of these molecules is very localised and they disappear, together with the wall material, as the invading hypha develops further (Figure 6). Moreover, this reaction appears to depend on the host plant involved ; for example, β (1-3) glucans are typically associated with the wall reaction of parenchyma cells to arbuscular mycorrhizal fungi in pea, tobacco and leek [3,14, unpublished results], but not in maize [15].

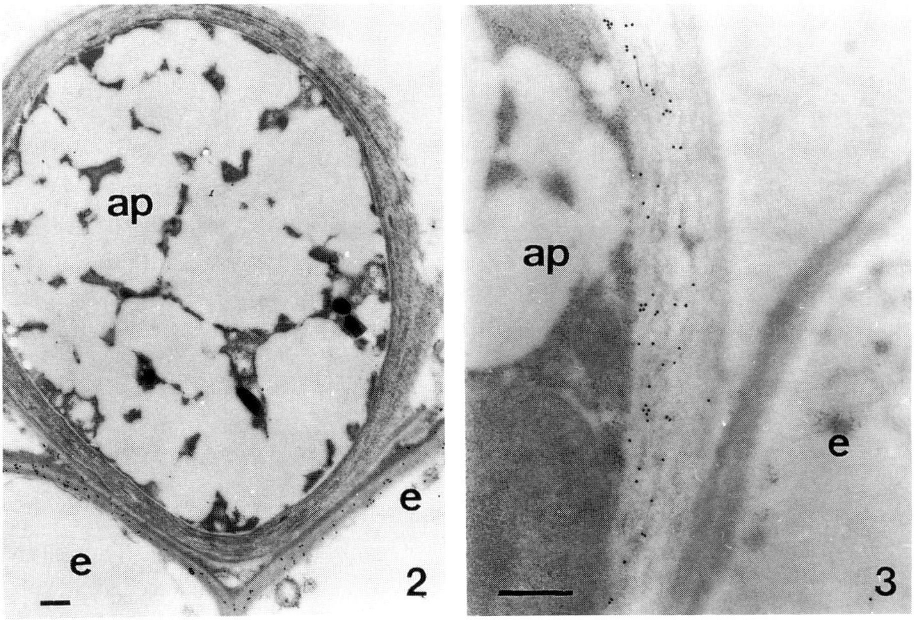

Figures 2 and 3. Immunolocalisation of pectin (2) and β (1-3) glucan chains (3) in pea roots colonised by *Glomus mosseae ;* no significant modifications occur in epidermal cells (e) below an appressorium (ap). Bar = 0.4 μm.

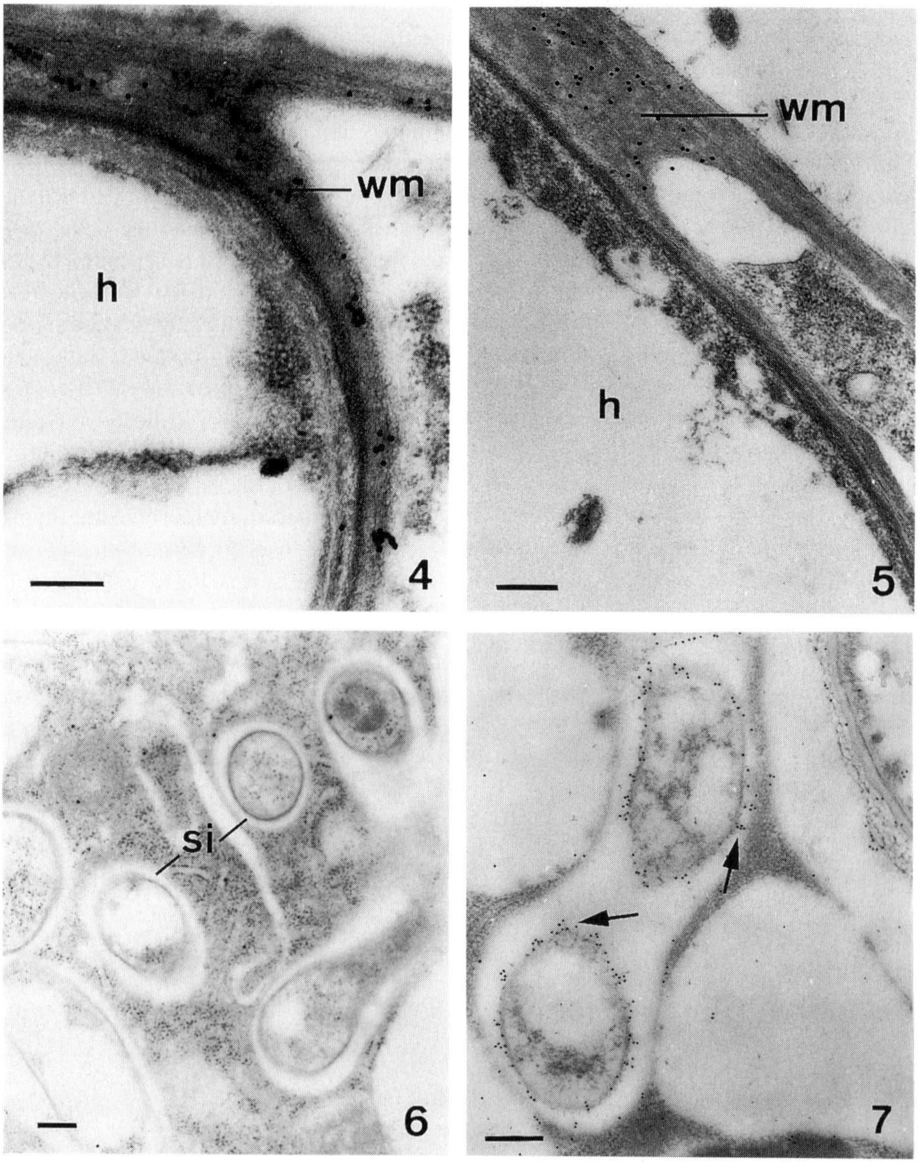

Figures 4-7. Roots of pea (4-6) and tobacco (7) colonised by *G. mosseae*. Synthesis of wall material (wm) containing pectin (4) and β (1-3) glucans (5) is induced as a hypha (h) penetrates parenchyma cortical cells. Wall material is absent from the symbiotic interface (si) (6) whilst PR-1 protein can be immunodetected there (arrow) (7). Bar = 0.2 μm.

Other defence-related molecules are localised in the interface between branching arbuscule hyphae and the surrounding plant membrane. Both pathogenesis-related PR-1 protein, which is considered to have antifungal activity [16], and hydroxyproline-rich glycoprotein, serologically homologous to that accumulating during pathogen interactions, have been detected in this compartment using antibodies (Figure 7) [14,15,17]. Accumulation of these molecules is considered to be a general plant response against pathogen attack and it does occur even in roots of pathogen-infected susceptible varieties [18,19]. Furthermore, *in situ* hybridisation has recently shown that transcripts for enzymes of the phenylpropanoid pathway, phenylalanine ammonium lyase and chalcone synthase, also accumulate within cells where arbuscules are developing [20]. However, it must be borne in mind that although *in situ* localisation of transcripts does indicate gene activation, it gives no information as to the site of final localization of the biosynthetic enzymes, their activity or utilisation of the products. No evidence for phenolics accumulation has been obtained by cytochemical staining [21]. The extremely weak autofluorescence under blue light of arbuscule-containing root tissue confirms this lack of phenolics accumulation (Figure 8), and contrasts with the strong autofluorescence observed when parenchyma cortical cells are colonised by a pathogenic fungus like *Phytophthora parasitica* (Figure 9).

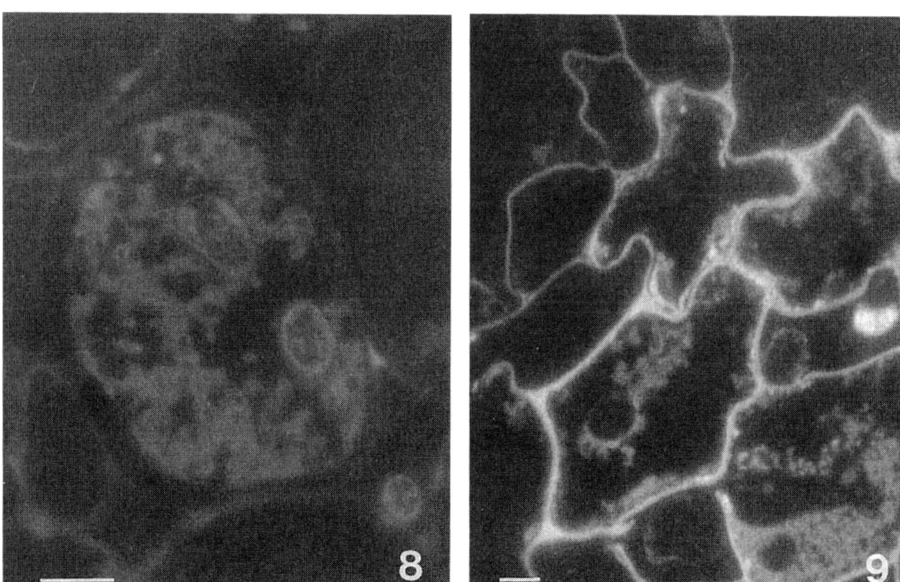

Figures 8 and 9. Cortical parenchyma of tomato roots containing arbuscules of *G. mosseae* only weakly fluoresces (8) compared to tissue infected by *Phytophthora parasitica* (9). Bar = 2.5 μm.

Peroxidases, which can be related to plant wall reinforcement, have not been found to be active in cells containing arbuscules using cytoenzymology techniques [22,23], and no significant alteration has been found to occur by immunolabelling in the distribution of chitinase in roots after mycorrhiza formation [24].

In all cases, when induction of defence-related molecules or genes does occur, it appears to be limited to arbuscule-containing cells and there is no generalised accumulation in infected tissues, contrary to that reported during pathogen attacks [18,25]. These cytological investigations confirm that during symbiotic interactions, arbuscular mycorrhizal fungi can activate a part of the plant's metabolic pathways associated with defence processes, but only in a weak and very localised fashion. Such localised activation could explain the low level of elicitation detected by quantitative analyses. In addition, arbuscules are ephemeral structures and the proportion of root containing actively developing infections decreases with plant age, which could also contribute to the apparent decrease in defence gene elicitation as the infection progresses [26,27]. At present, the reason for the low and very limited spatio-temporal activation of defence responses by arbuscular mycorrhizal fungi is not known. Production or activation of fungal elicitors may be related to fungal differentiation and limited to the developmental stage of the arbuscule. Alternatively, defence gene expression may represent a non-specific host response which is somehow suppressed, or maintained at a low level, by plant or fungal mechanisms specifically activated during symbiotic interactions.

3. Cell Responses To Arbuscular Mycorrhizal Fungi in Symbiotically-Defective Plant Mutants

Several mutants, isolated after chemical mutagenesis of pea and fababean, have been identified as having impaired symbiotic abilities vis-à-vis both arbuscular mycorrhizal fungi and *Rhizobium* [28], but not altered in their behaviour vis-à-vis root pathogens [7]. Some such mutants, called myc^{-1} nod^{-}, are able to elicit normal appressorium formation at the root surface but development of the mycorrhizal fungus is arrested at this stage (Figure 10) [29]. Such aborted infections are characterised by an abnormally thick reinforcement of the epidermal and hypodermal cell walls in contact with the fungal symbiont. Accumulation of heterogeneous paramural material, recalling some papilla deposits during defence responses to pathogens [30], is induced on the inner face of the plant walls (Figure 11) [29,31]. In resistant interactions with pathogens, alterations in host cell wall structures are among the cytologically most striking effects of microbial attack. Cytomolecular analyses, although not abundant, lend support to the belief that papillae and secondary wall structures contain a variety of fortifying components including callose, polyphenolic compounds, proteins or silicon which contribute to plant resistance [for references see 32].

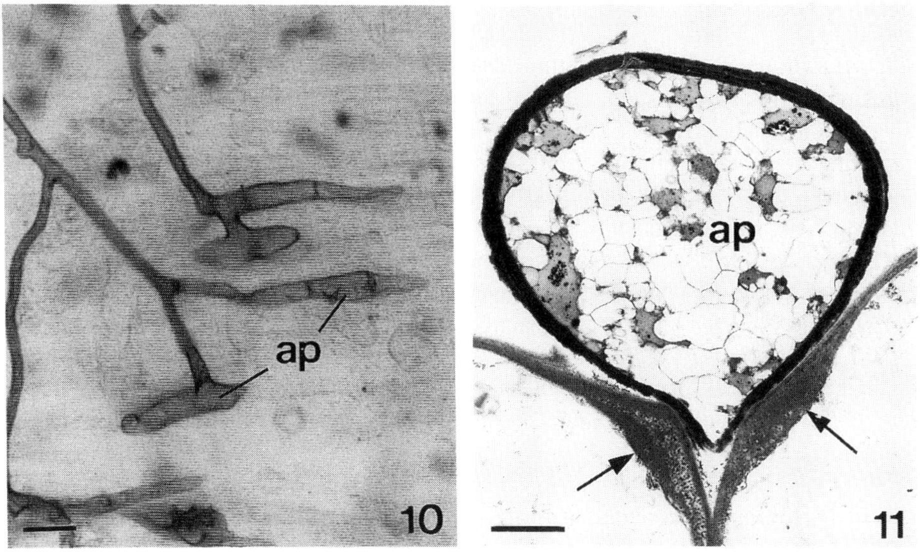

Figures 10 and 11. Myc^{-1} mutant pea roots showing appressorium (ap) formation by *G. mosseae* (10) and wall thickenings in underlying epidermal cells (arrow) (11). Bars = 20 μm (10) and 2 μm (11).

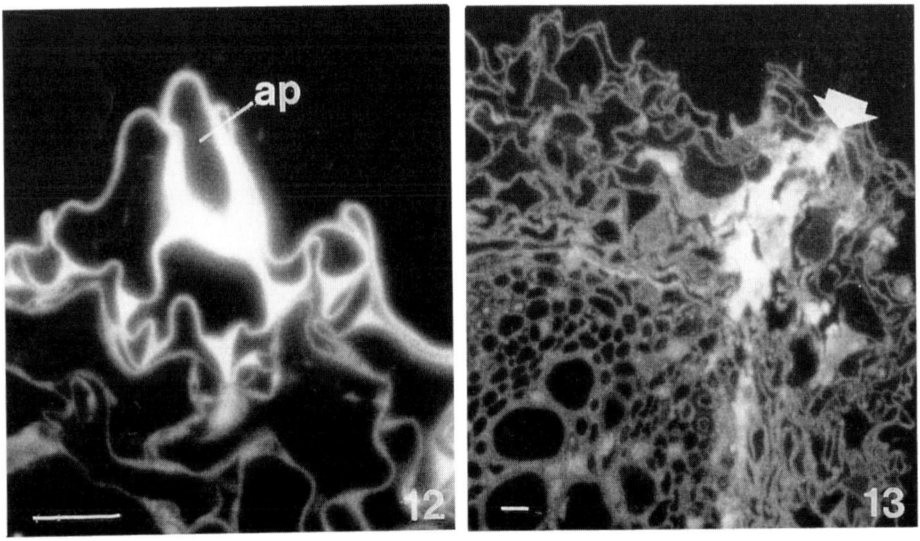

Figures 12 and 13. Strong autofluorescence of myc^{-1} pea root cells under an appressorium (ap) of *G. mosseae* (12), and in a pea root infected by *Chalara elegans* (arrow) (13). Bar = 10 μm.

In order to determine whether the wall structures observed in different myc^{-1} pea mutants could represent a defence response conferring resistance to arbuscular mycorrhizal fungi, their composition has been analysed by molecular cytology and compared to those structures elicited during induced resistance of pea roots to the fungal pathogen *C. elegans*. One feature of the secondary wall structures induced in the mutants is their strong autofluorescence under blue light (Figure 12) [31], comparable to that occurring in necrotic regions of *C. elegans*-infected roots (Figure 13). This has been consistently observed in three types of myc^{-1} mutants mutated for different genes [33]. Increase in autofluorescence indicates accumulation of phenolic compounds within the paramural deposits, and this is confirmed by a strong staining by toluidine blue and the nitroso reaction, whereas histochemical staining for lignin gives negative results [31]. Furthermore, pectins are only very weakly detected using antibodies but labeling can be enhanced by pretreatment of sections with hydrogen chloride. A probable interpretation for this is that compounds like phenolics are in fact inhibiting access of the antibody to the antigen [14].

Detailed analyses of one mutant, using immunocytochemical techniques, have shown that callose and PR-1 proteins are present in these secondary wall structures (Figures 14 and 16) [14,31]. These molecules are also typically associated with the wall deposits formed in pea cells reacting to *C. elegans*. (Figures 15 and 17). The wall reactions observed in myc^{-1} roots and induced by arbuscular mycorrhizal fungi therefore appear to be similar to those characteristic of the hypersensitive-like response in the incompatible interaction between pea roots and *C. elegans*. These secondary structures are clearly different from wall thickenings that have been described to occur in epidermal cells of leek in contact with appressoria of fungal symbionts. In this case, wall appositions do not contain callose and no strong autofluorescence has been observed [34]. Furthermore, contrary to the wall structures induced in the pea mutants, antibodies heavily label pectins in these wall reactions in leek roots (unpublished results) and hypodermal cell walls underlying appressoria do not show any thickening.

The important structural defence response in outer root cells associated with incompatibility between myc^{-1} roots and arbuscular mycorrhizal fungi contrasts with the very weak and localised defence reactions induced in parenchymal cells during normal mycorrhiza formation. This suggests that the expression of specific plant genes exerts a control over the expression of root defence responses early during symbiosis establishment. When these genes are mutated, they do not have this function anymore, leading to a strong defence response and no root colonisation by arbuscular mycorrhizal fungi. Whether products of symbiosis-related genes directly affect expression of defence genes or whether they act through activation of fungal-derived suppressors remains to be elucidated.

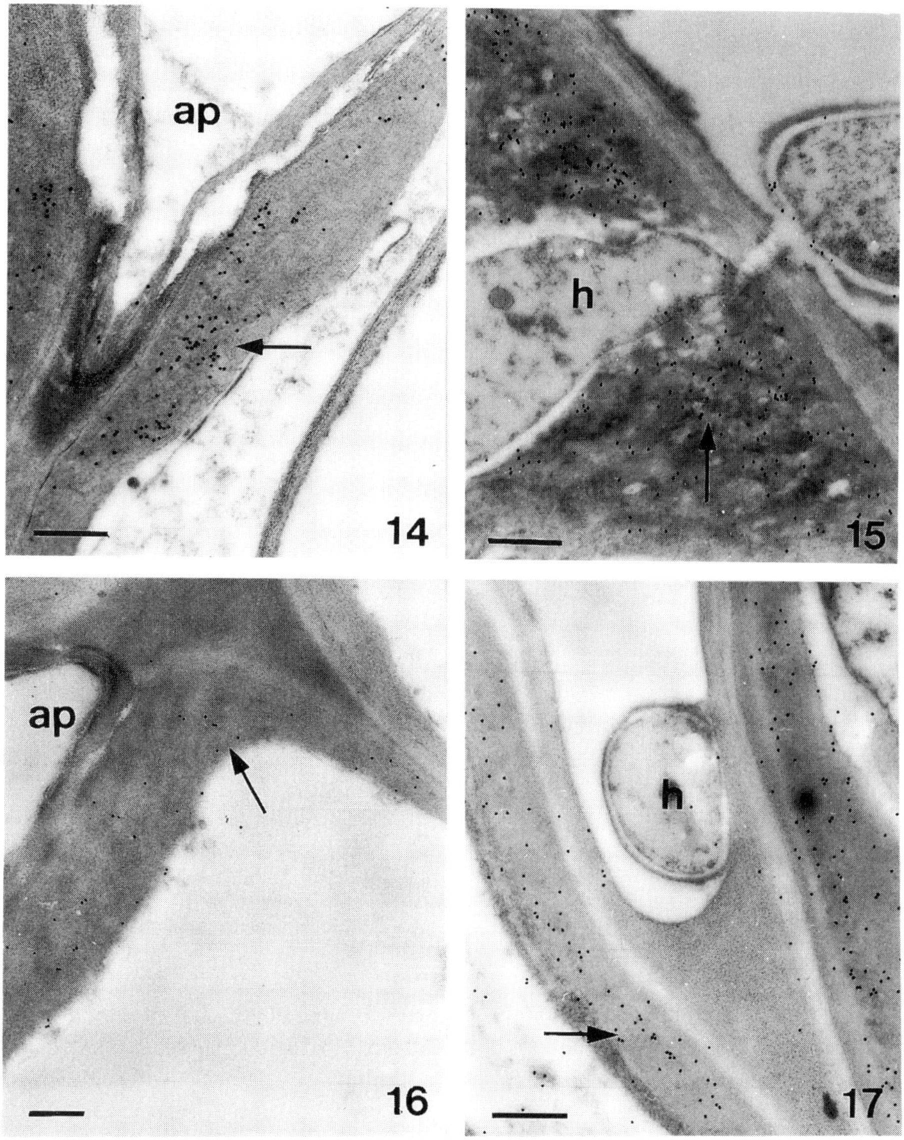

Figures 14-17. Immunolocalisation (arrows) of β (1-3) glucan chains (14,15) and PR-1 protein (16,17) in secondary wall structures induced in myc^{-1} mutant pea roots by an appressorium (ap) of *G. mosseae* (14,16) or by hyphae (h) of *C. elegans* (15,17). Bar = 0.4 μm.

4. Host Cell Responses To Nodulating Bacteria

The formation of nitrogen fixing root symbioses by bacteria of the genus *Rhizobium* or the actinobacterium *Frankia* and legume or non-legume plants, respectively, involves complex cellular interactions and differentiation of a highly organised organ, the nodule [for details see 35,36,37]. The infection process initiates with penetration of root hairs and the formation of transcellular tubular structures, the infection threads, in which the microorganisms migrate into the root tissue. Cytochemical and immunocytochemical techniques have shown that infection threads consist of an interface containing plant cell wall components like pectin (Figure 18) and glycoproteins (Figure 19) [8,38,39].

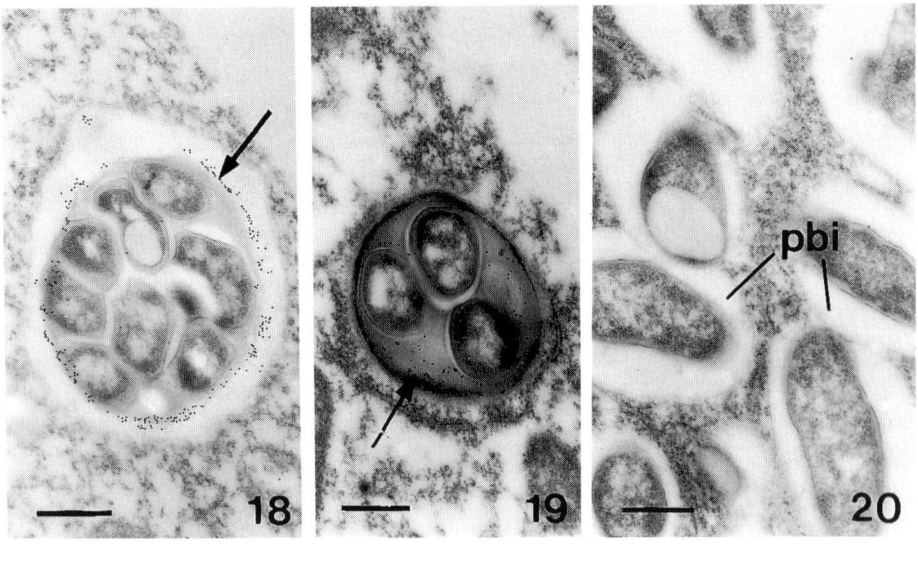

Figures 18-20. Immunolocalisation of pectin (18) and plant glycoprotein (19) in infection thread material surrounding bacteria (arrows). Neither molecule is detected in the peribacteroid interface (pbi) (20). Bar = 0.5 μm.

Infection threads penetrate cells of the nodule primordium, which originates from cortical cell divisions in advance of the extending infection thread [40]. Important morphological changes in the organisation of the infected plant cell occur as the microsymbionts proliferate. The latter become enclosed by a plant-derived perisymbiotic membrane and, in the case of *Frankia*, enzyme-gold affinity labelling has shown that host wall material persists in the symbiotic interface which forms [41]. This contrasts with legume nodules, where plant wall components like pectin cannot be detected in the interface surrounding the bacteroids as they differentiate from infection thread bacteria on release into the host cell (Figure 20) [for references see 35].

There have been few investigations of defence-related responses at the cellular or tissue level during the formation of effective nitrogen-fixing nodules. Immunolocalisation of chitinase or peroxidase in effective pea or soybean nodules [42,43], and *in situ* hybridisation of chalcone synthase transcripts [44] in effective pea nodules, have shown that such defence-related molecules accumulate predominantly in uninfected nodule tissue and not in the central infected region. It has been assumed that activation of defence-related metabolic pathways (mostly phytoalexins) early after *Rhizobium* inoculation of legumes represents a weak and non-specific plant response [4]. A similar interpretation has been made concerning lignin that has been cytochemically detected around intracellular *Frankia* in nodules of the non-legume *Casuarina* [45]. More attention has been paid to the role of inducible defence reactions in imbalanced interactions involving microbial mutants, and leading to the formation of pseudo- or ineffective nodules.

In the case of pseudonodules of legume species, the morphological differentiation of nodule-like structures is induced but neither infection threads nor bacteroids develop. In uninfected pseudonodules of alfalfa induced by a *Rhizobium* mutant deficient in an exopolysaccharide (EPS), cortical cell walls are abnormally thick and develop brown, necrotic areas with papillae-like wall structures. These are strongly autofluorescent under blue light and stain positively with sirofluor, indicating that they are encrusted with phenolics and callose [46]. These cytological observations confirm biochemical data of an induced plant defence response to the EPS mutant, and have led to the suggestion that surface EPS, or a related compound, acts a suppressor of the host plant's defence system to establish an effective symbiosis [46,47]. Similar plant defence responses occur during tissue and cell invasion of pea by lipopolysaccharide-defective (LPS) *Rhizobium* mutants. Such mutants are only partially successful in colonizing differentiated nodule tissue. Here, they form ineffective nodules with few infected cells and with infection threads giving rise to abnormal, thick « invasion structures » [8,44]. Infected cells of such ineffective nodules are strongly autofluorescent, indicating phenolics accumulation, and show *in situ* production of chalcone synthase transcripts. In addition, tissue and cell invasion by the LPS *Rhizobium* mutant is associated with increases in the infection thread matrix material, with secondary cell wall modifications and possible callose accumulation. It has been postulated from these different observations that LPS-defective bacteria trigger plant defence responses and that, like EPS, the essential role of the components of LPS at the bacterial cell surface is in somehow avoiding or preventing induction of defence reactions during symbiotic nodule interactions.

Only a small proportion of *Frankia* or *Rhizobium* infections in root hairs result in the formation of nodules, indicating an autoregulation of nodulation. In *Alnus*, arrested root-hair infection is associated with callose-containing wall appositions around *Frankia* hyphae where these penetrate the host cell, which appears senescent [48]. Similarly, in alfalfa, the arrest of infection threads induced by *Rhizobium* in subepidermal layers of the root tissue is accompanied by a process resembling that of host defence [49]. Autofluorescence, cytochemistry and immunocytochemistry have revealed the accumulation of phenolics, hydroxyproline-rich glycoprotein and chitinase in necrotic cells containing aborted infection threads in alfalfa roots. Consequently, host plants appear to control infection by wild-type symbiotic bacteria through the activation of defence mechanisms and so autoregulate nodulation. The molecular basis of the elicitation of such autoregulating mechanisms and the role, if any, of bacterial EPS and LPS remain to be elucidated.

5. Conclusions

Cytomolecular investigations indicate that defence-like reactions are not detectably elicited in response to early arbuscular mycorrhizal stages of appressorium formation and colonisation of outer root tissues in normal (myc^{+}) host roots. They are, however, activated in a weak and very localised manner during intracellular development of the endosymbiotic fungi in cells of the parenchymal cortex. Single gene mutations in pea induce root resistance to arbuscular mycorrhizal fungi, characterised by alterations in the plant response and the localised elicitation by appressoria of secondary host wall structures typical of defence. It seems likely that mutations have affected genes that are normally active upstream in the infection process, and that these represent symbiosis-related genes with some regulatory role in defence responses during initial infection steps. Low expression of defence mechanisms in later stages of arbuscule colonisation in parenchyma cells may be linked to a non-specific host response, or lack of active fungal elicitors at this morphogenetic stage where simplification in the molecular composition of the fungal wall reaches an extreme [50]. The role of wall components in eliciting defence responses by activating host receptors is a current concept in relation to the perception of pathogens by plants [for references see 32].

In contrast, plant defence responses do not appear to be activated in root or nodule cells where *Rhizobium* proliferates. The coincidence of enhanced defence responses with reduced cell and tissue invasion by bacterial mutants devoid of EPS or LPS components, is consistent with the idea that surface molecules of the microorganisms are somehow involved in preventing plant defence. It has been proposed that EPS and LPS may be synchronised in suppressing plant defence responses during early and later stages of the symbiosis, respectively [43]. However, the fact that activation of plant defence can already occur with wild-type Rhizobia in host roots means that mechanisms involved in their suppression must be more complex.

In conclusion, in endocellular root symbioses like arbuscular mycorrhiza and nodules, the delicately balanced interactions imply that the microsymbiont proliferates within host tissues in a controlled fashion. Cytological investigations indicate that defence reactions typical of pathogen infections do not play a major role in this control, since such host responses are either not observed or extremely localised in root cells. The presence or the deposition of host wall material in the path of the invading microorganism may be a mechanism by which surface contact between the symbiotic partners is restricted until the microorganism has transformed its cell wall components so as not to elicit a resistance reaction by the host. Furthermore, interactions involving symbiotically-defective host plant or microbial mutants trigger strong defence responses in host cells, suggesting that active prevention of these is a common key event in root symbioses.

Acknowledgements

The authors thank C. Arnould and D. Dubois for excellent technical assistance, and C. Picard for invaluable secretarial help. Antibodies were from Biosupplies (Mab/ β(1-3) glucan) or kindly provided by J.P. Knox (JIM5/pectin), N.J. Brewin (MAC265/glycoprotein) and J.P. Carr (MAb/PR-1 protein).

References

1. Cullimore, J.V., and Bennett, M.J.: Nitrogen assimilation in the legume root nodule : current status of the molecular biology of the plant enzymes, *Canadian Journal of Microbiology* **38** (1992), 461-466.
2. Bethlenfalvay, G., and Linderman, R.G.: *Mycorrhizae in Sustainable Agriculture*, ASA Special Publication, Madison, WI, 1992.
3. Gianinazzi-Pearson, V.: Morphofunctional compatibility in interactions between roots and arbuscular endomycorrhizal fungi : molecular mechanisms, genes and gene expression, in K. Kohmoto, R.P. Singh and U.S. Singh (eds.), *Pathogenesis and Host Specificity in Plant Diseases*, Pergamon, Elsevier Science Ltd, Oxford, vol. II (1995), pp. 251-263.
4. McKhann, H.I., and Hirsch, A.M.: Does *Rhizobium* avoid the host reponse ?, in J.L. Dangl (ed.), *Bacterial Pathogenisis of Plants and Animals*, Springer-Verlag, Berlin (1994), pp. 139-162.
5. Gianinazzi-Pearson, V., Gianinazzi, S., and Brewin, N.J.: Immunocytochemical localisation of antigenic sites in the perisymbiotic membrane of vesicular-arbuscular endomycorrhiza using monoclonal antibodies reacting against the peribacteroid membrane of nodules, in P. Nardon, V. Gianinazzi-Pearson, A.M. Grenier, L. Margulis and D.C. Smith (eds.), *Endocytobiology IV*, INRA, Paris (1990), pp. 127-131.
6. Mellor, R.B.: Bacteroids in the *Rhizobium*-legume symbiosis inhabit a plant internal lytic compartment : implications for other microbial endosymbioses, *Journal of Experimental Botany* **40** (1989), 831-839.
7. Gianinazzi-Pearson, V., Gollotte, A., Dumas-Gaudot, E., Franken, P. and Gianinazzi, S.: Gene expression and molecular modifications associated with plant responses to infection by arbuscular mycorrhizal fungi, in M. Daniels (ed.), *Advances in Molecular Genetics of Plant-Microbe Interactions*, Kluwer Academic Publishers, Dordrecht (1994), pp. 179-186.
8. Perotto, S., Brewin, N.J., and Kannenberg, E.L.: Cytological evidence for a host defense response that reduces cell and tissue invasion in pea nodules by lipopolysaccharide-defective mutants of *Rhizobium leguminosarum* strain 3841, *Molecular Plant-Microbe Interactions* **7** (1994), 99-112.
9. Newman, E.I., and Reddell, P.: The distribution of mycorrhizas among families of vascular plants, *New Phytologist* **106** (1987), 745-751.
10. Pirozynski, K.A., and Dalpe, Y.: Geological history of the Glomaceae with particular reference to mycorrhizal symbiosis, *Symbiosis* **7** (1989), 1-36.
11. Remy, W., Taylor, T.N., Hass, H., and Kerp, H.: Four hundred-million-year-old vesicular arbuscular mycorrhizae, *Proceedings of the National Academy of Sciences USA* **91** (1994), 11841-11843.
12. Bonfante, P., and Perotto, S.: Strategies of arbuscular mycorrhizal fungi when infecting host plants, *New Phytologist* **130** (1995), 3-21.

13. Gollotte, A., Gianinazzi-Pearson, V., and Gianinazzi, S.: Immunodetection of infection thread glycoprotein and arabinogalactan protein in wild type *Pisum sativum* (L.) or an isogenic mycorrhiza-resistant mutant interacting with *Glomus mosseae, Symbiosis* **18** (1995), 69-85.
14. Gollotte, A., Gianinazzi-Pearson, V., and Gianinazzi, S.: Etude immunocytochimique des interfaces plante-champignon endomycorhizien à arbuscules chez des pois isogéniques myc$^+$ ou résistants à l'endomycorhization, *Acta Botanica Gallica* **141** (1994), 449-454.
15. Balestrini, R., Romera, C., Puigdomenech, P., and Bonfante, P.: Location of a cell wall hydroxyproline-rich glycoprotein, cellulose and β-1,3 glucans in apical and differentiated regions of maize mycorrhizal roots, *Planta* **195** (1994), 201-209.
16. Alexander, D., Goodman, R.M., Gut-Rella, M., Glascock, C., Weymann, K., Friedrich, L., Maddox, D., Ahl-Goy, P., Luntz, T., Ward, E., and Ryals, J.: Increased tolerance to two oomycete pathogens in transgenic tobacco expressing pathogenesis-related protein 1a, *Proceedings of the National Academy of Sciences USA* **90** (1993), 7327-7331.
17. Gianinazzi-Pearson, V., Tahiri-Alaoui, A., Antoniw, J.F., Gianinazzi, S., and Dumas, E.: Weak expression of the pathogenesis related PR-b1 gene and localization of related protein during symbiotic endomycorrhizal interactions in tobacco roots, *Endocytobiosis and Cell Research* **8** (1992), 177-185.
18. Benhamou, N., Mazau, D., and Esquerré-Turgayé, M.T.: Immunocytochemical localization of hydroxyproline-rich glycoproteins in tomato root cells infected by *Fusarium oxysporium* f.sp *radicis-lycopersici* : study of a compatible interaction, *Phytopathology* **80** (1990), 163-173.
19. Tahiri-Alaoui, A., Dumas-Gaudot, E., Gianinazzi, S., and Antoniw, J.F.: Expression of the PR-$b_{1..}$ gene in roots of two *Nicotiana* species and their amphidiploid hybrid infected with virulent and avirulent races of *Chalara elegans*, *Plant Pathology* **42** (1993), 728-736.
20. Harrison, M.J., and Dixon, R.A.: Spatial patterns of expression of flavonoid/isoflavonoid pathway genes during interactions between roots of *Medicago truncatula* and the mycorrhizal fungus *Glomus versiforme*, *The Plant Journal* **6** (1994), 9-20.
21. Codignola, A., Verotta, L., Spanu, P, Maffei, M., Scannerini, S., and Bonfante-Fasolo, P.: Cell wall bound phenols in roots of vesicular-arbuscular mycorrhizal plants, *New Phytologist* **112** (1989), 221-228.
22. Spanu, P., and Bonfante-Fasolo, P.: Cell wall bound peroxidase activity in roots of mycorrhizal *Allium porrum*, *New Phytologist* **109** (1988), 119-124.
23. Gianinazzi, S., and Gianinazzi-Pearson, V.: Cytology, histochemistry and immunocytochemistry as tools for studying structure and function in endomycorrhiza, *Methods in Microbiology* **24** (1992), 109-139.
24. Spanu, P., Boller, J., Ludwig, A., Wiemken, A., Faccio, A., and Bonfante-Fasolo, P.: Chitinase in roots of mycorrhizal *Allium porum* : regulation and localization, *Planta* **177** (1989), 447-455.
25. Tahiri-Alaoui, A., Dumas-Gaudot, E., and Gianinazzi, S.: Immunocytochemical localization of pathogenesis-related PR-1 proteins in tobacco root tissues infected *in vitro* by the black root rot fungus *Chalara elegans*, *Physiological and Molecular Plant Pathology* **42** (1993), 69-82.
26. Harrison, M.J., and Dixon, R.A.: Isoflavonoid accumulation and expression of defense gene transcripts during the establishment of vesicular-arbuscular mycorrhizal associations in roots of *Medicago truncatula*, *Molecular Plant Microbe Interactions* **6** (1994), 643-654.
27. Volpin, H., Elkind, Y., Okon, Y., and Kapulnik, Y.: A vesicular arbuscular mycorrhizal fungus (*Glomus intraradix*) induces a defense response in alfalfa roots, *Plant Physiology* **104** (1994), 683-689.
28. Duc, G., Trouvelot, A., Gianinazzi-Pearson, V., and Gianinazzi, S.: First report of non-mycorrhizal plant mutants (myc$^-$) obtained in pea (*Pisum sativum* L.) and fababean (*Vicia faba* L.), *Plant Science* **60** (1989), 215-222.
29. Gianinazzi-Pearson, V., Gianinazzi, S., Guillemin, J.P., Trouvelot, A., and Duc, G.: Genetic and cellular analysis of the resistance to vesicular-arbuscular (VA) mycorrhizal fungi in pea mutants, in H. Hennecke and D.P.S. Verma (eds.), *Advances in Molecular Genetics of Plant-Microbe Interactions*, Kluwer Academic Publishers, Dordrecht (1991), pp. 336-342.
30. Aist, J.R.: Papillae and related wound plugs of plant cells, *Annual Review of Phytopathology* **14** (1976), 145-163.

31. Gollotte, A., Gianinazzi-Pearson, V., Giovannetti, M., Sbrana, C., Avio, L., and Gianinazzi, S.: Cellular localization and cytochemical probing of resistance reactions to arbuscular mycorrhizal fungi in a 'locus a' mutant of *Pisum sativum* (L.), *Planta* **191** (1993), 112-122.

32. Collinge, D.B., Gregersen, P.L., and Thordal-Christensen, H.: The induction of gene expression in response to pathogenic microbes, in A.S. Basra (ed.), *Mechanisms of Plant Growth and Improved Productivity : Modern Approaches and Perspectives*, Marcel Dekker, New York (1994), pp. 391-433.

33. Gollotte, A., Lemoine, M.C., and Gianinazzi-Pearson, V.: Morphofunctional integration and cellular compatibility between endomycorrhizal symbionts, in K.G. Mukerji (ed.), *Concepts in Mycorrhizal Research*, Kluwer Academic Publishers, Dordrecht (in press).

34. Garriock, T.L., Peterson, R.L., and Ackerley, C.A.: Early stages in colonization of *Allium porrum* (leek) roots by the vesicular-arbuscular mycorrhizal fungus, *Glomus versiforme, New Phytologist* **112** (1989), 85-92.

35. Brewin, N.: Development of the legume root nodule, *Annual Review of Cell Biology* **7** (1991), 191-226.

36. Baker, D.D., and Mullin, B.C.: Actinorrhizal symbioses, in G. Stacey, R.H. Bunis and H.J. Evaro (eds.), *Biological Nitrogen Fixation*, Chapman and Hall, New York (1992), pp. 229-292.

37. Hirsch, A.M.: Developmental biology of legume nodulation, *New Phytologist* **122** (1992), 211-237.

38. Liu, Q., and Berry, A.M.: Localization and characterization of pectin polysaccharides in roots and root nodules of *Ceanothus* spp. during intercellular infection by *Frankia*, *Protoplasma* **163** (1991), 93-101.

39. VandenBosch, K.A., Bradley, D.J., Knox, J.P., Perotto, S., Butcher, G.W., and Brewin, N.J.: Common components of the infection thread matrix and the intercellular space identified by immunocytochemical analysis of pea nodules and uninfected roots, *EMBO Journal* **8** (1989), 335-342.

40. Newcomb, W., Sippel, D., and Peterson, R.L.: The early morphogenesis of *Glycine max* and *Pisum sativum* root nodules, *Canadian Journal of Botany* **57** (1979), 2603-2616.

41. Berg, R.H.: Cellulose and xylans in the interface capsule in symbiotic cells of actinorrhizae, *Protoplasma* **159** (1990), 35-43.

42. Staehelin, C., Müller, J., Mellor, R.B., Wiemken, A., and Boller, T.: Chitinase and peroxidase in effective (fix^+) and ineffective (fix^-) soybean nodules, *Planta* **187** (1992), 295-300.

43. Parniske, M., Schmidt, P.E., Kosch, K., and Müller, P.: Plant defense responses of host plants with determinate nodules induced by EPS-defective *exoB* mutants of *Bradyrhizobium japonicum*, *Molecular Plant-Microbe Interactions* **7** (1994), 631-638.

44. Yang, W.-C., Canter Cremers, H.C.J., Hogendijk, P., Katinakis, P., Wijffelman, C.A., Franssen, H., van Kammen, A., and Bisseling, T.: *In situ* localization of chalcone synthase mRNA in pea root nodule development, *The Plant Journal* **2** (1992), 143-151.

45. Berg, R.H., and McDowell, L.: Cytochemistry of the wall of infected cells of *Casuarina* actinorrhizae, *Canadian Journal of Botany* **66** (1988), 2038-2047.

46. Niehaus, K., Kapp, D., and Pühler, A.: Plant defence and delayed infection of alfalfa pseudonodules induced by an exopolysaccharide (EPS 1)-deficient *Rhizobium meliloti* mutant, *Planta* **190** (1993), 415-425.

47. Niehaus, K., Kapp, D., Lorenzen, J., Meyer-Gattermann, P., Sieben, S., and Pühler, A.: Plant defence in alfalfa pseudonodules induced by an exopolysaccharide (EPS 1)-deficient symbiont (*Rhizobium meliloti*), *Acta Horticulturae* **381** (1994), 258-264.

48. Berry, A.M., and McCully, M.E.: Callose-containing deposits in relation to root-hair infections of *Alnus rubra* by *Frankia*, *Canadian Journal of Botany* **68** (1989), 798-802.

49. Vasse, J., de Billy, F., and Truchet, G.: Abortion of infection during the *Rhizobium meliloti*-alfalfa symbiotic interaction is accompanied by a hypersensitive reaction, *The Plant Journal* **4** (1993), 555-566.

50. Lemoine, M.C., Gollotte, A., and Gianinazzi-Pearson, V.: Localization of β(1-3) glucan in walls of the endomycorrhizal fungi *Glomus mosseae* (Nicol. & Gerd.) Gerd. & Trappe and *Acaulospora laevis* Gerd. & Trappe during colonization of host roots, *New Phytologist* **129** (1995), 97-105.

HISTOLOGY AND CYTOCHEMISTRY OF INTERACTIONS BETWEEN PLANTS AND XANTHOMONADS

B. BOHER, I. BROWN*, M. NICOLE, K. KPEMOUA, V. VERDIER, U. BONAS**, J.F. DANIEL, J.P. GEIGER and J. MANSFIELD*
ORSTOM, Laboratoire de Phytopathologie, BP 5045, 34032, Montpellier, France
**University of London, Wye College, Ashford, England, TN25 5AH*
*** CNRS, Institut des Sciences Végétales, 91198, Gif/Yvette, France*

1. Introduction

Most of the bacteria in the genus *Xanthomonas* are plant pathogens. The genus includes six species and no less than 140 different pathovars in the species *X. campestris* [17]. The parasitic development of xanthomonads includes three stages:

- an epiphytic stage when the bacteria adhere to the surface of aerial organs and multiply without causing symptoms,
- an intercellular parasitic stage within parenchyma, causing necrotic spots on aerial organs and
- a vascular and systemic parasitic stage inside the xylem which is often associated with a more or less complete colonization of the plant.

Many pathovars develop primarily in intercellular spaces, with only restricted vascular development, including X. *campestris* pv. *cassavae*, pv. *citri*, pv. *glycines*, pv. *malvacearum*, pv. *pruni*, pv. *vesicatoria*, *X. fragariae* and X. *oryzae (X. c.)* pv. *oryzicola*. In these *Xanthomonas*, penetration into vessels and systemic development are rarely observed. Others, including *X. c.* pv. *campestris*, pv. *graminis*, pv. *manihotis*, pv. *pelargonii* and *X. o.* pv. *oryzae* are characterized by systemic colonization of the plant. A great variety of symptoms such as necrotic spots, wilting, defoliation and dieback are produced. Disease development is often associated with contamination of vegetative planting material.

This chapter deals with histological and cytological aspects of the various

M. Nicole and V. Gianinazzi-Pearson (eds.), Histology, Ultrastructure and Molecular Cytology of Plant-Microorganism Interactions, 193-210.

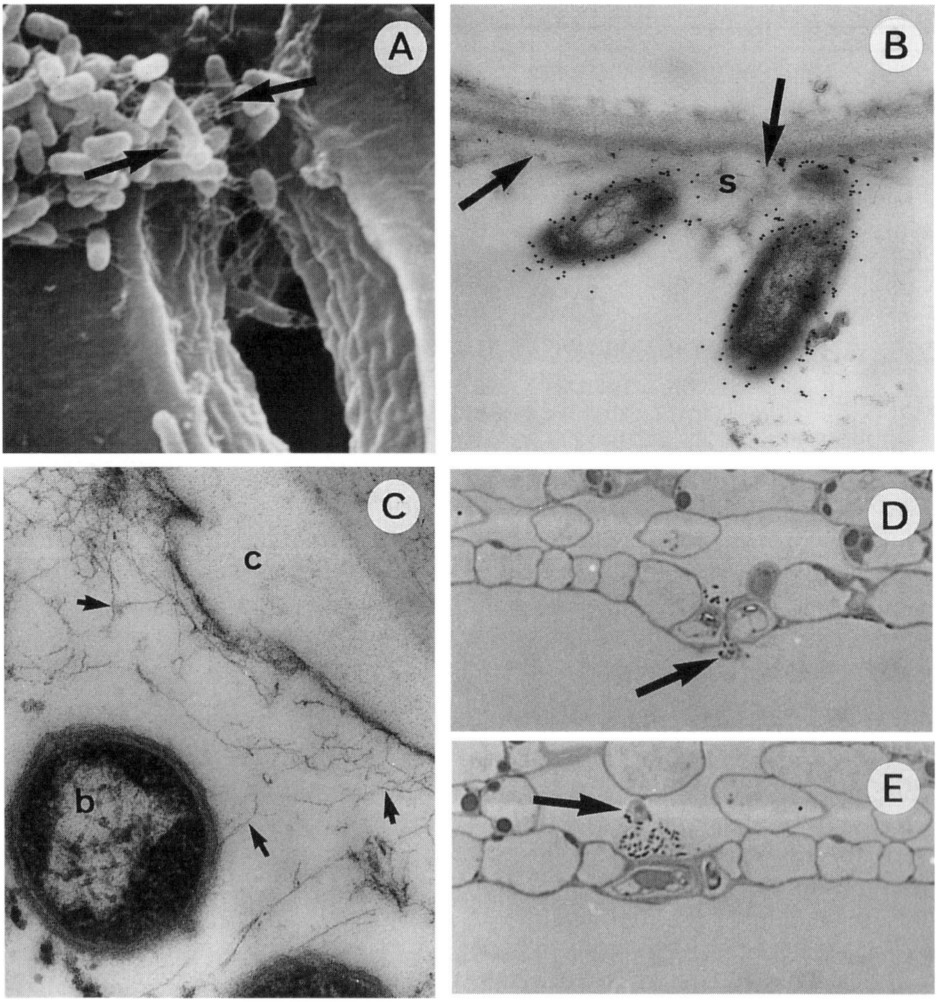

Figure 1A : Scanning electron micrograph of a cassava leaf surface colonized by *Xanthomonas campestris* pv. *manihotis*. Bacteria are located on the lower epidermis close to a stomatal aperture. A fibrillar sheath surrounds the bacterial colony (arrows) (x 7600).

Figure 1B : Transmission electron micrograph (TEM) showing the surfaces of *Xanthomonas campestris* pv. *manihotis*. decorated with gold particles after labelling with polyclonal antibodies raised against whole killed bacteria. Note that the bacterial sheath (s) adheres to the plant cuticule (arrows) (x 32000).

Figure 1C : *Xanthomonas campestris* pv. cassavae on cassava leaf; a fibrillar network (arrows) is seen between the bacteria (b) and the epidermal cuticle (c) (x 77000).

Figure 1D, E : Light microscope observation of a *Xanthomonas campestris* pv. *manihotis*. colony (arrow) during penetration through the stomatal pore (1D) and located in the substomatal cavity at the lower surface of a cassava leaf (1E) (1D : x 760; 1E : x 760).

stages of *Xanthomonas*-induced pathogenesis, including compatible interactions and the resistance of plants to colonization.

2. Epiphytic Development and Penetration into Host Tissues

Many xanthomonads are able to survive and multiply at the leaf surface without inducing symptoms. This epiphytic phase is important for the production of primary inoculum, dispersal and survival. Immunofluorescence microscopy applied directly to the leaf, or to a nitrocellulose print which traps bacteria on the surface, is particularly useful for the observation of the parasite on the epidermis. Colonies are often embedded in a matrix whose fibrillar texture is visible with scanning electron (Figure 1A) or transmission electron microscopy(Figure 1B, 1C) [2, 36]. In the *X. c.* pv. *citri*/citrus interaction, the presence of the polysaccharide extracellular matrix is necessary for bacterial adhesion to the epidermal surface; a host-produced agglutinin also seems to be involved in this process [32]. The microcolonies of several species are preferentially localized in the vicinity of stomata (Figure 1A, 1D) [10], suggesting either an electrostatic interaction between the negatively charged bacterial bodies and the positively charged stomatal cells [1], or the occurrence of chemotaxis [15]. Fimbriae are proteinic filiform appendages which are not common in *Xanthomonas* [30]. However, they are produced in large numbers in *X. c.* pv. *hyacinthi* and they specifically adhere to stomatal cells [10]. It seems that fimbriae are present also in *X. c.* pv. *vesicatoria* and that their production is induced by host compounds [27].

Since *Xanthomonas* species are unable to penetrate the epidermal cuticle, they reach host tissues through natural openings, stomata and hydathodes, or wounds. In cassava, stomatal pores become plugged by microcolonies and the extracellular matrix of *X. c.* pv. *manihotis* (Figure 1D); these plugs may be sucked inside the substomatal cavities during gas exchange (Figure 1E). High stomatal density [26] and large stomatal openings are also factors that may favour infection [24].

Hydathodes localized at the margins of leaves are structures similar to stomata, but with a larger opening. They play an important role in the infection of rice by *X. o.* pv. *oryzae* [25] and cabbage by *X. c.* pv. *campestris* [6]. Guttation occurs at the hydathodes and the exuded liquid may be reabsorbed, thus facilitating a direct contamination of xylem vessels.

Infection through wounds places the parasite directly into contact with the host intercellular milieu. For example, *X. c.* pv. *vesicatoria* can penetrate through break points that frequently appear at the base of the trichomes of the tomato leaf [35]. With time, entry through wounds may be limited by

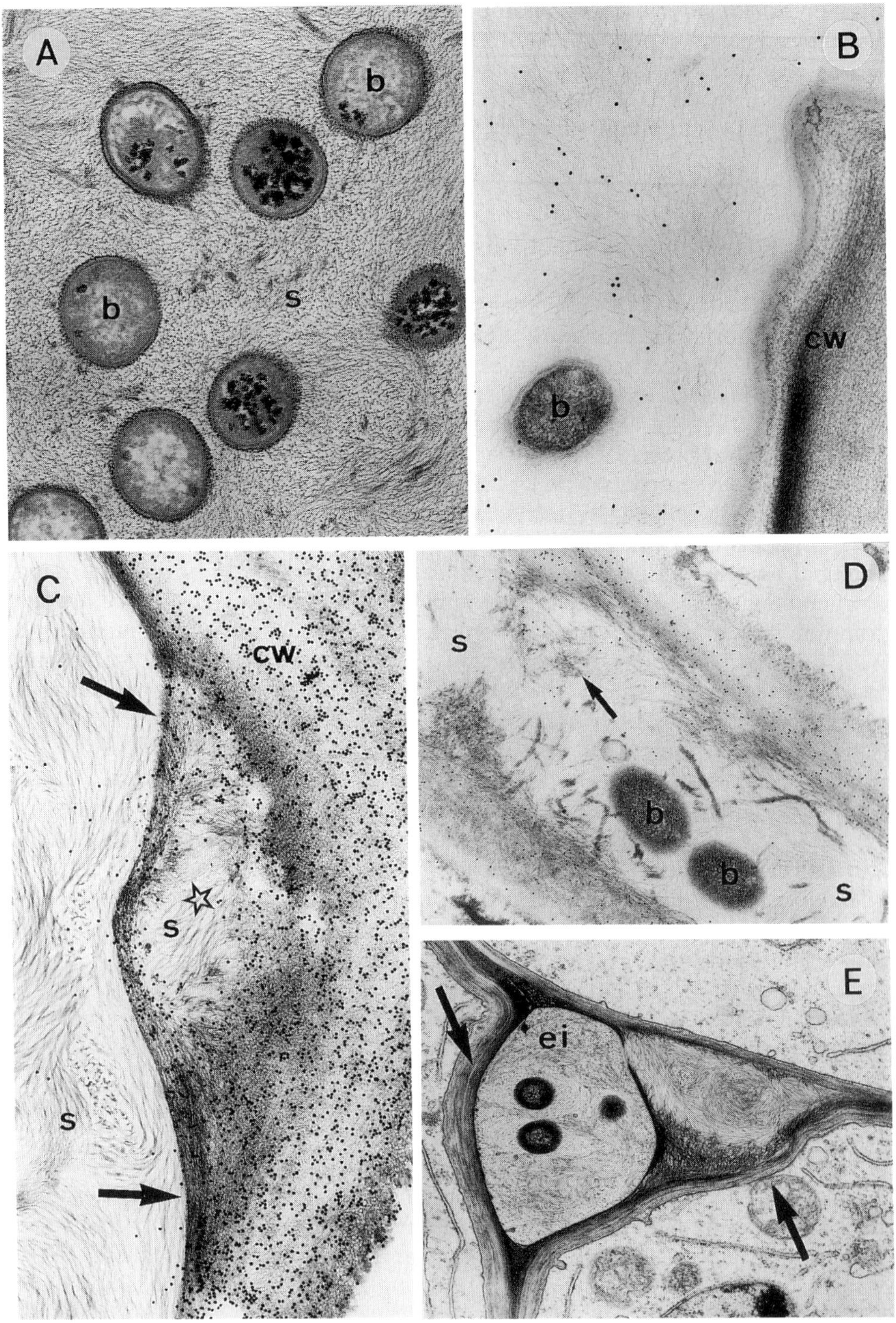
A
b
s
b
B
cw
b
C
cw
s
s
D
s
b
b
s
E
ei

hyperplasia and hypertrophy of mesophyll cells close to the wounding site which is often associated with callose deposits, suberization and lignification of walls [9, 21]. *X. c.* pv. *pruni* penetrates into peach trees through leaf scars. These are normally protected against infection by the dense abscission layer that forms before the leaf falls. Only the scars of leaves prematurely torn away by the wind can constitute entry points for bacteria [14].

3. Intercellular Development in Parenchyma

Following penetration, two types of interaction can be observed : compatibility in which bacteria multiply in the intercellular spaces of the leaf parenchyma and incompatibility characterized by the activation of plant responses which restrict bacterial multiplication.

3.1. COMPATIBLE INTERACTIONS

In the intercellular spaces bacteria are often separated from each other and from the walls by a matrix which is pink-stained by toluidine blue and has the appearance of a fibrillar network under the electron microscope. Although the polysaccharidic nature of the matrix is demonstrated by ruthenium red or the PATAg reaction (Figure 2A), its origin remains controversial. A bacterial origin was demonstrated in *X. c.* pv. *vesicatoria* and *X. c.* pv. *manihotis* [3, 7] using monoclonal antibodies which are specific to the lateral chains of xanthan, an exopolysaccharide (EPS) of *X. c.* pv. *campestris* (Figure 2B) [16]. On the other

---

Figure 2A : PATAg reaction for polysaccharide detection. TEM showing that the extracellular sheath (s) of *Xanthomonas campestris* pv. *manihotis* (b) reacts positively to this test (x 32000).

Figure 2B : The use of monoclonal antibody XB3 revealed that the exopolysaccharides of *Xanthomonas campestris* pv. *cassavae* (b) are evenly labelled. No labelling is seen over the cassava plant cell wall (cw) (x 33000).

Figure 2C : TEM showing the close association (arrows) of *Xanthomonas campestris* pv. *malvacearum* exopolysaccharides (s) with the cell wall of a cotton mesophyll cell (cw). The plant cell wall is labelled with an exoglucanase complexed to colloidal gold for ß-1,4-glucan localization. The plant cell wall displays degraded areas at the junction with the bacterial sheath (*) (x 28000).

Figure 2D : The bacterial sheath (s) of *Xanthomonas campestris* pv. *manihotis* (b) closely associated with the middle lamella of cassava mesophyll cells contains fragments of material labelled with the monoclonal antipectin antibody JIM 5 (arrows) (x 23000).

Figure 2E : Bacterial strands of *Xanthomonas campestris* pv. *manihotis* are visible in intercellular spaces (ei) of the vascular parenchyma of a cassava stem, causing host cell wall tearing (arrows). Note the occurrence of abundant exopolysaccharides (x 10000).

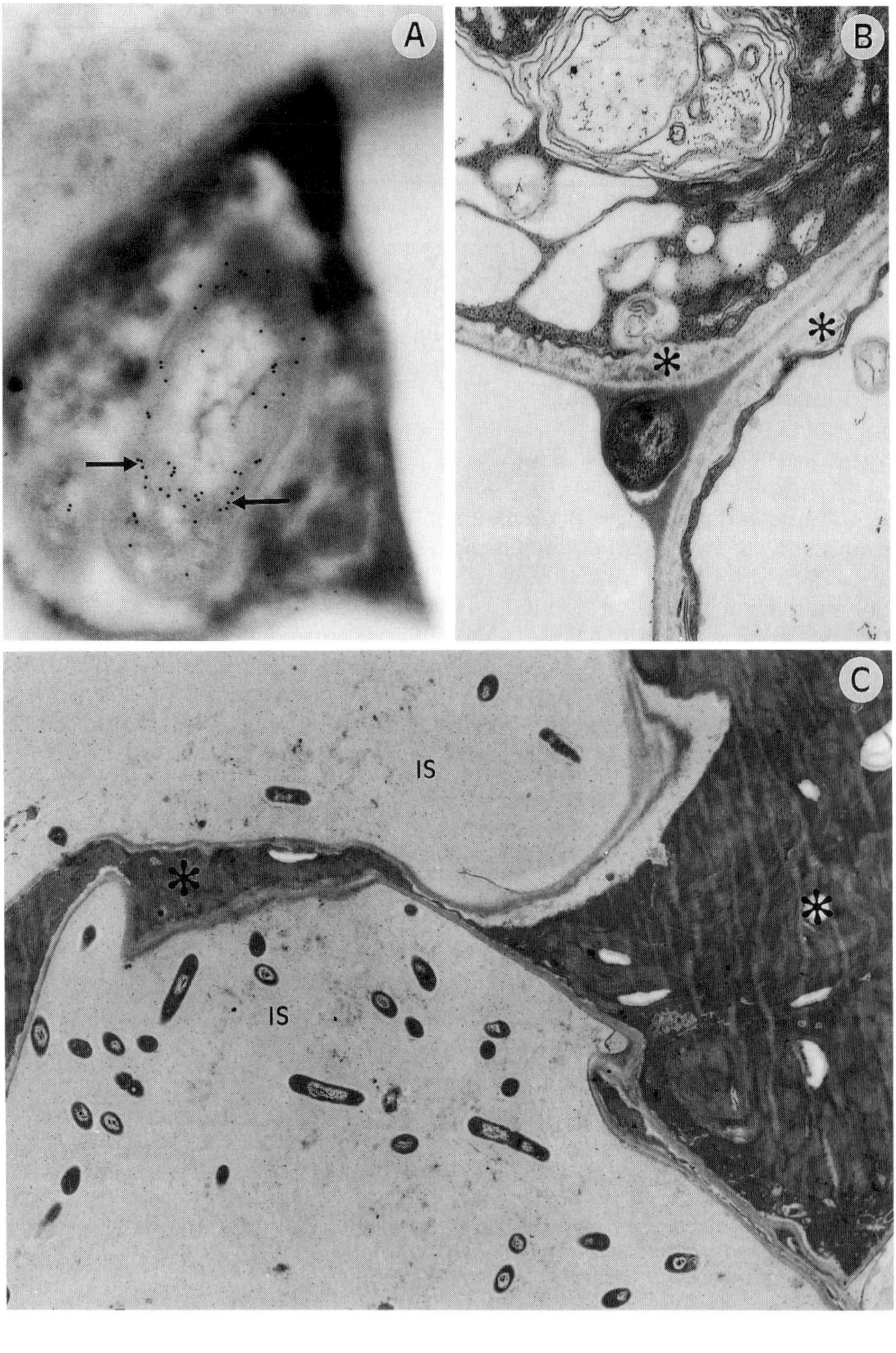
A
B
*
*
C
IS
IS
*
*

hand, in *X. o.* pv. *oryzae*/rice interaction, this matrix was labelled by an EPS specific polyclonal antiserum only at the immediate vicinity of the bacteria suggesting plant origin for more distant fibrils [39]. After infiltration using purified EPS from *X. c.* pv. *pelargonii* [28] images of a fibrillar structure identical to that observed in infected tissues were obtained. In cassava, gold-probes targeting wall compounds such as cellulose, esterified or non esterified pectin, and xyloglucans do not recognize this matrix, but they demonstrate the occurrence of wall degradation products within EPS, as also evidenced with *X. c.* pv. *malvacearum* (Figure 2C) [3]. *In vitro*, several, but not all xanthomonads are known to produce hydrolytic enzymes such as cellulase, pectinase, xylanase, and proteinase [29], which can assist in the degradation of the host walls. The degradation of the middle lamella and walls during pathogenesis has been observed by light and electron microscopy [8, 20, 33, 38, 42]. The dissolution of the middle lamella causes the separation of cells and facilitates the intercellular progress of bacterial colonies (Figure 2D). Rapid intercellular progression observed in the spongy parenchyma is slowed down in paravascular parenchyma, xylem and phloem. However, the pathogen can spread intercellularly in these tissues in the form of strands (Figure 2E). *X. c.* pv. *glycines* and *X. c.* pv. *vignicola* are exceptional in that they cause the formation of pustules on the host leaves. These pustules are mostly caused by cellular hypertrophy resulting from the presence of the bacteria [19].

3.2. INCOMPATIBLE INTERACTIONS.

3.2.1. *The hypersensitive reaction.*

The rapid necrosis of plant cells at inoculation sites is characteristic of the hypersensitive reaction (HR) in non-host plants and resistant cultivars. In cotton, the occurrence of the HR controlled by single genes for resistance to *X.c.* pv. *malvacearum* has been linked with the subsequent accumulation of

---

Figure 3A : Immunocytochemical localisation of AvrBs3 protein in cells of *Xanthomonas campestris*pv. *vesicatoria* strain 85-10::*hrpE*75(pD36) 48h after inoculation into ECW-30R. The bacteria are encapsulated onto the plant cell wall and the labelled AvrBs3 (arrows) is restricted to the cytoplasm (x42000).

Figure 3B : Early stages of cytoplasmic collapse caused by *Xanthomonas campestris* pv. *vesicatoria* 85-10::*hrpE*75 harbouring pD36 which allows over-expression of the avirulence gene *avrBs3*. Inoculation with this *hrp* mutant results in a patchy macroscopic HR after 48h in ECW-30R. Note that paramural deposits (asterisks) are induced as well as cytoplasmic collapse (x22000).

Figure 3C : Final stage of the HR in ECW-30R, 48 h after inoculation with *Xanthomonas campestris* pv. *vesicatoria* 85-10 (pL3XV1-6) which expresses *avrBs3*. The mesophyll cell has collapsed, its cytoplasm is autolyzed and electron dense. Surrounding bacteria remain dispersed in EPS within the intercellular space (IS) and are dividing (arrow) but show some vesiculation indicative of abnormal development (x 5400).

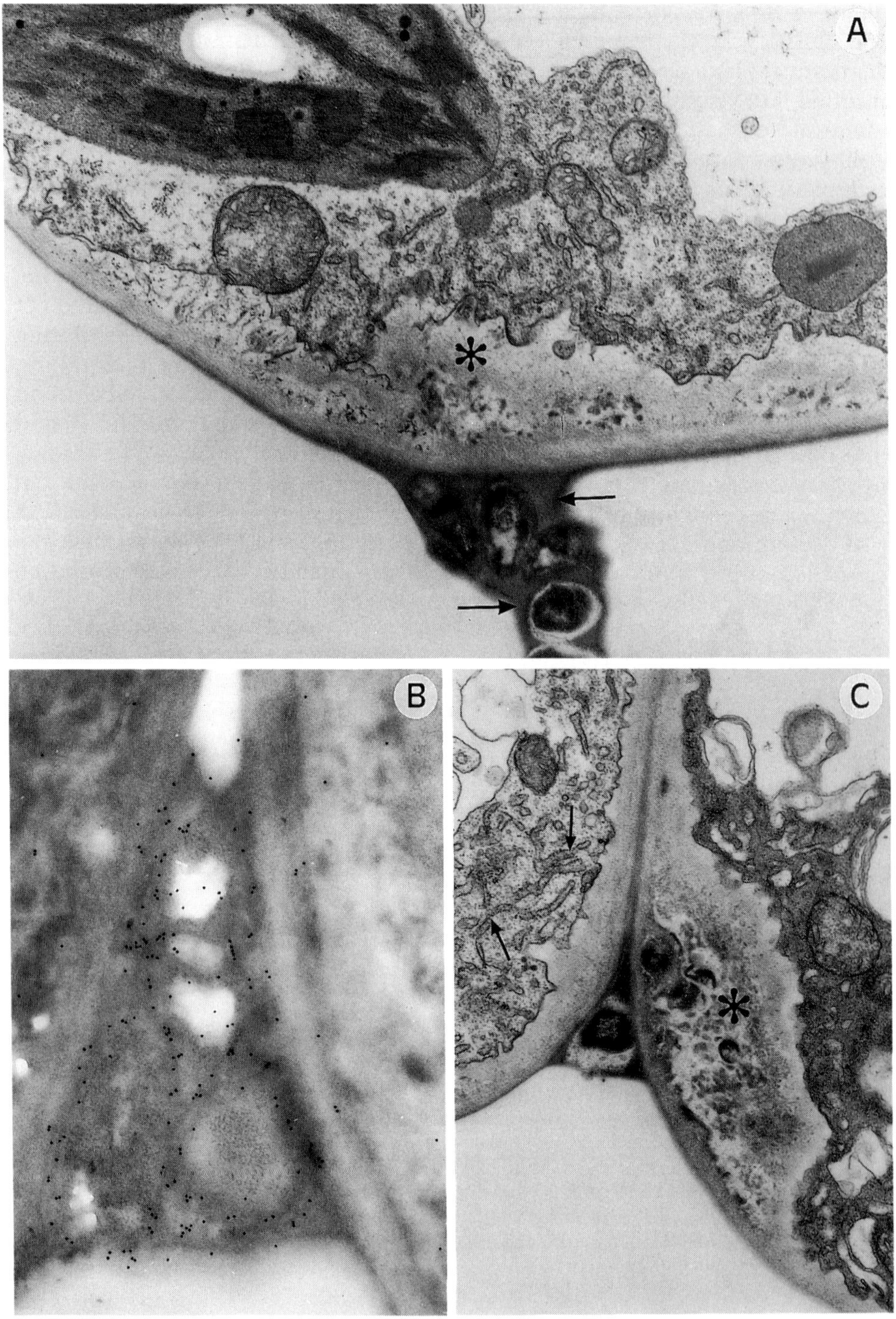
A
B
C

phytoalexins identified as sesquiterpenoid lacinilene derivatives [13]. The accumulation of the phytoalexins within cells undergoing the HR has been demonstrated by fluorescence microscopy utilizing the striking yellow-green autofluoresence of the compounds under UV light [12].

In certain resistant cotton varieties specific attachment of *X. c.* pv. *malvacearum* has been reported to be required before the HR [8]. Similar specificity at the cell surface was not observed in pepper leaves undergoing the HR in cv ECW-30R which is controlled by the *Bs3* gene for resistance to *X.c.* pv. *vesicatoria* strains with the matching avirulence gene *avrBs3* [7]. The *avrBs3* protein has been located by immunocytochemistry within the cytoplasm of *X.c.* pv. *vesicatoria* (Figure 3A), indicating that it probably does not act directly as an elicitor of the HR. The earliest signs of the onset of the HR in pepper cells is a localized vesiculation of the cytoplasm adjacent to the bacterial colony (Figure 3B). This response is rapidly followed by general decompartmentalization and cytoplasmic collapse (Figure 3C).

The protein product of the avirulence gene *avrXa10* from *X o..* pv. *oryzae* has also been immunolocalized in the bacterial cytoplasm [40]. The HR in rice cultivars with the *Xa10* gene for resistance to bacterial blight has been associated with increases in peroxidase activities [22]. Antibodies raised to a synthesized peptide specific to an induced cationic peroxidase (PO-C1) have been used to localize the enzyme during infection of rice leaves [41]. Immunogold labelling showed that PO-C1 accumulated within the apoplast of mesophyll cells and within the cell walls and vessel lumen of xylem elements in plants undergoing the incompatible interaction. Localization of peroxidase was closely correlated with sites of lignin deposition within tissues undergoing the HR [22].

Figure 4A : Paramural deposit in a mesophyll cell of pepper cv. ECW-30R induced by *Xanthomonas campestris* pv. *vesicatoria* 85-10::*hrpD*140, 24 h after inoculation. Note that bacteria are embedded in amorphous material (arrows) and that the paramural deposit (asterisk) contains electron-dense vesicles within an opaque matrix (x15000).

Figure 4B : Immunocytochemical localisation of HRGPs around cells of *Xanthomonas campestris* pv. *vesicatoria* 85-10::*hrpD*140, 48 h after inoculation into ECW-30R. The antibody, raised to tomato extensin, reveals that the amorphous material surrounding and apparently encapsulating *hrp* mutants is rich in the cell wall protein (x42000).

Figure 4C : Large paramural deposits produced in mesophyll cells of ECW-30R in response to inoculation with T55, a saprophytic strain of *Xanthomonas campestris* lacking the entire *hrp* cluster. Note the layered structure of the paramural deposit (asterisk) and the proliferation of endoplasmic reticulum (arrows). (x15000).

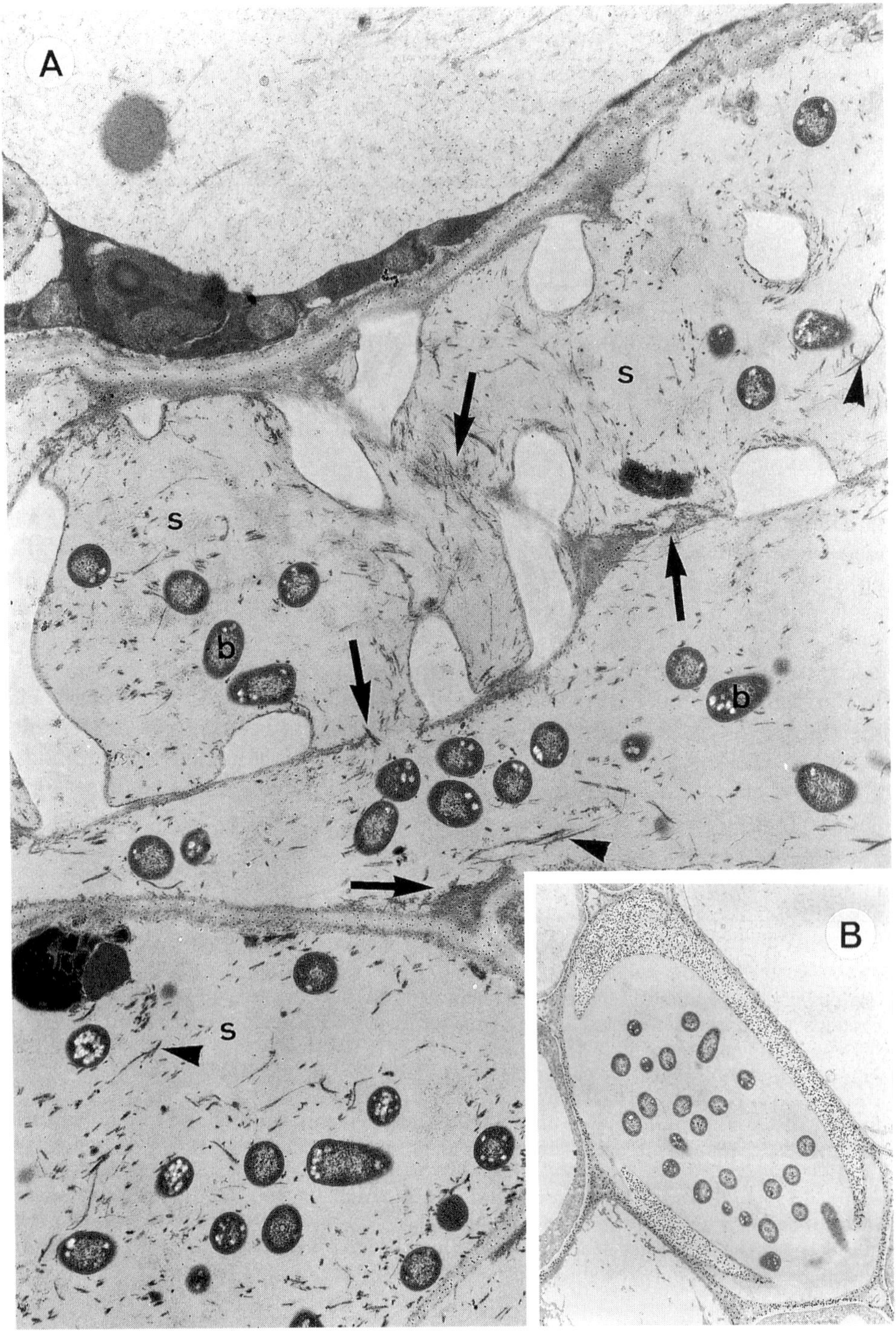
A
s
s
b
b
s
B

3.2.2. *Responses to* hrp *mutants and saprophytic strains.*
The *hrp* genes within xanthomonads determine basic pathogenicity to the host and ability to cause the HR in resistant plants [4]. Strains with mutations in *hrp* loci fail to multiply in the plant. Microscopical studies have revealed that pepper mesophyll cells undergo localized reactions to *hrp* mutants of *X. c.* pv. *vesicatoria* and also the saprophytic X. *campestris* strain T55 although no macroscopic symptoms develop [Brown, unpublished]. Deposition of large paramural papillae containing phenolics, callose and hydroxyproline rich glycoproteins (HRGPs) is characteristically observed next to bacteria which are encapsulated within an amorphous matrix (containing HRGPs) on the plant cell wall (Figure 4A, B and C). The rapid and localized responses to *hrp* mutants may be suppressed by pathogenic strains of *X. c.* pv. *vesicatoria.*

Deposition of papillae and lignification are thought to retard fungal growth through plant cell walls. However, with pathovars of *Xanthomonas* that grow within the intercellular spaces it is unlikely that plant cell wall alterations directly restrict multiplication. Factors associated with peroxidase activity and lignification, such as the accumulation of phenolic free radicals and active oxygen species, may well lead to generation of antibacterial conditions around invading bacteria. Deposition of HRGPs as illustrated in Figure 4B, is also likely to involve peroxidases in oxidative cross-linking reactions [5]. Clearly, there are many potential mechanisms that might restrict bacterial multiplication within the intercellular spaces in resistant plants. In this respect the most important factor to be considered is the localization of responses at sites of colony development either within parenchyma or vascular tissues as described below.

4. Development in Xylem Vessels

Only a few pathovars are able to develop systematically in the xylem and rapidly colonize organs far from the original point of infection. Penetration into xylem vessels may be passive through hydathodes, e.g., in *X. c.* pv. *campestris* [6] and X. *o.* pv. *oryzae* [25, 31], or active as in *X. c.* pv. *manihotis,* after dissolution of the middle lamella and the primary wall (Figures 5A,B) [3]. In

---

Figure 5A : TEM of xylem cells in a cassava leaf infected with *Xanthomonas campestris* pv. *manihotis* (b). Primary cell walls and middle lamellae of vessels are labelled for pectin localization. The bacterial fibrillar sheath (s) is closely associated with the degraded portions of the plant cell walls (arrows). Strands of unlabelled material occur within the bacterial sheath (arrowheads) (x 12000).

Figure 5B : Numerous cells of *Xanthomonas campestris* pv. *manihotis* are located within the vessel lumen in a cassava stem. The plant cell walls are labelled with an exoglucanase complexed to colloidal gold for ß-1,4-glucan localization (x 7000).

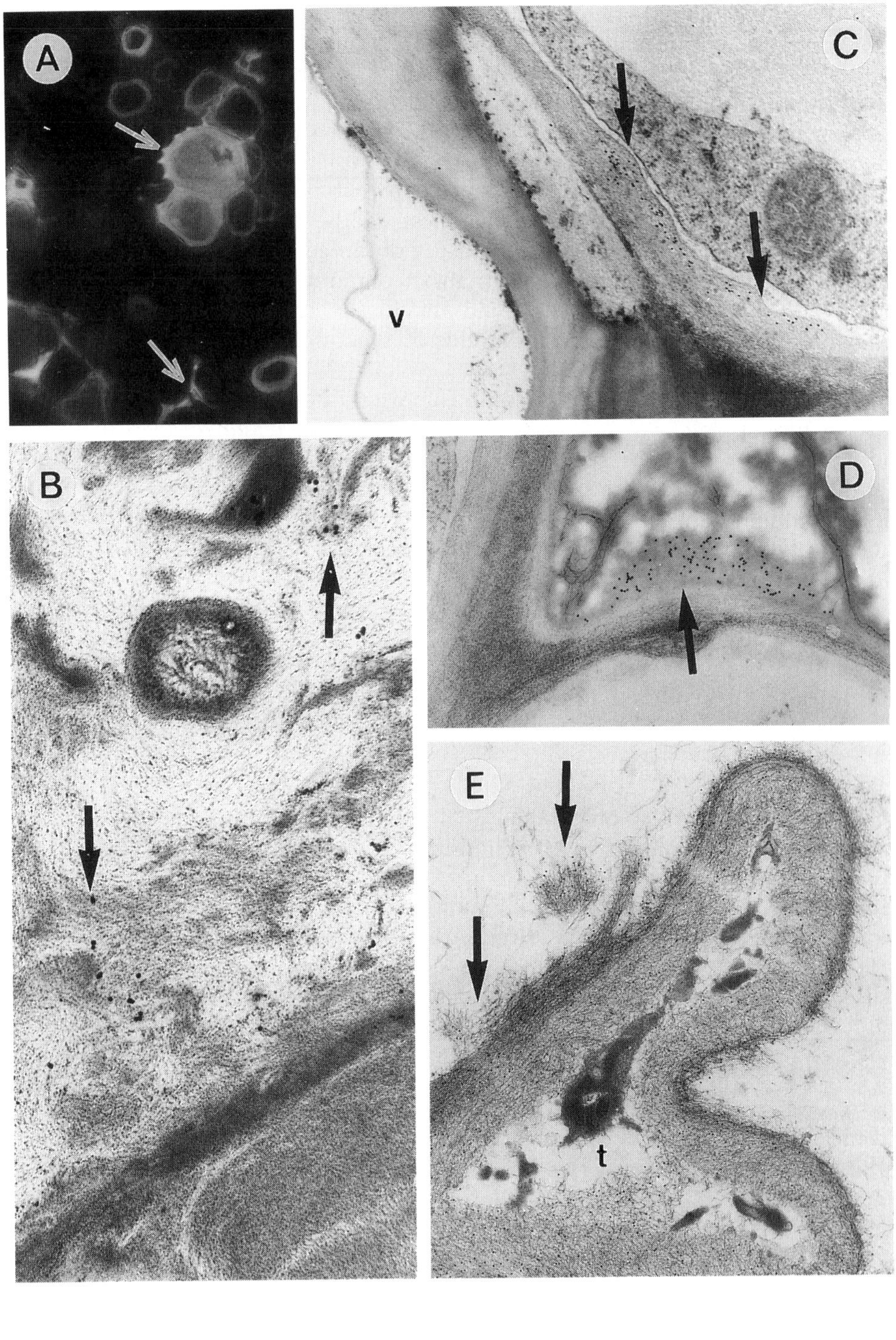
A
B
C
v
D
E
t

vessels, bacteria are associated with a fibrillar matrix which fills the lumen. In *X. o.* pv. *oryzae*, this matrix is less electron-dense in the immediate vicinity of the bacteria and it has been reported not to be recognized by ConA, WGA, and anti-EPS and anti-LPS antisera [18, 39].

On the other hand, the matrix in xylem vessels of cassava is identical to that found within intercellular spaces and it reacts positively with a monoclonal antibody that recognizes a xanthan gum epitope [3]. *X. populi* is restricted to poplar protoxylem [11] and *X. c.* pv. *pruni* cannot move transversely from one vessel to the next in the peach tree [14]. However, *X. c.* pv. *manihotis*, *X. c.* pv. *pelargonii* and *X. c.* pv. *campestris* are able to move to nearby protoxylem and metaxylem vessels [37, 38].

During compatible interactions, plant defense responses may slow down intercellular and systemic development. In cassava and cotton infected by *X. c.* pv. *manihotis* and *X. c.* pv. *malvacearum* respectively, increased autofluorescence of walls and cells close to the parasite is a common reaction (Figure 6A). In areas of fluorescence, vacuoles, the cytoplasm and periplasmic spaces of the cells often contain osmiophilic compounds, as seen with electron microscopy. The fluorescent compounds impregnate cell walls, the bacterial extracellular matrix, and bacteria that are present in the intercellular spaces. The phenolic nature of these fluorescent compounds can be demonstrated by laccase conjugated to colloidal gold (Figure 6B).

In cassava, in addition to accumulation of autofluorescent phenolics, there is probably a reinforcement of walls through a lignification process revealed by the red colouration of tissues after phloroglucinol treatment. In phloem sieve tubes (Figure 6D) and contact cells of xylem vessels (Figure 6C), infection also results in callose deposition on walls or in paramural areas. In resistant cassava

---

Figure 6A : Observations under UV illumination of sections of a cassava stem infected by *Xanthomonas campestris* pv. *manihotis* show the enhanced fluorescence of the cell walls in infected phloem and xylem (arrows) (x 100).

Figure 6B : Cytochemical localization of phenol compounds with a gold-conjugated laccase revealed that labelling is observed over degraded cell wall portions (arrows) of cassava xylem infected by *Xanthomonas campestris* pv. *manihotis* (x 42000).

Figure 6C, D : Immunogold localization of ß-1,3-glucans indicated that callose accumulates in paramural areas of vascular companion cells (6C, arrows) and in intercellular spaces of phloem cells (6D, arrow) of cassava stems infected by *Xanthomonas campestris* pv. *manihotis* (v : vessel) (6C: x 24000; 6D: x 25000).

Figure 6E : TEM showing immunogold localization of pectin. The tylosis (t) produces labelled portions of material (arrows) that accumulate in the vessel lumen of a cassava xylem infected by *Xanthomonas campestris* pv. *manihotis* (x 24000).

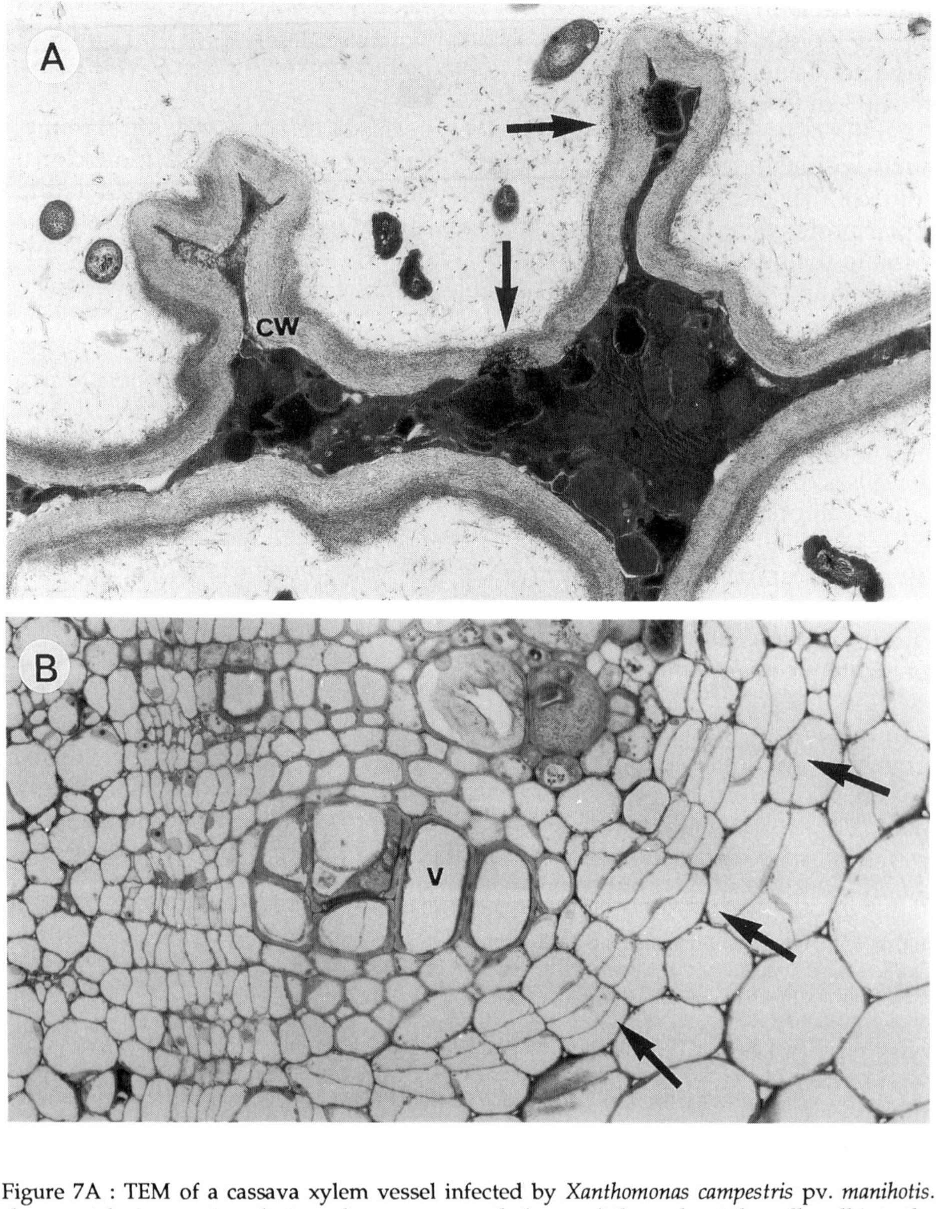

Figure 7A : TEM of a cassava xylem vessel infected by *Xanthomonas campestris* pv. *manihotis*. shows a tylosis secreting electron-dense compounds (arrows) throughout the cell wall into the vessel lumen (cw). Bacterial cells close to this tylosis are shrunken (x 12500).

Figure 7B : An area of hyperplasia (arrows) in the vicinity of an infected vascular bundle (v) in a stem infected by *Xanthomonas campestris* pv. *manihotis*. (x 250).

varieties, the various cellular reactions are stronger and appear earlier than in susceptible varieties.

Within infected xylem vessels of *Pelargonium*, the electron-dense material coating secondary walls and pits displays an autofluorescence and reacts to phloroglucinol [37, 23]. In rice infected by *X. o.* pv. *oryzae* [34], electron-dense deposits originating from companion cells were found to block the vessel lumen. Callose, pectin and lignin were cytochemically identified in xylem vessel plugs in cassava infected by *X. c.* pv. *manihotis* (Figure 6E).

In resistant cassava, parenchyma cells adjacent to xylem vessels react rapidly to the presence of the parasite. Their ultrastructure indicates an intense metabolical activity associated with the production of osmiophilic compounds, which are also found in nearby vessels. Tyloses contribute to the obstruction of vessels and they are often observed in the stele of wounded plants. These responses are typical reactions of host plants to invasion by xanthomonads. In cassava, tyloses appear in both resistant and susceptible varieties. However, in the resistant reaction, some tyloses show a specific differentiation associated with the secretion of phenolic compounds which may be bactericidal and slow down bacterial multiplication (Figure 7A). Cell division often occurs in geranium and cassava around sites of vascular colonization and leads to the formation of suberized barriers around infections [23, 37] (Figure 7B).

Late in the infection of its host *X. c.* pv. *manihotis* can induce the formation of large lysis pockets in the phloem, where the latex produced by the bursting of laticifers accumulates, together with bacteria and EPS. Rupturing of the girdle of phloem fibers causes exudation which is responsible for dissemination of bacteria.

5. Conclusion

Electron microscopy of interactions between plants and species of *Xanthomonas* allows definition of the structural and temporal framework within which recognition and response occur in the challenged plant. The recent application of immunocytochemical and advanced histochemical techniques described here for the localization of key determinants of pathogenicity, such as components of EPS and the products of avirulence genes, continues to provide new insights into cellular reactions. As parallel molecular studies advance they will provide new probes to examine interactions occurring at critical micro-sites within infected tissue. Close integration of ultrastructural and biochemical studies will be needed to develop a full understanding of bacterial pathogenesis and the subtle co-ordination of the plant's response occurring at the sub-cellular level.

6. References

1. Alippi, A.M. (1992) Histopatologia de hojas de tomate inoculadas con *Xanthomonas campestris* pv. *vesicatoria*. *Agronomie* **12**, 115-122.
2. Bashan, Y. and Okon Y. (1985) Internal and external infections of fruits and seeds of peppers by *Xanthomonas campestris* pv. *vesicatoria*. *Can. J. Bot.* **64**, 2865-2871.
3. Boher, B., Kpemoua, K., Nicole, M., Luisetti, J. and Geiger, J.P. (1995) Ultrastructure of interactions between cassava and *Xanthomonas campestris* pv. *manihotis* : Cytochemistry of cellulose and pectin degradation in a susceptible cultivar. *Phytopathology* **85**, in press.
4. Bonas, U. (1994) *hrp* genes of phytopathogenic bacteria. In J.L. Dangl (ed.) Bacterial pathogenesis of plants and animals: Molecular and cellular mechanisms. *Curr. Top. Microbiol. Immunol.* **192** Springer-Verlag, Berlin, Heidelberg pp. 79-98.
5. Bradley, D.J., Kjellbom, P. and Lamb, C.J. (1992) Elicitor- and wound-induced oxidative cross-linking of a proline-rich plant cell wall protein: a novel, rapid defense response. *Cell* **70**, 21-30.
6. Bretschneider, K.E., Gonella, M.P. and Robeson, D.J. (1989) A comparative light and electron microscopal study of compatible and incompatible interactions between *Xanthomonas campestris* pv. *campestris* and cabbage (*Brassica oleracea*). *Physiol. Mol. Plant Pathol.* **34**, 285-297.
7. Brown, I., Mansfield, J., Irlam, I., Conrads-Stauch, J. and Bonas, U. (1993) Ultrastructure of interactions between *Xanthomonas campestris* pv. *vesicatoria* and pepper, including immunocytochemical localization of extracellular polysaccharides and the *AvrBs3* protein. *Mol. Plant-Microbe Interact.* **3**, 376-386.
8. Cason, E.T., Richardson, P.E., Essenberg, M.K., Brinkerhoff, L.A., Johnson, E.M. and Venere, R.J. (1978) Ultrastructural cell wall alterations in immune cotton leaves inoculated with *Xanthomonas malvacearum*. *Phytopathology* **68**, 1015-1021.
9. Dienelt, M.M. and Lawson, R.H. (1989) Histopathology of *Xanthomonas campestris* pv. *citri* from Florida and Mexico in wound-inoculated detached leaves of *Citrus aurantifolia* : Transmission electron microscopy. *Phytopathology* **79**, 336-348.
10. Doorn van, J., Boonekamp P.M. and Oudega, B. (1994) Partial characterization of fimbriae of *Xanthomonas campestris* pv. *hyacynthi*. *Mol. Plant-Microbe Interact.* **7**, 334-344.
11. Ebrahim-Nesbat, F., von Tiedemann S., Albrecht, S., Heitefuss, R. and Hüttermann, A. (1990) Electron microscopical studies of poplar clones inoculated with *Xanthomonas* populi subsp. *populi*. *Eur. J. For. Pathol.* **20**, 367-375.
12. Essenberg, M. and Pierce, M.L. (1994) Role of phytoalexins in resistance of cotton to bacterial blight. In D.D. Bills and S.A. King (eds.) Biotechnology and Plant Protection, Bacterial Pathogenesis and Disease Resistance. World Scientific, Singapore, pp. 303-314.
13. Essenberg, M., Pierce, M.L., Hamilton, B., Cover, E.C., Scholes, V.E. and Richardson, P.E. (1992) Development of fluorescent, hypersensitively necrotic cells containing phytoalexins adjacent to colonies of *Xanthomonas campestris* pv. *malvacearum* in cotton leaves. *Physiol. Mol. Plant Pathol.* **41**, 85-95.
14. Feliciano, A. and Daines, R.H. (1970) Factors influencing ingress of *Xanthomonas pruni* through peach leaf scars and subsequent development of spring cankers. *Phytopathology* **60**, 1720-1726.
15. Feng, T.Y. and Kuo, T.T. (1975) Bacterial leaf blight of rice plants. VI. Chemotactic responses of *Xanthomonas oryzae* to water droplets exudated from water pores on the leaf of rice plants. *Acad. Sin. Inst. Bot. Bull.* (Taipei) **16**, 126-136.
16. Haaheim, L.R., Kleppe, G. and Sutherland, I.W. (1989) Monoclonal antibodies reacting with the exopolysaccharide xanthan from *Xanthomonas campestris*. *J. Gen. Microbiol.* **135**, 605-612.
17. Hayward, A.C. (1993) The hosts of *Xanthomonas*, in *Xanthomonas* J.G. Swings and E.L.

Civerolo (eds.), Chapman & Hall, pp. 1-17.

18. Horino, O., Watabe, M. and Fuagai, M. (1993) Electron microscopic examination of fibrillar materials and bacterial lipopolysaccharide observed in xylem vessels of rice leaves infected with *Xanthomonas oryzae* pv. *oryzae*. *Ann. Phytopathol. Soc.* Japan **59**, 188-191

19. Jones, S.B. and Fett, W.F. (1987) Bacterial pustule disease of soybean : microscopy of pustule development in a susceptible cultivar. *Phytopathology* **77**, 266-274.

20. Koisumi, M. (1979) Ultrastructural changes in susceptible and resistant plants of citrus following artificial inoculation with *Xanthomonas citri* (Hasse) Dowson. *Ann. Phytopathol. Soc.* Japan **45**, 635-644.

21. Koisumi, M. (1983) Relationship between wound-healing process of citrus leaf tissues and successful infection through wounds by *Xanthomonas campestris* pv. *citri* (Hasse) Dye. *Ann. Phytopathol. Soc.* Japan **49**, 352-360.

22. Leach, J.E., Guo, A., Reimers, P., Choi, S.H., Hopkins, C.M. and White, F.F. (1994) Physiology of resistant interactions between *Xanthomonas oryzae* pv. *oryzae* and rice. In C.I. Kado, J.H. Crosa, eds, Molecular Mechanisms of Bacterial Virulence. Kluwer Academic Publishers, Dordrecht, The Netherlands, pp. 551-560.

23. Lemattre, M. (1963) Centres de prolifération et de nécrose induits par *Xanthomonas pelargonii* (Brown) Starr et Burk. dans les tissus de *Pelargonium zonale* : détection, nature, genèse. *C. R. Acad. Sc.* Série III, Paris **256**, 4494-4497.

24. Matthee, F.N. and Daines, R.H. (1969) The influence of nutrition on susceptibility of peach foliage to water congestion and infection by *Xanthomonas pruni*. *Phytopathology* **59**, 285-287.

25. Mew, T.W., Mew I-pin, C. and Huang, J.C. (1984) Scanning electron microscopy of virulent and avirulent strains of *Xanthomonas campestris* pv. *oryzae* on rice leaves. *Phytopathology* **74**, 635-641.

26. Ramos, L.J. and Volin, R.B. (1987) Role of stomatal opening and frequency on infection of *Lycopersicon* spp. by *Xanthomonas campestris* pv. *vesicatoria Phytopathology* **77**, 1311-1317.

27. Romantschuk, M. (1993) Fimbriae (pilus) mediated attachment of *Pseudomonas syringae*, *Xanthomonas campestris* and *Erwinia rhapontici* to plant surfaces, in C.I. Kado and J. Crosa (eds.), Molecular Mechanisms of Bacterial Virulence, Kluwer Academic Publishers, Dordrecht.

27. Rudolph, K. (1993) Infection of the plant by *Xanthomonas*, in *Xanthomonas*, J.G. Swings and E.L. Civerolo (eds.), Chapman & Hall, pp. 193-264.

29. Rudolph, K., Ebrahim-Nesbat, F., Mendgen, K. and Thiele, C. (1987) Cytological observations of leaf spot-causing bacteria in susceptible and resistant hosts, in E.L. Civerolo, A. Collmer, R.E. Davis and G.E. Gillaspie (eds.), Proceedings of the Sixth International Conference on Plant Pathogenic Bacteria, Maryland, 2-7 june 1985, Martinus Nijhoff Publishers, Dordrecht, pp.604-612.

30. Stemmer, W.C. and Sequeira, L. (1987) Fimbriae of phytopathogenic and symbiotic bacteria. *Phytopathology* **77**, 1633-1639.

31. Tabei, H. (1967) Anatomical studies of rice plant affected with bacterial leaf blight, with special reference to stomatal infection at the coleoptile and the foliage leaf sheath of rice seedling. *Ann. Phytopathol. Soc.* Japan **33**, 12-16.

32. Takahashi, T. and Doke, N. (1985) Purification and partial characterization of an agglutinin in citrus leaves against extracellular polysaccharides of *Xanthomonas campestris* pv. *citri*. *Physiol. Plant Pathol.* **27**, 1-13.

33. Thiers, H.D. and Blank, L.M. (1951) A histological study of bacterial blight of cotton. *Phytopathology* **41**, 499-510.

34. Tsuno, K. and Wakimoto, S. (1989) Ultrastructural study on the compatibility between rice leaf vessel system and various bacteria in Z. Klement (ed.) Plant Pathogenic Bacteria, Akadémiai Kiado, Budapest, 45-50.

35. Vakili, N.G. (1967) Importance of wounds in bacterial spot (*Xanthomonas vesicatoria*) of

tomatoes in the field. *Phytopathology* **57**, 1099-1103.

36. Verdier, V., Schmit, J. and Lemattre, M. (1990) Etude en microscopie à balayage de l'installation de deux souches de *Xanthomonas campestris* pv. *manihotis* sur feuilles de vitroplants de manioc. *Agronomie* **2**, 93-102.
37. Wainwright, S.H. and Nelson, P.E. (1972) Histopathology of *Pelargonium* species infected with *Xanthomonas pelargonii*. *Phytopathology* **62**, 1337-1347.
38. Wallis, F.M., Rijkenberg, F.H.J., Joubert, J.J. and Martin, M.M. (1973) Ultrastructural histopathology of cabbage leaves infected with *Xanthomonas campestris*. *Physiol. Plant Pathol.* **3**, 371-378.
39. Watabe, M., Yamagushi M., Kitamura S. and Horino, O. (1993) Immunohistochemical studies on localization of the extracellular polysaccharide produced by *Xanthomonas oryzae* pv. *oryzae* in infected rice leaves. *Can. J. Microbiol.* **39**, 1120-1126.
40. Young, S.A., White, F.F., Hopkins, C.M. and Leach, J.E. (1994) *AvrXa10* protein is in the cytoplasm of *Xanthomonas oryzae* pv. *oryzae*. *Mol. Plant-Microbe Interact.* **7**, 799-804.
41. Young, S.A., Guo, A., Guikema, J.A., White, F.F. and Leach, J.E. (1995) Rice cationic peroxidase accumulates in xylem vessels during incompatible interactions with *Xanthomonas oryzae* pv. *oryzae*. *Plant Physiol.* **107**, 1333-1341.
42. Zumoff, C.H. and Dickey, R.S. (1987) Histopathology of *Syngonium podophyllum* artificially inoculated with *Xanthomonas campestris* pv. *syngonii*. *Phytopathology* **77**, 1263-1268.

COMPARTMENTALIZATION IN TREES: NEW FINDINGS DURING THE STUDY OF DUTCH ELM DISEASE

D. RIOUX
Natural Resources Canada, Canadian Forest Service - Quebec Region, 1055 du P.E.P.S., P.O. Box 3800, Sainte-Foy, Quebec G1V 4C7, Canada

1. Introduction

Compartmentalization processes have been described as one of the most important mechanisms that explains tree resistance to various diseases (21). These processes are based on the formation of anatomical barriers that limit the extent of injured tissues in the xylem as well as in the bark. Even if the first observations of an unusual tissue between infected and sound wood were made as early as 1887 (9), the importance of the "walling off" processes has been particularly recognized since 1977, when the CODIT (compartmentalization of decay in trees) model was proposed as a defense mechanism against wood decay (22). This model involves four "walls". The first three, also called reaction zones, are associated with the xylem present at the moment of infection. Wall 1 limits the upward movement of pathogens by the occlusion of the conducting elements following the formation of structures such as tyloses and gums (gels). Wall 2 restricts the inward spread by means of the thick walls of fibers comprised in latewood. Ray cells form wall 3 and limit the tangential colonization. Finally, the strongest barrier, wall 4 (barrier zone), is formed by the cambium after the onset of infection. Although this concept was originally postulated to explain the restriction of colonization by decay-causing fungi, it has since been associated with other types of tree diseases, including wilt diseases (8, 23, 25).

In the bark, it appears that similar processes occurred to hinder the progression of canker-causing microorganisms (5). Briefly, an impervious lignified (IL) and suberized layer of cells forms around the necrotic tissues, followed a few hours later by the differentiation, immediately internally to IL, of a new periderm called the "necrophylactic periderm", which is almost entirely composed of suberized cells. When the defense seems adequate, the necrophylactic periderm is usually in continuity with the normal periderm at the periphery of the injured area. Details on the defensive barriers that occur in the bark can be found in the following review papers (3, 4, 16).

The Dutch elm disease (DED) is still a serious wilt disease in Europe as well as in North America. Millions of elms have been killed by this disease. In order to improve our knowledge of DED, we have conducted studies using the approach of inoculating small branches of nonhost trees and shrubs to observe how they react to the presence of *Ophiostoma ulmi* (the DED pathogen) and to be able to compare with the histopathological changes of a susceptible host, *Ulmus americana*. The purpose of this chapter is to present our principal histochemical and cytochemical results, which may shed new light on the spatial organization and the chemical composition of many anatomical changes associated with compartmentalization processes in trees.

M. Nicole and V. Gianinazzi-Pearson (eds.), Histology, Ultrastructure and Molecular Cytology of Plant-Microorganism Interactions, 211-225.

2. Materials and Methods

This section summarily describes the main procedures followed to obtain the results presented in this chapter. For *Ulmus americana* L.,*Prunus pensylvanica* L.f. and *Populus balsamifera* L., the trees were inoculated in the field or in a nursery with aggressive or nonaggressive isolates of *Ophiostoma ulmi* (Buism.) Nannf. Samples were also collected from naturally infected elms.

For histochemical tests, samples were embedded in Tissue-Tek O.C.T. Compound at -15°C and the sections (about 20 µm) were obtained with a Histostat cryomicrotome (Reichert). Phloroglucinol-HCl was used to identify lignified structures. In addition, when lignin is stained with this solution, it loses its capacity to autofluoresce while that of suberin remains intact and can then be easily detected under violet light illumination (Biggs 1984).

For standard light and transmission electron microscopy (TEM), samples were fixed in 2% glutaraldehyde in 0.1M sodium cacodylate buffer (pH 7.2) and postfixed in 1% OsO_4 in the same buffer. After dehydration with ethanol, samples were embedded in Epon 812. For light microscopy, sections (0.5-1 µm) were stained with toluidine blue and safranine. For electron microscopy, sections (90 nm) were contrasted with uranyl acetate and lead citrate.

The cytochemical tests included the use of an exoglucanase (kindly provided by Dr. C. Breuil, Forintek Canada) having affinity for ß(1-4)-D-glucans, the monomers of cellulose molecules. Two monoclonal antibodies (kindly provided by Dr. K. Roberts, John Innes Institute, UK) recognizing un-esterified (JIM5) and esterified (JIM7) pectin epitopes were also used to detect pectic compounds.

A complete description of the different tests used in our studies may be obtained by consulting the references given in the text.

3. Barrier Zone Formation (BZF), Wall 4

The most common reaction of many nonhost trees to the inoculation of *O. ulmi* was BZF (18). BZF was particularly spectacular in *Prunus pensylvanica* and *Populus balsamifera*, where it occurs about 15 days post-inoculation at 2.5 cm above the inoculation point (Figs. 1 and 2). By comparison, in *Ulmus americana*, BZF was usually absent when the branch died or when present, barrier zones took 30 days to form and even then, they were often discontinuous (Fig. 3).

Following these first observations, histochemical tests were carried out principally to characterize the chemical nature of these barrier zone cell walls (19). Staining with phloroglucinol-HCl showed that lignin was apparently more abundant in barrier zones than elsewhere in healthy and affected xylem tissues. With similarly stained cryo-made sections, the detection of suberized cell wall layers was very easy under violet epifluorescence (Fig. 4), according to the method described by Biggs (1).

Figures 1-2. Continuous barrier zones (between arrowheads) surround the invading xylem in *P. pensylvanica* (Fig. 1) and *P. balsamifera* (Fig. 2). The xylem between the cambium (Fig. 1; arrow) and the barrier contains fewer vessel elements that have smaller diameters than usual. Bars = 300 µm.
Figure 3. Discontinuities are shown (arrows) in this barrier zone of *U. americana*. Bar = 200 µm.

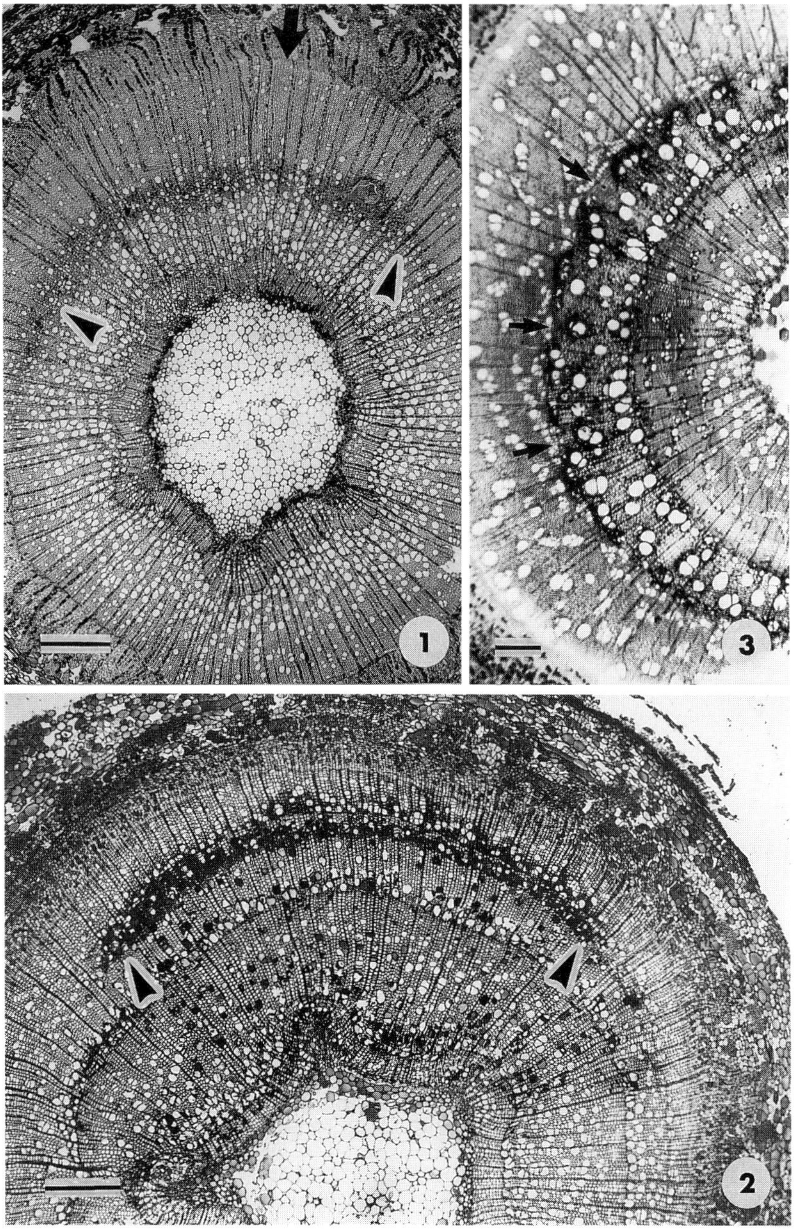
1
3
2

Due to the effectiveness of this test, it was possible to observe only a few suberized cells forming discontinuous barrier zones in the elm, which otherwise would have gone unnoticed (Fig. 5). This difficulty to observe such discontinuous barriers is perhaps the reason why in previous studies it was reported that BZF was rare or absent in small branches of the elms studied (8, 23). Occasionally, *O. ulmi* cells were observed between such discontinuous barriers and the cambial zone, whereas when the barriers were continuous in the nonhosts and in the elm (a rare event in the latter), the fungus was never seen crossing this sheath (19). When the fungus is close to the cambium, it will probably affect this meristem by means of toxins and enzymes. If most of the cambium cells are affected to the point that they are no longer capable of regenerating new usual cells or barrier zone cells, the tree, or a large part of it, will inevitably die. When the barriers were continuous, normal production of xylem resumed after BZF though the vessel elements were often of a smaller diameter than expected when compared with controls (Fig. 1).

It is hard to overemphasize the importance of BZF as a tree defense mechanism. Naturally, if the physiological mechanisms involved in BZF were known, it would perhaps be possible to find ways to stimulate them in order to improve tree protection against many invading microorganisms. BZF might find another practical application in the study of the fungicides injected to protect or cure trees of wilt diseases. Knowing that the distribution of such fungicides is an important feature being taken into consideration each time such products are evaluated, barrier zones might be used for studying the penetration of fungicides into refractory walls composed of significant amounts of suberin and lignin. Among the reagents we used to study BZF, there was $KMnO_4$, which is known to react with lignin (10). Briefly, the method consisted in putting samples in 2% $KMnO_4$ for 4 h to study lignin distribution in microscopy (19). Surprisingly, we realized that the solution did not penetrate the continuous barrier zones, probably due to the presence of lignified and suberized walls. Since this solution oxidizes almost all the tissues it encounters, the natural fluorescence of the unpenetrated tissues was easilydetected under blue or violet illumination (Fig. 6). Contrary to the situation in the nonhosts, in the elm we found small groups of invaded vessels entirely encompassed by impermeable cells (Fig. 7). As we detect only scattered suberized cells above these vessels but strong reactions for lignin all around them, it seems that lignification processes are responsible for the lack of penetration of the $KMnO_4$ solution. In another study not related to BZF, it was clearly demonstrated that intensely lignified walls did not permit the passage of such a $KMnO_4$ solution (15). When diseased elms are injected with fungicides, the vessel occlusion probably precludes a good distribution of the active ingredient. One may also think that the impermeable walls around the colonized vessels might also prevent the passage of the fungicide into these vessel elements. A possible avenue of research in the control of Dutch elm disease might be to inoculate a wilt-causing agent less virulent than *O. ulmi* in small branches of *U. americana* to promote BZF.

Figures 4-5. Sections stained with phloroglucinol-HCl and observed under u.v. illumination. **Figure 4.** The autofluorescence of suberin is easily detected in fibres of this barrier zone of *P. balsamifera*. Some vessel elements included in or adjacent to this barrier are occluded by suberized tyloses (arrowheads). Bar = 40 µm. **Figure 5.** Two groups of suberized cells (arrowheads) forming a discontinuous barrier zone in *U. americana*. Bar = 120 µm. **Figures 6-7.** Sections treated with $KMnO_4$ and observed under u.v. illumination. **Figure 6.** In *P. balsamifera*, the invaded xylem and some phloem fibres (bottom left) were not penetrated by $KMnO_4$ and they show an obvious autofluorescence. Bar = 200 µm. **Figure 7.** In *U. americana*, a group of invaded vessels (arrow) is completely surrounded by cells not oxidized by $KMnO_4$. C, contamination. Bar = 100 µm.

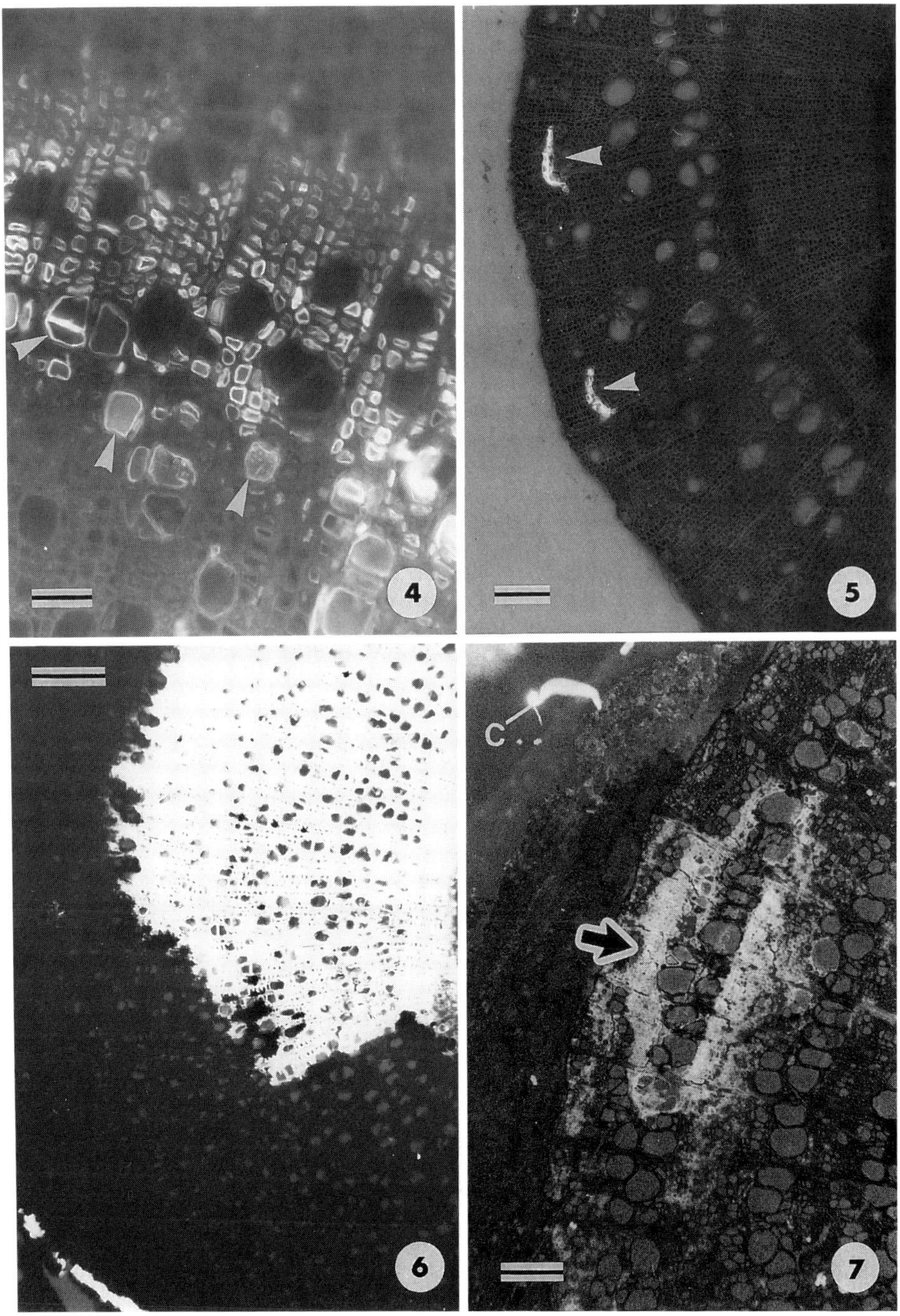
4
5
6
C
7

Such barrier zones with impermeable walls might then be used macroscopically with a fungicide to evaluate the penetration facility of the different solvents used with the active ingredient to find the one allowing a better distribution of the product.

4. Reaction Zones: Wall 3

The reaction zone that hinders the tangential spread of pathogens was reported to be formed essentially by xylem ray cells (22). These cells would react to the infection by accumulating, for instance, fungitoxic substances. As xylem rays have a limited height that varies according to tree species, the wall 3 reaction zone is considered discontinuous. If it is, why does the pathogen not succeed in crossing this zone?

The implication of suberin in reaction zones, particularly wall 3, has been strongly suggested (14). Biggs (2) also has reported the presence of suberin in xylem parenchyma cells following wounding of different tree species. Together with the plugging of vessels by amorphous material and tyloses, he suggested that these reactions contribute to the effectiveness of walls 1, 2 and 3. However, as the distribution of these structures is often random, he mentioned that they cannot form continuous barriers.

In studying the reactions of *P. pensylvanica* and *P. balsamifera* to the inoculation of *O. ulmi*, it was clear that wall 3 successfully limits the colonization of this pathogen. The presence, in *P. pensylvanica*, of many gels (Fig. 8), and in *P. balsamifera*, of numerous tyloses (Fig. 9), seemingly was important for the effectiveness of this reaction zone. Some of these occluding structures were also observed in the invaded xylem but usually they were not concentrated in a particular area. As shown in Figure 9, the tyloses of wall 3 in *P. balsamifera* were suberized contrarily to those of the invaded xylem, except the ones found adjacent to or included in barrier zones (Fig. 4). By comparison, in the host *U. americana*, such suberized tyloses were formed randomly (Fig. 10), which may explain why the pathogen frequently can spread easily in a tangential manner in infected susceptible elms. Until now, we have not been able to obtain good radial longitudinal sections to see whether these structures in the nonhosts, together with the reacting ray cells, may give a continuous configuration to the wall 3 reaction zone. However, their formation should undoubtedly improve the continuity and, consequently, the effectiveness of this zone as a defense anatomical barrier.

The occlusion of vessel elements by tyloses and gels is apparently a consequence of xylem vessel dysfunction and not a cause of it (27). Their formation might seal such vessel elements and thus impede longitudinal air movement within the tree. But when tyloses expand within a vessel, they have difficulty reaching and blocking completely such areas as near the rim of vessel elements, pit chambers and in void spaces in front of the junction of the vessel with at least two other cells (at corner-like areas of vessels). To seal the vessel completely, the tyloses secrete pectic molecules, as revealed by immunogold labelling at a junction between the vessel and other xylem cells (Fig. 11). On the other hand, in places where the tylosis primary wall is intimately in contact with the vessel wall, this pectic layer may be very thin and hardly discernible (Fig. 12). The labelling in the last figure was done with an exoglucanase conjugated to colloidal gold and shows that the tylosis wall in this sample of *P. balsamifera* also contains a suberized lamellar layer

---

Figure 8. Gels are concentrated in the wall 3 region (between brackets) on the left margin of this barrier zone (arrowhead) of *P. pensylvanica*. Bar = 150 µm. **Figures 9-10.** Sections stained with phloroglucinol-HCl and observed under u.v. illumination. **Figure 9.** In *P. balsamifera*, numerous suberized tyloses are shown located in wall 3 (arrowhead). B, barrier zone. Bar = 200 µm. **Figure 10.** Suberized tyloses in *U. americana* that are formed at random in the xylem. Bar = 150 µm.

8

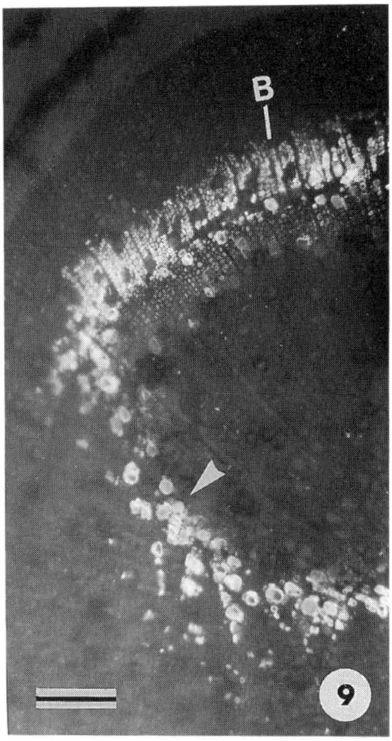
B
9

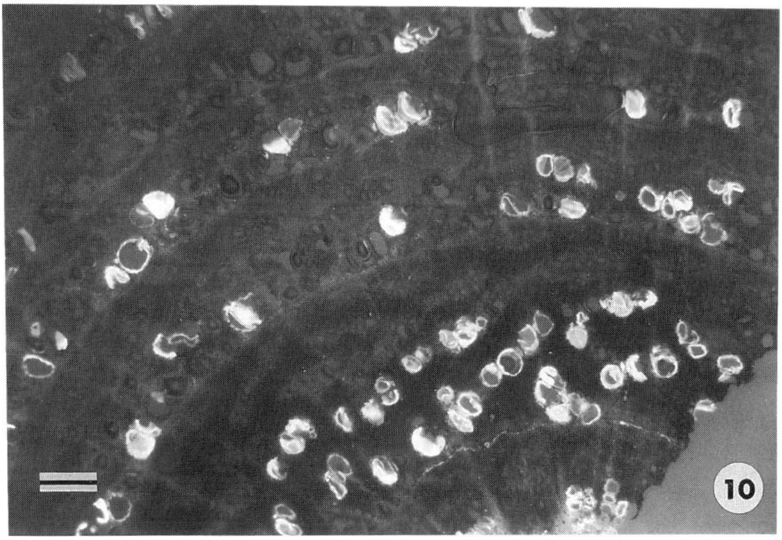
10

localized between two cellulosic layers. In the hundreds of suberized tyloses studied, this external layer often had the appearance of a middle lamella (17).
With recent immunogold labelling, we found out that a part, at least, of the material involved in gel formation could show a great similarity to the pectic material secreted by many tyloses. Until now, positive results in that sense were obtained in the three species of trees studied, that is *Hevea brasiliensis*, *U. americana* and *P. pensylvanica*. The last was the most intensively studied and was also the species that gave the most obvious immunogold labelling showing that pectic compounds are secreted by xylem parenchyma cells through the pit membrane into the vessel element. In addition, most of the time it appears that the pectic compounds migrate a certain distance from the protective layer, which is frequently thicker in front of the vessel element, before being secreted (Fig. 13). This could happen concomitantly with the retraction of the plasmalemma membrane (Fig. 14). As with the tyloses of *P. balsamifera*, the pectic compounds generally had a fibrillar appearance (Fig. 14).
The tylosis primary wall through which the pectin is secreted is generally considered to be an extension of the protective layer (7, 11, 12, 13). Considering this fact, it is no surprise that a parenchyma cell may secrete pectic compounds in a vessel element via this protective layer. This idea was proposed for the first time by Bonsen and Kucera (7). Our immunolabelling results confirm without a doubt that these two phenomena show similarities concerning the secretion mechanism of pectic molecules to completely obstruct the vessel elements.

5. A New Reaction Zone Similar to Wall 2

As described by Shigo and Marx (22), wall 2 is formed by already existing cells, essentially by fibers of latewood having thick cell walls impeding the movement of pathogens. During the study of defense reactions of *P. balsamifera*, we observed that an unusual suberized band of cells is formed in the perimedullary area only in front of the invaded xylem (Fig. 15). This band of suberized cells originated from the dedifferentiation of the parenchyma cells present in the perimedullary zone. Although these parenchyma cells are covered by xylem and bark tissues, it seems that they remain alive for a certain number of years. For instance, in two- to three-year-old branches of *P.* x *canadensis*, these cells were still capable of photosynthetic activities (26). Ultrastructural observations have shown that the suberized layers frequently had an obvious lamellar structure (Fig. 16). Labelling with the exoglucanase also evidenced that a succession of suberized and cellulosic layers might occur in such cells (Fig. 17). This succession was at times also observed within the walls of suberized tyloses in *P. balsamifera* (17).

Figures 11-14. TEM observations. **Figures 11-12.** Suberized tyloses of *P. balsamifera*. Bars = 0.3 µm. **Figure 11.** The labelling for pectin (JIM5 antibody) is intense over the outer wall layer of the tylosis that is particularly thick at the junction of this vessel with two fibres. Many gold particles are also present over the tylosis primary wall (P). S, suberized layer; V, vessel secondary wall. **Figure 12.** A section treated with the exoglucanase. The external wall layer is hardly discernible (arrow) in this portion of the tylosis. The lamellar suberized layer (S) is located between two cellulosic layers. **Figures 13-14.** In *P. pensylvanica*, pectic compounds that appear to be secreted by the xylem parenchyma cells are detected with the JIM5 antibody. **Figure 13.** Before being secreted, pectin appears to move within the protective layer (arrow). Bar = 0.5 µm. **Figure 14.** The retraction of the plasmalemma (arrow) is evident in this parenchyma cell. The fibrillar appearance of pectin can be seen particularly between the protective layer (curved arrow) and the plasmalemma. Bar = 0.4 µm.

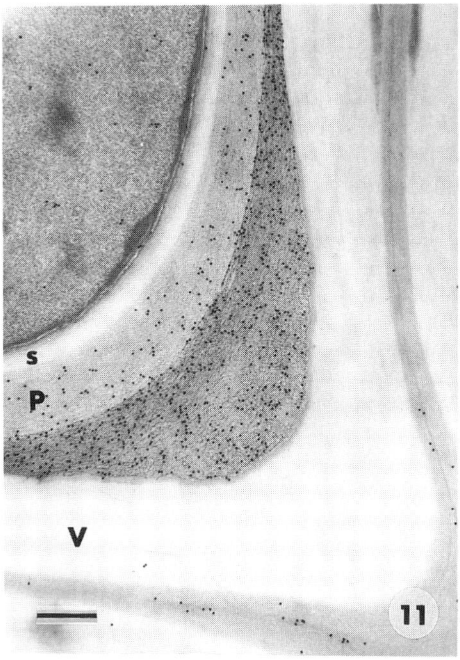
s
P
V
11

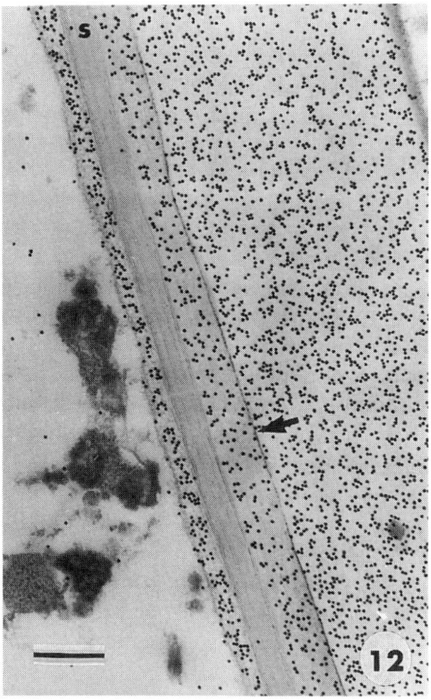
s
12

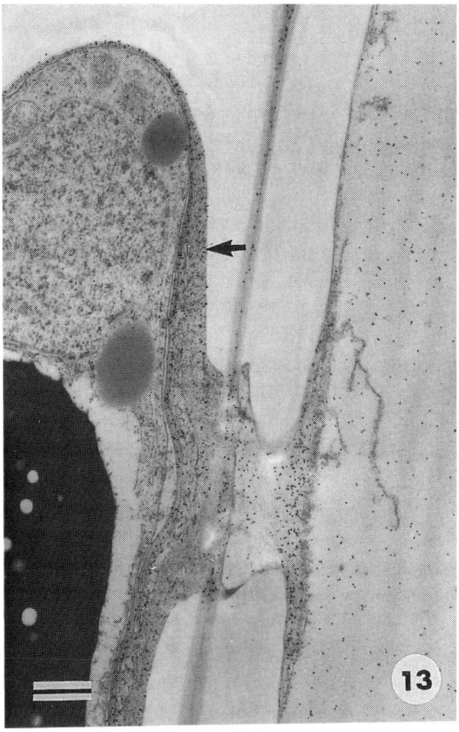
13

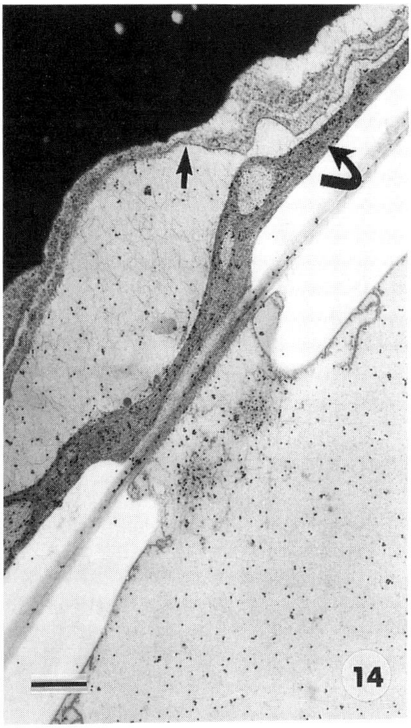
14

However, so far, the significance for the plant wall of alternate suberized and cellulosic layers is difficult to understand. It might be possible that the suberized layers confer impermeability to the walls while cellulosic layers are formed to strengthen such walls. Plasmodesmata were at times observed crossing the suberized layers, which may explain how these cells can survive the initial deposition of such impermeable wall layers.

6. Ultrastructure of Suberized Wall Layers

In previous studies, we reported that the intensity of the autofluorescence of suberized layers in light microscopy seemed positively correlated with the facility to observe a lamellar structure in transmission electron microscopy (19, 20). However, at that time, many amorphous wall layers were discarded as being suberized because they looked so white that we thought they represented only a void space, for instance as a consequence of the retraction of the plasmalemma. Such a unequally developed layer can be seen in a tylosis of *P. balsamifera* (Fig. 18). We have observed so many similar layers that we are now convinced that they represent suberized entities. In addition, it is possible at times to observe plasmodesmata through them that indicate that they are not void spaces, as can be seen between two suberized ray cells of a barrier zone in *P. pensylvanica* (Fig. 19). On the other hand, in this species, it was at times possible to observe very intense lamellation of an additional wall layer formed in small vessels included in a barrier zone (Fig. 20) but this lamellation gave only moderate autofluorescence under violet light excitation. At least with these tree species, all the results strongly point out that there is no link between the intensity of the autofluorescence and the lamellar structure of the suberized wall layers.

7. Compartmentalization vs Dessication of Xylem Tissues

Different putative roles have been associated with compartmentalized processes. Various studies suggest the following functions: 1) to limit the movement of microorganisms; 2) to prevent the diffusion of toxins or enzymes produced by these microorganisms; 3) to increase the quantity of phytoalexins in the invaded xylem only, avoiding damage to healthy tissues of the tree; 4) to limit the availability of nutrient elements for the microorganisms; 5) to prevent the desiccation of adjacent tissues. Concerning the last possibility, recent studies strongly suggest that the tree compartmentalizes primarily as a defense against the presence of air within its tissues (see review 6) rather than limiting the colonization of microorganisms. According to this work, healthy xylem tissues would be inadequate to promote the growth of fungi because their water content is too high. Following wounding or other injuring processes, a certain amount of air may penetrate the tissues accompanied by pathogens. These microorganisms are now in a medium that can

Figures 15-17. In *P. balsamifera*, observations related to the suberized band formed in the perimedullary area in front of the invaded xylem. **Figure 15.** Section stained with phloroglucinol-HCl and observed under u.v. illumination. Detection of suberin in this reaction zone formed between the pith (P) and the invaded xylem (IX). Bar = 150 µm. **Figures 16-17.** TEM observations. **Figure 16.** The lamellation of the suberized layers (S) is obvious in these two cells. The internal layer (C) is probably composed of cellulose. Bar = 0.2 µm. **Figure 17.** Exoglucanase-gold labelling. The cell on the left is completely ccluded by a succession of suberized and cellulosic layers. S, suberized layers. Bar = 0.3 µm.

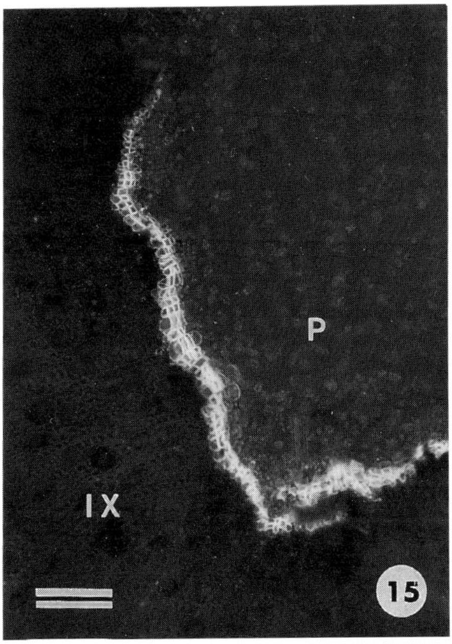
P
IX
15

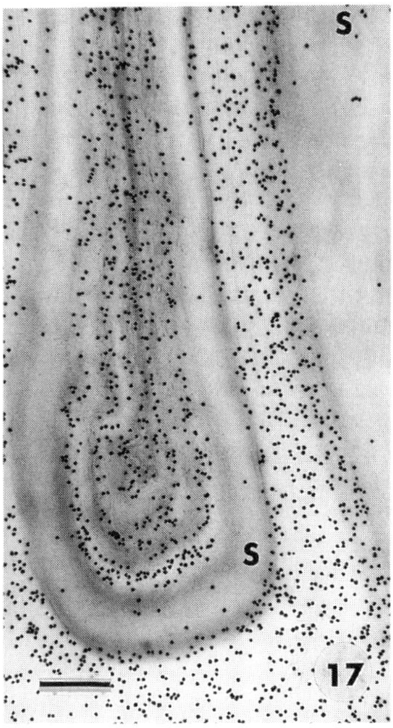
S
S
17

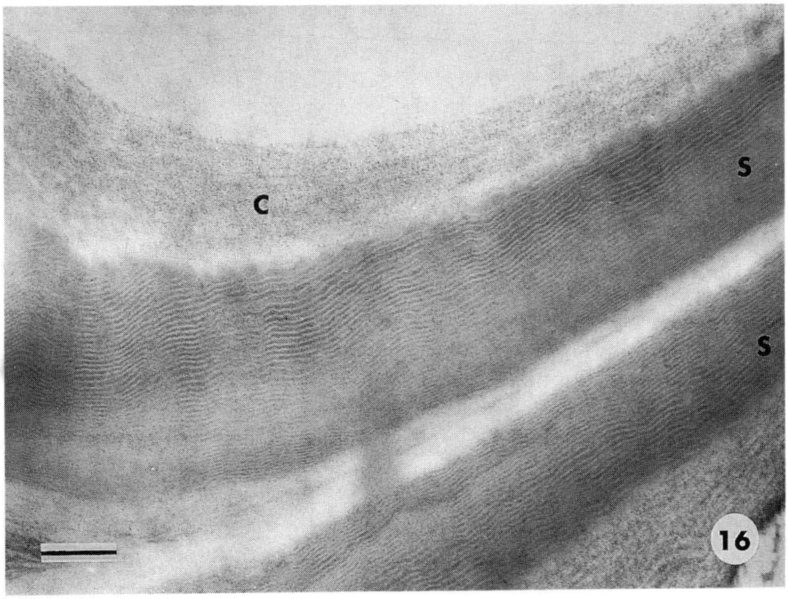
S
C
S
16

favor growth. The spread of these microorganisms is made possible when these fungi degrade walls to allow air to cause, among other problems, embolism of adjacent conducting xylem elements. In addition, it has been shown that keeping wood sections under water saturation does not permit degradation of wood by microorganisms already present in the wood, while decreasing the water content during drying, but naturally not below a critical level, triggered the degradation of the wood.
In line with these results, other branches of *P. balsamifera* were infected with different inocula of *O. ulmi* in the summer of 1994. If only a small quantity of spores was inoculated, the reactions of the tree were hardly discernible in microscopy many days after inoculation, at periods of time when BZF usually can be observed. If the concentration of propagules, especially hyphae, is increased, then the tree reacts intensely by forming the different barriers associated with compartmentalization. As reported in the present study, these barriers are also strongly impregnated by compounds such as suberin. Although these represent only preliminary investigations that should be repeated in the near future, we can nevertheless say that at this stage it seems that if the vessels become plugged and thus embolized, the tree will react rapidly by forming structures (tyloses, barrier zones) that eventually will be suberized to render them impermeable to air movement and at the same time more resistant to attack of microorganisms.

8. Conclusion

This chapter has shown that investigations into the defense mechanisms of host and nonhost trees following the inoculation of a particular wilt-causing agent, *O. ulmi*, can broaden our knowledge of the compartmentalization processes that seem to be a ubiquitous defense response of trees. It is foreseeable that future studies will reveal new and interesting facts about the formation of the different barriers involved in such walling off processes. However, it is highly desirable that several of these studies will bear on the physiological processes involved in the production of these anatomical changes. It might thus be possible to know more about the hormones involved and in the particular case of BZF, to obtain additional information about the differentiation of xylem tissues, a topic of general interest to many research groups around the world.
Besides this acquisition of knowledge, some of these data might help in research to stimulate the defense reactions of trees. Together with the normal production of phytoalexins, the triggering of compartmentalization processes, by means of hormonal treatment for instance, might give the tree just enough time to win the battle with the intruder, as often the issue seems to rest on a precarious equilibrium between the conditions that favor the tree or the microorganism. One of the ultimate goals of all these studies would be that one day it could be possible, by traditional breeding techniques or by using the latest developments in biotechnology, to transfer genes responsible for good compartmentalization responses to our most beautiful or useful tree species.

Figures 18-20. TEM observations of suberized layers; *P. balsamifera* (Figs. 18-19); *P. pensylvanica* (Fig. 20). **Figure 18.** Exoglucanase-gold labelling. The white suberized layer (arrow) is obvious on the left but very thin or absent on the right portion of this tylosis. This layer looks like a void space caused by the retraction of the plasmalemma. V, vessel secondary wall; curved arrow, tylosis primary wall. Bar = 0.6 µm. **Figure 19.** Plasmodesmata are shown crossing two white suberized layers (S) in these barrier zone cells. Bar = 0.3 µm. **Figure 20.** An additional suberized layer with a clear lamellar structure occurs between remnants of cytoplasm (arrow) and the vessel secondary wall (V). This vessel was found within a barrier zone. Bar = 0.4 µm.

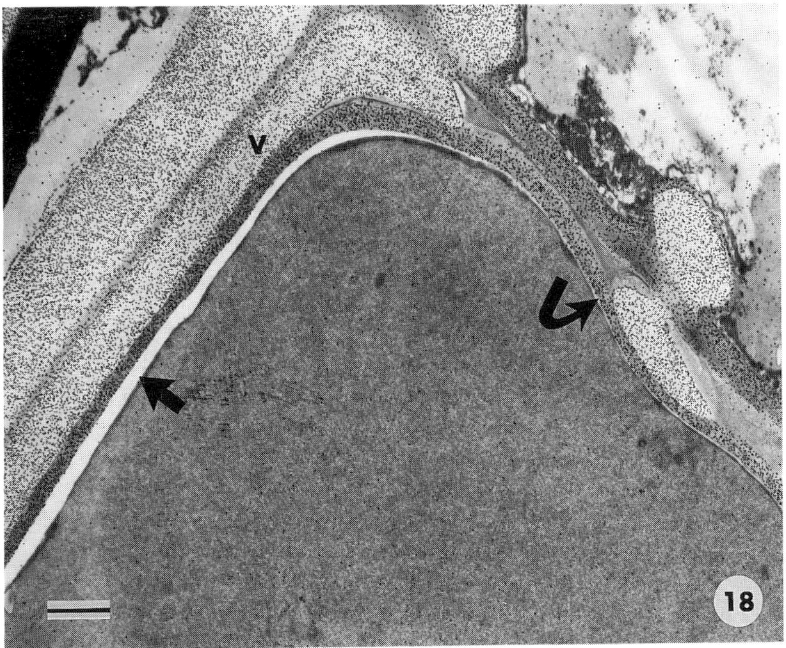
V
18

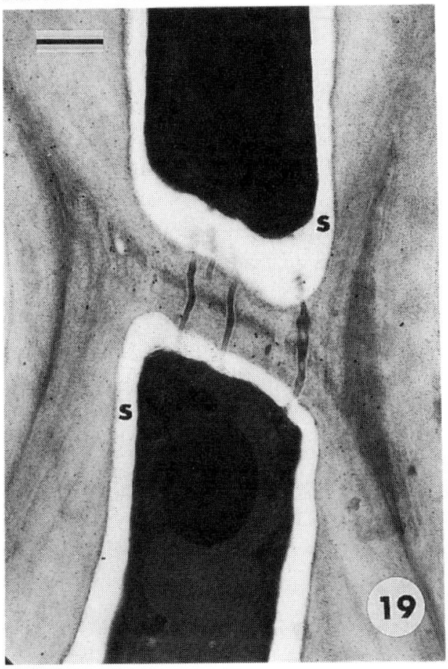
s
s
19

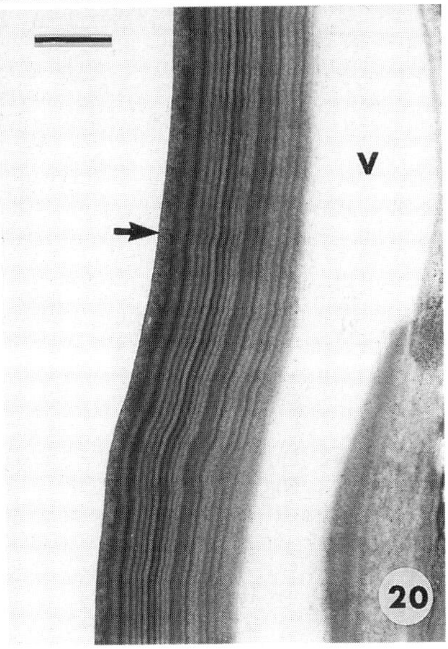
V
20

9. References

1. Biggs, A.R. (1984) Intracellular suberin: occurrence and detection in tree bark, *IAWA Bull. n.s.* **5,** 243-248.
2. Biggs, A.R. (1987) Occurrence and location of suberin in wound reaction zones in xylem of 17 tree species, *Phytopathology* **77,** 718-725.
3. Biggs, A.R. (1992a) Anatomical and physiological responses of bark tissues to mechanical injury, in R.A. Blanchette and A.R. Biggs (eds.), *Defense Mechanisms of Woody Plants against Fungi*, Springer-Verlag, Berlin, pp. 13-40.
4. Biggs, A.R. (1992b) Responses of angiosperm bark tissues to fungi causing cankers and canker rots, in R.A. Blanchette and A.R. Biggs (eds.), *Defense Mechanisms of Woody Plants against Fungi*, Springer-Verlag, Berlin, pp. 41-61.
5. Biggs, A.R., Merril, W., and Davis, D.D. (1984) Discussion: response of bark tissues to injury and infection, *Can. J. For. Res.* **14,** 351-356.
6. Boddy, L. (1992) Microenvironmental aspects of xylem defenses to wood decay fungi, in R.A. Blanchette and A.R. Biggs (eds.), *Defense Mechanisms of Woody Plants against Fungi*, Springer-Verlag, Berlin, pp. 96-132.
7. Bonsen, K.J.M. and Kucera, L.J. (1990) Vessel occlusions in plants: morphological, functional and evolutionary aspects, *IAWA Bull. n.s.* **11,** 393-399.
8. Bonsen, K.J.M., Scheffer, R.J., and Elgersma, D.M. (1985) Barrier zone formation as a resistance mechanism of elms to Dutch elm disease, *IAWA Bull. n.s.* **6,** 71-77.
9. Hartig, R. (1894) *Text-book of the Diseases of Trees*, MacMillan and Co., London.
10. Hepler, P.K., Fosket, D.E., and Newcomb, E.H. (1970) Lignification during secondary wall formation in *Coleus*: an electron microscopic study, *Am. J. Bot.* **57,** 85-96.
11. Meyer, R.W. and Côté, W.A. (1968) Formation of the protective layer and its role in tylosis development, *Wood Sci. Technol.* **2,** 84-94.
12. Murmanis, L. (1975) Formation of tyloses in felled *Quercus rubra* L., *Wood Sci. Technol.* **9,** 3-14.
13. Ouellette, G.B. (1980) Occurrence of tyloses and their ultrastructural differentiation from similarly configured structures in American elm infected by *Ceratocystis ulmi*, *Can. J. Bot.* **58,** 1056-1073.
14. Pearce, R.B. and Holloway, P.J. (1984) Suberin in the sapwood of oak (*Quercus robur* L.): its composition from a compartmentalization barrier and its occurrence in tyloses in undecayed wood, *Physiol. Plant Pathol.* **24,** 71-81.
15. Rioux, D. (1994) Anatomy and ultrastructure of pith fleck-like tissues in some Rosaceae tree species, *IAWA J.* **15,** 65-73.
16. Rioux, D. and Biggs, A.R. (1994) Cell wall changes in host and nonhost systems: microscopic aspects, in O. Petrini and G.B. Ouellette (eds.), *Host Wall Alterations by Parasitic Fungi*, APS Press, St. Paul, Minnesota, pp. 31-44.
17. Rioux, D., Chamberland, H., Simard, M., and Ouellette, G.B. (1995) Suberized tyloses in trees: an ultrastructural and cytochemical study, *Planta* **196,** 125-140.
18. Rioux, D. and Ouellette, G.B. (1989) Light microscope observations of histological changes induced by *Ophiostoma ulmi* in various nonhost trees and shrubs, *Can. J. Bot.* **67,** 2335-2351.
19. Rioux, D. and Ouellette, G.B. (1991a) Barrier zone formation in host and nonhost trees inoculated with *Ophiostoma ulmi*. I. Anatomy and histochemistry, *Can. J. Bot.* **69,** 2055-2073.
20. Rioux, D. and Ouellette, G.B. (1991b) Barrier zone formation in host and nonhost trees inoculated with *Ophiostoma ulmi*. II. Ultrastructure, *Can. J. Bot.* **69,** 2074-2083.
21. Shigo, A.L. (1984) Compartmentalization: a conceptual framework for understanding how trees grow and defend themselves, *Annu. Rev. Phytopathol.* **22,** 189-214.
22. Shigo, A.L. and Marx, H.G. (1977) Compartmentalization of decay in trees, *Agric. Inf. Bull. USDA For. Serv. No:* 405.
23. Shigo, A.L. and Tippett, J.T. (1981) Compartmentalization of American elm tissues infected by *Ceratocystis ulmi*, *Plant Dis.* **65,** 715-718.
24. Tainter, F.H. and Fraedrich, S.W. (1986) Compartmentalization of *Ceratocystis fagacearum* in Turkey oak in South Carolina, *Phytopathology* **76,** 698-701.
25. Tippett, J.T. and Shigo, A.L. (1981) Barrier zone formation: a mechanism of tree defense against vascular pathogens, *IAWA Bull.* **2,** 163-168.

26. van Cleve, B., Forreiter, C., Sauter, J.J., and Apel, K. (1993) Pith cells of poplar contain photosynthetically active chloroplasts, *Planta* **189,** 70-73.
27. Zimmermann, M.H. (1983) *Xylem Structure and the Ascent of Sap.* Springer, Berlin Heidelberg New York.

VIRUS OF PLANT TRYPANOSOMES (*Phytomonas* spp.)

M. DOLLET, S. MARCHE*, D. GARGANI, E. MULLER, T. BALTZ*
Unité de Recherche Commune Virologie CIRAD - Laboratoire de Phytovirologie des Régions Chaudes CIRAD-ORSTOM - BP 5035- 34032 Montpellier Cedex 1, France.
**Laboratoire d'Immunologie et Parasitologie Moléculaire, Université Bordeaux II, 33076 Bordeaux, France.*

1. Introduction

Trypanosomes have been known to exist in the laticifers of latex-bearing plants since 1909 (15). The pathogenicity of these organisms in plants has been the subject of much discussion, but it can now be said that no plant diseases exist with which intralaticiferous trypanosomes are specifically associated (5). On the other hand, intraphloemic trypanosomes are specifically associated with decay in cultivated plants in South America: coffee phloem necrosis (25), coconut Hartrot (21), oil palm Marchitez (8), and in the Caribbean: decay in *Alpinia purpurata* (P. Hunt, M. Dollet, 1991, not published).

Whether they be intralaticiferous or intraphloemic, these organisms have the same morphology: fine (0.5 to 1.5 μ in diameter), and elongated (12 to 20 μ), with their anterior end terminating in a flagellum. They are of the so-called "promastigote" form.

It is therefore virtually impossible to distinguish between these organisms, apart from their cytological location within a plant. This is all the more awkward in that as soon as they were discovered, plant trypanosomes were classified in a single arbitrary genus, created especially for them: *Phytomonas* (9), whilst the morphology of *Herpetomonas* or *Leptomonas*, which are insect trypanosomes, is very similar. Moreover, as plant trypanosomes spend part of their cycle inside insect vectors (5) (16), it is impossible to distinguish within the insect between those trypanosomes that are parasites of the insects by which they are being harboured, and the plant trypanosomes - "*Phytomonas*" - that the insects transmit. This situation has become even more confused over recent years, with the discovery of trypanosomatids that develop in various fruits and seeds, in which they are deposited by insects that come to feed (4) (13). Such trypanosomes do not multiply in the other parts of the plant.

M. Nicole and V. Gianinazzi-Pearson (eds.), Histology, Ultrastructure and Molecular Cytology of Plant-Microorganism Interactions, 227-236.

One of the main concerns for teams working on trypanosomes is to find the best methods, techniques and criteria for characterizing trypanosomes found in plants. One of these studies led to the discovery of virus-like double-stranded RNA in the group of intraphloemic trypanosomes (18). In this article, we describe the virus-like structures found in these organisms and discuss their role and use as characterization and classification criteria.

2. Trypanosomes taxonomy and characterization

2.1. REMINDER

Trypanosomatids are flagellate protozoans belonging to the Kinetoplastidae family.

Kingdom	Protista
Class	Zoomastigophora
Order	Kinetoplastida
Family	Trypanosomatidae

Crithidia Leptomonas Herpetomonas Blastocrithidia Phytomonas Trypanosoma Leishmania Endotrypanum Rhynchoidomonas.

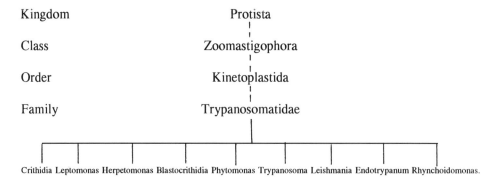

Kinetoplastidae have a single mitochondrion in which thousands of DNA minicircles and a few maxicircles are massed in the anterior end of the organism: kinetoplast (*Figure 1*).

The most common trypanosome species, of the *Trypanosoma* genus, is the causal agent of sleeping sickness, *T. gambiense*.

2.2. ULTRASTRUCTURAL CHARACTERISTICS

The flagellate protozoans found in plants all have the same ultrastructural characteristics (23) (7) (13). The organism is surrounded by a pellicular membrane line with microtubules; the cytoplasm contains a nucleus, a more or less branched mitochondrion containing a substantial fibrillar mass of DNA - kinetoplast - endoplasmic reticulum networks, ribosomes, vesicles and various dense bodies (*Figure 1 and 2*). A limited study of isolates from latex plants revealed differences in ultrastructural organization :

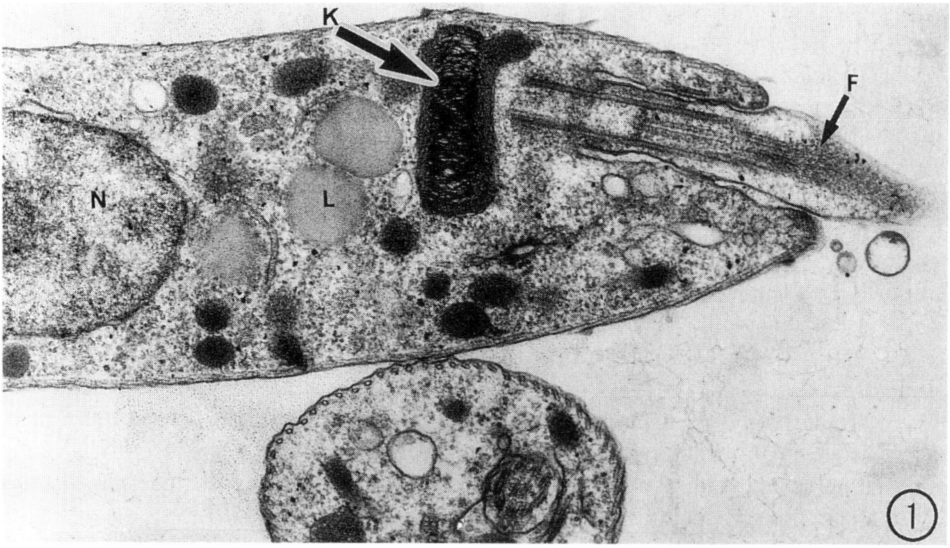

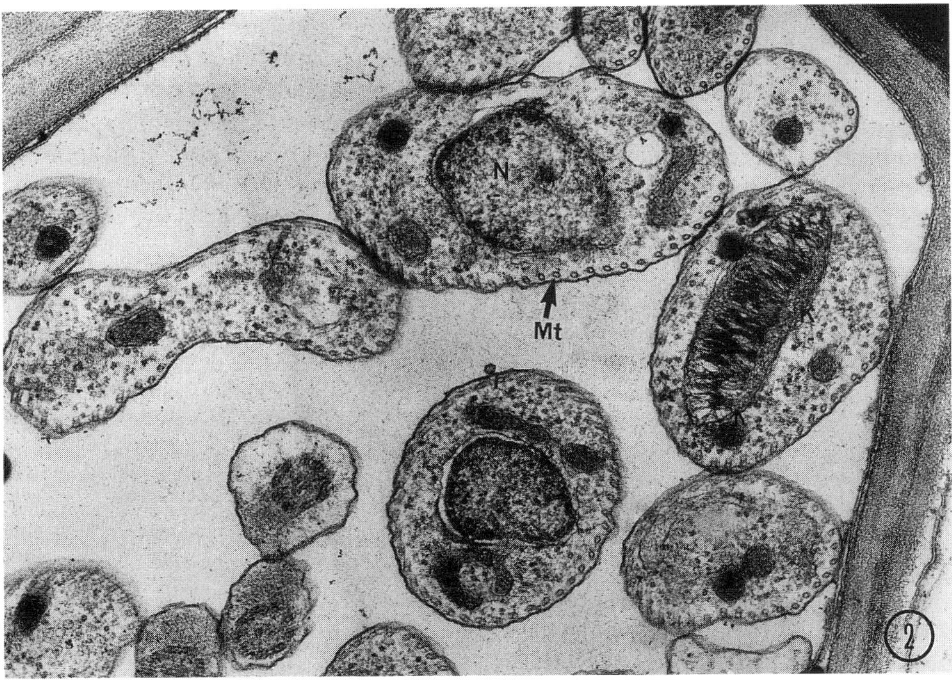

Figure 1 : Longitudinal section of the anterior end of a phloem-restricted trypanosomatid associated with Hartrot disease of coconut in French Guiana. The kinetoplast (K) lies in the mitochondrial matrix at the base of the flagellum (F). N, nucleus; L, lipid globule. (X 36000)

Figure 2 : Transverse section of a sieve tube of *Alpinia purpurata* affected by a decay in Grenada (Windward islands). Mt, microtubules ; N, nucleus ; K, kinetoplast (X 47500).

arrangement of the endoplasmic reticulum - sub-pellicular or central, lamellar or showing a peculiar branching pattern, more or less loose network of DNA minicircles in the kinetoplast, etc. (3). However, a long and expensive ultrastructural study is hardly feasible as a routine characterization or classification technique. Apart from the work involved in taking cross-sections and carrying out observations, fixation methods need to be developed for each isolate and will differ slightly from one to another (3). Moreover, the culture medium has a considerable effect on the morphology and growth of organisms and it may well be necessary to observe a very large number of cross-sections, carried out on organisms cultured under different conditions, to define ultrastructural characterization criteria.

2.3. OTHER CHARACTERIZATION METHODS

There are currently no methods available that can be used alone to characterize plant trypanosomes (6). Alongside the biological criteria, such as cellular location, specific association or not with a pathological syndrome, the medium used for their *in vitro* culture and serology, various molecular biology techniques have led to substantial steps forward in Trypanosome characterization. Studies of isozymes (12) (20) and kinetoplastic DNA (1) (2), have revealed the existence of considerable heterogeneity in plant trypanosomes. The results obtained with these techniques tally with biological observations, namely that there are probably three types of plant trypanosomes: intraphloemic trypanosomes associated with decay in cultivated plants in South and Central America, intralaticiferous trypanosomes in latex-bearing plants, spread throughout all the continents, and trypanosomes associated with fruits and seeds. A recent study of 18S ribosomal RNA confirmed latex trypanosomes/phloem trypanosomes separation, even if the organisms are isolated on the same continent, at the same geographical site (e.g. an intraphloemic isolate from a diseased coconut palm in French Guiana and an intralaticiferous isolate from a euphorbia growing at the foot of the coconut palm are very different from each other) (12), (19).

3. Virus-like particles

3.1. ds RNA

During studies on ribosomal RNA, a double-stranded RNA (ds RNA) with a molecular weight of 4.7 kb was discovered in intraphloemic trypanosomes associated with decay in oil palm and coconut in South America (18) and with decay in *Alpinia* in the Caribbean (17). However, two intraphloemic isolates from South America do not have this ds RNA. The virus of *Trichomonas vaginalis* can disappear after several months' culture. It is therefore reasonable to assume that the two intraphloemic isolates without ds RNA have lost their virus.

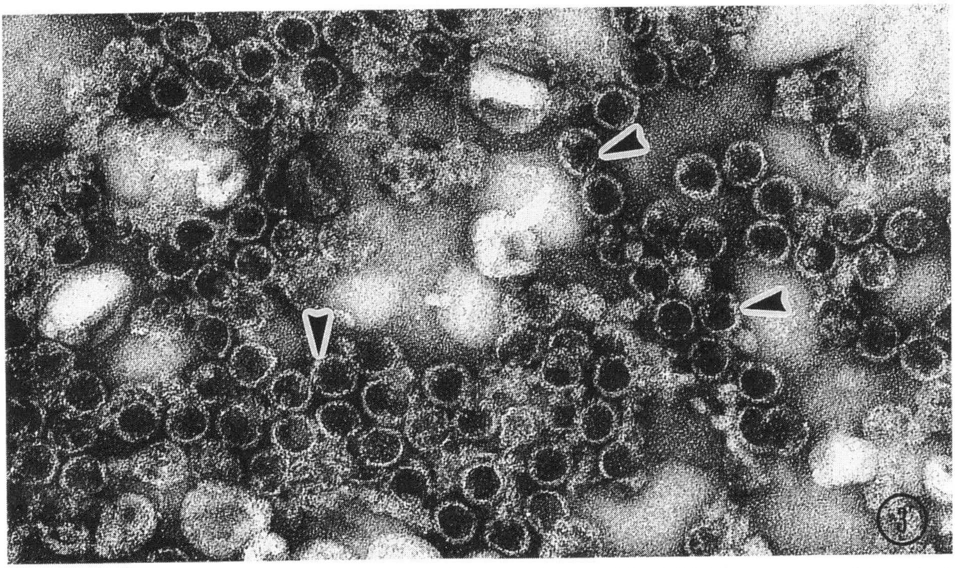

Figure 3 : Virus-like particles (arrows) observed in a 15-30 % sucrose gradient containing ds RNA from trypanosomatids associated with Hartrot disease. Samples were mixed v/v with aqueous Bacitracin (10 mg ml^{-1}) and stained with 1,5 % aqueous uranyl acetate. (X 185000)

This ds RNA exists in the cytoplasmic fraction of the parasite, in which there are also 34 ± 2 nm spherical virus-like particles (*Figure 3*). Most of these particles are penetrated by staining, as already seen with other protozoan viruses with ds RNA (14) (24). Primary culture of intraphloemic trypanosomes requires feeder cells - insect cells - (6). The origin of the virus could well have been in these cells, but an examination of the nucleic acids in these cells did not reveal the existence of this characteristic 4.7 kb band (18).

ds RNA viruses have very different hosts ranging from plants to vertebrates and including insects, bacteria, yeasts and protozoans. Most of these viruses are infectious and cause serious pathological syndromes. The viruses of protozoans are few in number and little is known about their biology (22) (26). Only a virus of *Giardia lamblia* and of *Leishmania braziliensis* have been characterized and classed within the Totiviridae group (10).

3.2. ULTRASTRUCTURAL STUDY

We carried out an ultrastructural study of an intraphloemic isolate associated with coconut Hartrot in French Guiana, in order to totally refute the hypothesis of external contamination. On its 164th transfer, the axenic culture - a cloned isolate - was fixed with 2% glutaraldehyde in a 0.1 M sodium cacodylate buffer, pH 7.2. After post-fixation with 1% O_sO_4, the samples were dehydrated in an acetone gradient then embedded in EPON.

To fix the organisms after centrifugation of the culture medium, we either fixed the pellet directly, or fixed the organisms in a mass, in agarose or Bacto agar. The best results were obtained with direct fixation.

A cytological examination of the fixed organisms did not reveal any cytopathological abnormality. The cross-sections revealed normal cytoplasmic structures for Kinetoplastidae and the pellicular membrane was usually intact. On the other hand, other spherical particles 32-36 nm in diameter were seen to exist in certain vesicles (*Figure 4*). These particles therefore had approximately the same diameter as the particles seen with negative staining (*Figure 5*) sedimenting with the ds RNA. To our knowledge, no structure of this type, spherical and regular in diameter, electron dense and intravesicular, had ever been described in trypanosomatids, and we believe that they are therefore viral particles associated with intraphloemic isolates.

4. Discussion

Virus-like particles are associated with intraphloemic trypanosome isolates from diseased plants in South America and the Caribbean. These particles are quite easily detected by negative staining. They sediment at the same level as a 4.7 kb doubled-stranded RNA. A study of the partial sequence of this RNA revealed the existence of patterns, specific to protozoan and yeast viruses RNA dependent RNA polymerase, and strong protein sequence homologies with a virus of *Saccharomyces cerevisiae* (17). They are therefore probably ds RNA trypanosome viruses. Compared to the ds RNA concentration found with electrophoresis, and the particle concentration seen with negative staining, the low concentration of virus-like particles observed *in situ* remains unexplained. It also needs to be said that the publications describing the existence of trypanosomes in plant sieve tubes (7) (8) (21) have never mentioned the existence of virus-like particles. We recently took another look at samples of oil palm affected by Marchitez in Peru and of *Alpinia purpurata* affected by decay in Grenada, to try and find virus-like particles in the intraphloemic trypanosomes, but without success.

It may be that not all ds RNA are encapsidated, as is the case with Chestnut Blight fungus for example (11). It would therefore be interesting to search for the nucleotide sequence of the ds RNA by *in situ* molecular hybridization. The biological role of these particles remains to be seen. Axenic organism culture and the possibility of isolating these particles in a sucrose gradient should enable the infection of isolates without ds RNA by electroporation or some other technique yet to be defined, but the fact that these virus-like particles have only been found in intraphloemic isolates associated with decay already provides a further criterion for plant trypanosomes classification.

Figure 4 : Longitudinal section of a cultured trypanosomatid associated with Hartrot disease of coconut. The arrow shows a virus-like particle, in a vesicle, in the cytoplasm of the organism. M, mitochondrion ; ER, endoplasmic reticulum ; Mt, microtubules. (X 47000)

Figure 4B: Higher magnification of the virus-like particle (X 80000).

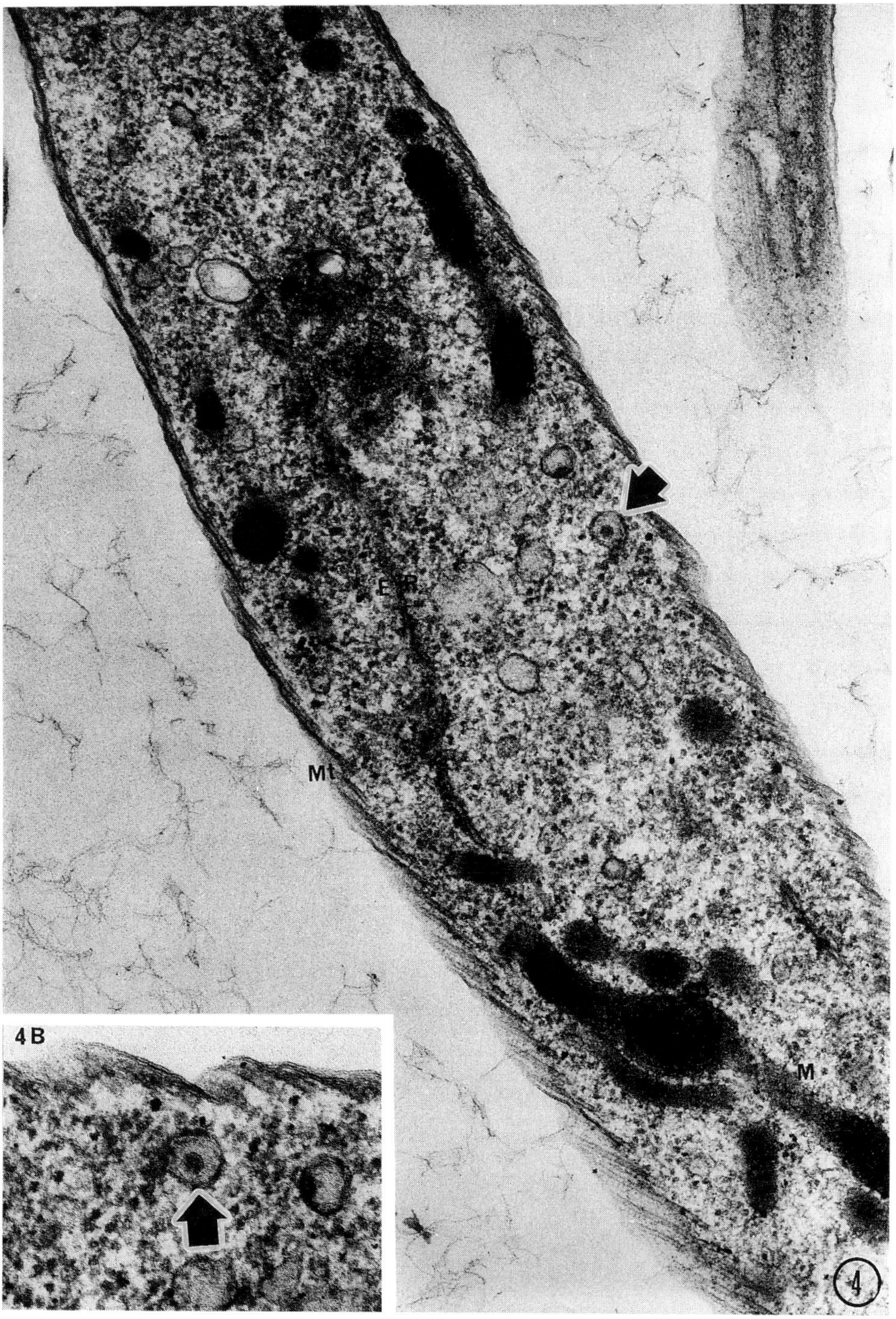

ER
Mt
M
4B
4

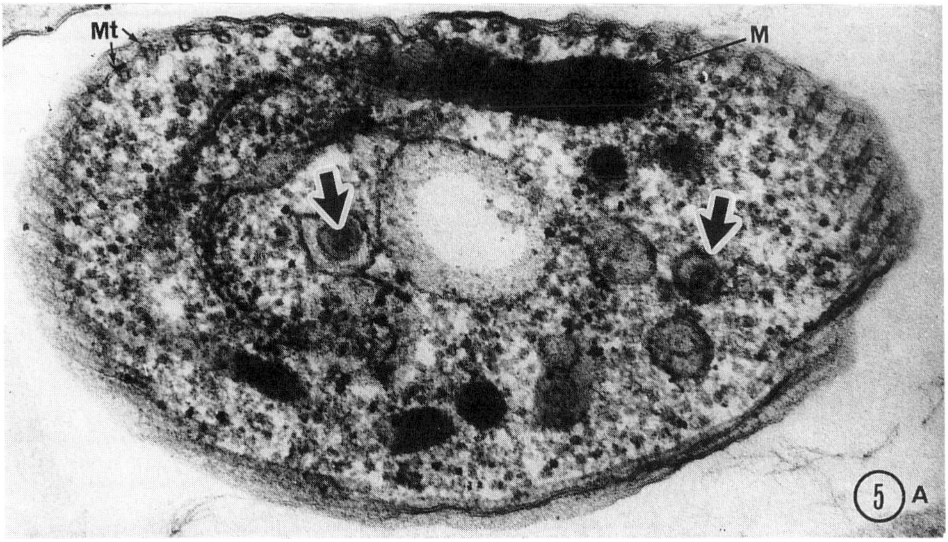

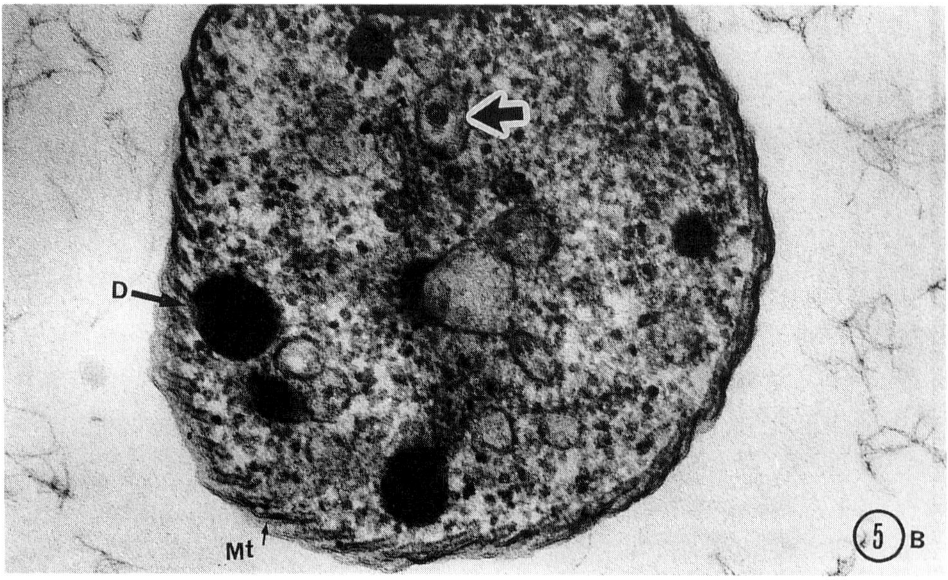

Figures 5A-5B : Transverse section of cultured organisms associated with Hartrot of coconut. Arrows show virus-like particles in vesicles. M, mitochondrion ; D, dense bodies ; Mt, microtubules.

(5A : X 80000) (5B : X 80000)

5. References

1. Ahomadegbe, J.C., Dollet, M., Coulaud, D., Gargani, D., and Riou, G. (1990) Kinetoplast DNA permits characterization of pathogenic plant trypanosomes of economic importance, *Biol. Cell*, **70**, 167-176.
2. Ahomadegbe, J.C., Dollet, M., and Riou, G. (1992) Kinetoplast DNA from plant trypanosomatids responsible for the Hartrot disease in coconut trees, *Biol. Cell.*, **74**, 273-279.
3. Attias, M., Roitman, I., Plessmann, Camargo C., Dollet, M., and De Souza, W. (1988) Comparative analysis of the fine structure of four isolates of Trypanosomatids of the genus *Phytomonas*, *J. Protozool.,* **35**, 3, 365-370.
4. Conchon, I., Campaner, M., Sbravate, C., and Camargo, E.P. (1989) Trypanosomatids, other than *Phytomonas* spp., isolated and cultured from fruit, *J. Protozool.*, **36**, 412-414.
5. Dollet, M. (1984) Plant diseases caused by flagellate protozoa (*Phytomonas*), *Annu. Rev. Phytopathol.*, **22** , 115-132.
6. Dollet, M. (1994) Identification and characterization of pest organisms; plant trypanosomes case study. In: *The identification and characterization of Pest organisms*, Ed. by D.L. Hawksworth. CAB Intal and the Systemic Association, 415-426.
7. Dollet, M. (1991) Infection of plants by flagellate protozoa (*Phytomonas* spp., Trypanosomatidae), In: *Electron Microscopy of Plant Pathogens,* K.Mendgen & D.E.Lesemann, Eds, Springer-Verlag.
8. Dollet, M., Giannotti, J., and Ollagnier, M. (1977) Observation de protozoaires flagellés dans les tubes criblés de palmiers à huile malades, *C.R. Acad. Sci. Paris,* Série D, **284** , 643-645.
9. Donovan, C. (1909) Kala-azar in Madras, especially with regard to its connexion with the dog and the bug (*Conorrhinus*), *Lancet* **177**, 1495-1496.
10. Francki, R.I.B., Fauquet, C.M., Knudson, D.L., Brown, F. (1991) Classification and nomenclature of viruses, *Fifth report of the International committee on taxonomy of viruses*, *Archives of virology supplementum* 2. Springer verlag New-York.
11. Ghabrial S.A. (1994) New developments in fungal virology, *Adv. Virus Res.*, **43**, 303-346.
12. Guerrini, F., Segur, C., Gargani, D., Tibayrenc, M., and Dollet, M. (1992) An isoenzyme analysis of the genus *Phytomonas* : genetic, taxonomic and epidemiologic significance, *J. Protozool.*, **39**, 4, 516-524.
13. Jankevicius, J.V., Jankevicius, M., Campaner, M., Conchon, I., Maeda, L.A., Teixeira, M.M.G., Freymuller, E., and Camargo, E.P. (1989) Life cycle and culturing of *Phytomonas serpens* (Gibbs), a trypanosomatid parasite of tomatoes, *J. Protozool.*, **36**, 3, 265-271.
14. Khoshnan, A., and Alderete, J.F. (1993) Multiple double-stranded RNA segments are associated with virus particles infecting *Trichomonas vaginalis*, *J. Virol.*, **67**, 12, 6950-6955.
15. Lafont, A. (1909) Sur la présence d'un parasite de la classe des flagellés dans le latex de l'*Euphorbia pilulifera*, *C.R. Soc. Biol.*, **66**, 1011-1013.
16. Louise, C., Dollet, M., and Mariau, D. (1986) Recherches sur le Hartrot du cocotier, maladie à *Phytomonas* (Trypanosomatidae) et sur son vecteur *Lincus* sp. (Pentatomidae) en Guyane, *Oléagineux*, **41**, 10, 437-449.
17. Marché, S. (1995) Caractérisation et identification des *Phytomonas* : Trypanosomatides de plantes, Thèse Université Bordeaux II.
18. Marché, S., Roth, C., Manohar, S.K., Dollet M., and Baltz T. (1993) RNA virus-like Particles in Pathogenic Plant Trypanosomatids, *Mol. Biochem. Parasitol.* **57**, 2, 261-268.
19. Marché, S., Roth, C., Philippe, H., Dollet, M., and Baltz, T. (1995) Characterization and detection of plant trypanosomatids by sequence analysis of small submit ribosomal RNA gene, *Mol. Biochem. Parasitol.,* **71**, 15-26.
20. Muller, E, Gargani, D., Schaeffer, V., Stevens, J., Fernandez-Becerra C., Sanchez-Moreno, M., Dollet, M. (1994) Variability in the phloem restricted plant trypanosomes (*Phytomonas* spp) associated with wilts of cultivated crops, *Eur. J. Plant Pathol.*, **100**, 425-434.

21. Parthasarathy, M.V., Van Slobbe, W.G. , and Soudant, C. (1976) Trypanosomatid flagellate in the phloem of diseased coconut palm, *Science*, **192**, 1346-1348.
22. Patterson, J.L. (1990) Viruses of protozoan parasites, *Exp. Parasitol.*, **70**, 111-113.
23. Paulin, J.J. and McGhee, R.B. (1971) An ultrastructural study of the Trypanosmatid, *Phytomonas elmassiani*, from the milkweed, *Asclepias syriaca*, *J. Parasitol.*, **57**, 1279-1287.
24. Roditi, I., Wyler, T., Smith, N., and Braun, R. (1994) Virus-like particles in *Eimeria nieschulzi* are associated with multiple RNA segments, *Mol. Biochem. Parasitol.*, **63**, 275-282.
25. Stahel, G. (1931) Zur kenntnis der Siebröhrenkrankheit (Phloemnekrose) des Kaffeebaumes in Surinam, I, *Phytopath. Z.*, **4**, 65-82.
26. Wang, A.L., and Wang, C.C. (1991) Viruses of the protozoa, *Annu. Rev. Microbiol.*, **45**, 251-263.

PLANT CELL MODIFICATIONS BY PARASITIC NEMATODES

W.M. ROBERTSON
Scottish Crop Research Institute
Invergowrie
Dundee DD2 5DA U.K.

1. Introduction

Plant parasitic nematodes have a wide range of feeding habits. Amongst the simplest is that of *Aporcelaimus* spp. which have a short tooth which they use to puncture the walls of root epidermal cells prior to sucking out their contents. Amongst the most sophisticated are various sedentary root endoparasites such as *Meloidogyne* and *Globodera* which, at their feeding sites, at or close to the stele, modify the cells around their heads so that they provide a constant and abundant supply of food.

In between these extremes there are a wide range of feeding behaviours. Trichodrid nematodes graze epidermal cells at the root-tip and, like *Aporcelaimus*, have a solid stylet. However, they create a feeding tube, made from oesphageal gland exudate, through the cell wall. They also inject a more fluid exudate through this feeding tube which causes the contents of the plant cell to accumulate around the point of penetration (Wyss, 1971). *Aporcelaimus* and trichodorids are exceptional as most plant parasitic nematode species have hollow stylets which performs the same functions as the trichodorid feeding tube and they can inject secretions from the oesophageal glands in the nematode feeding apparatus. The stylet, as in *Aporcelaimus,* is connected via a cuticle lined tube to a pump which is used to withdraw the contents of the cell under attack. Nematodes with short stylets , such as *Tylenchus* spp. feed only on epidermal cells, but others with longer stylets such as *Xiphinema* and *Longidorus* spp. can feed several cells deep within the root where they also induce changes in host cells, which increases their nutritional value to the feeding nematode.

Increasingly complex host-nematode interactions are often accompanied by a narrowing of the nematodes` host range. This is particularly true for cyst nematodes (*Globodera* and *Heterodera*) and some species of root-knot nematodes (*Meloidogyne*). However, most *Longidorus* and *Xiphinema* spp. have wide host ranges, and some polyploid, parthenogenetic species of *Meloidogyne* can parasitize several thousand species of plants.

Undoubtedly it is those nematodes which induce plant cell modifications that are the most challenging as these are also those against which plant resistance occurs most frequently. Hence, it is on these that this review will concentrate, comparing the changes induced in plants by nematodes. A recent review of root nematode parasitic strategies will provide a broader perspective (Sijmons *et al.*, 1994).

M. Nicole and V. Gianinazzi-Pearson (eds.), Histology, Ultrastructure and Molecular Cytology of Plant-Microorganism Interactions, 237-244.

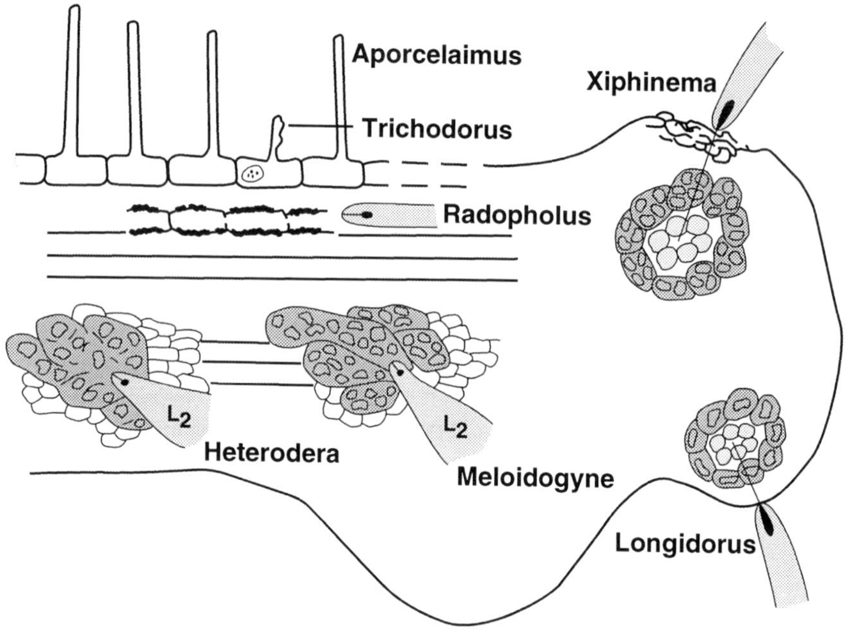

Figure 1. Diagram of a parasitized root-tip showing the main feeding sites of the ectoparasites (*Longidorus, Xiphinema, Aporcelaimus and Trichodorus* spp.) and the endoparasites (*Meloidogyne and Heterodera*).

2. Biology

2.1 ENDOPARASITIC NEMATODES - *Heterodera, Globodera* and *Meloidogyne*

Most nematodes usually pass through four juvenile stages before becoming adult but plant parasitic nematodes mostly undergo the first moult in the egg. In cyst (*Heterodera-Globodera* spp) and root-knot (*Meloidogyne* spp.) nematodes the second stage juveniles hatch from the egg and are the invasive stage. Hatching of cyst nematodes is often in response to specific chemicals exudating from the host roots. Once hatched, the juveniles can find the host roots from several centimetres. The zone of elongation is the preferred site of entry into the root. Cyst nematodes cause substantial cell death as they move intracellularly through the cortical cells to their feeding site immediately adjacent to the endodermis. In contrast, juvenile root-knot nematodes usually move inter-cellularly and therefore cause less damage during invasion. During invasion they move within the root towards the root-tip. Once beyond the region of differentiation of the endodermis they turn back up the root and enter the vascular tissue to locate a feeding site without damaging the endodermal cell layer (Sijmons *et al.*, 1994). This method of invasion is thought to cause less damage to the host roots than that of the cyst nematodes and confers the advantage of attenuating the production of signals likely to be received by the plant defence recognition mechanisms. It may be that this behaviour contributes to the root-knot nematode's ability to parasitize a wide range of plant species compared to the much more limited host range of *Heterodera* and *Globodera* spp..

Plant cell modification induced by *Meloidogyne* spp. occurs once the invading nematode has settled and selected a group of cells. It breaks through each of the cell walls with its stylet without damaging the plasmalemma. A plug of solid material which appears to emanate from the oesophageal and amphidial glands supports the cell wall where the stylet breaks through. The stylet is then projected a few microns into the cell with the cell membrane draped around it and a fluid secretion from gland cells associated with the feeding apparatus is exuded into the cells. This secretion is thought to induce cell enlargement and repeated, synchronous nuclear division. Subsequent cell wall thickening takes place with a large increase in surface area to allow for greatly increased flow of solutes into each giant cell (Jones, 1981). A further secretion is exuded from the stylet and precipitates in the cytoplasm in the form of a coiled tube which acts as a filter, enabling the nematode to remove cytosol without damaging the cytoplasm. Once the nematode has established the feeding site it becomes sedentary and loses the body wall muscle. In addition to inducing giant cell formation, the surrounding cells are stimulated

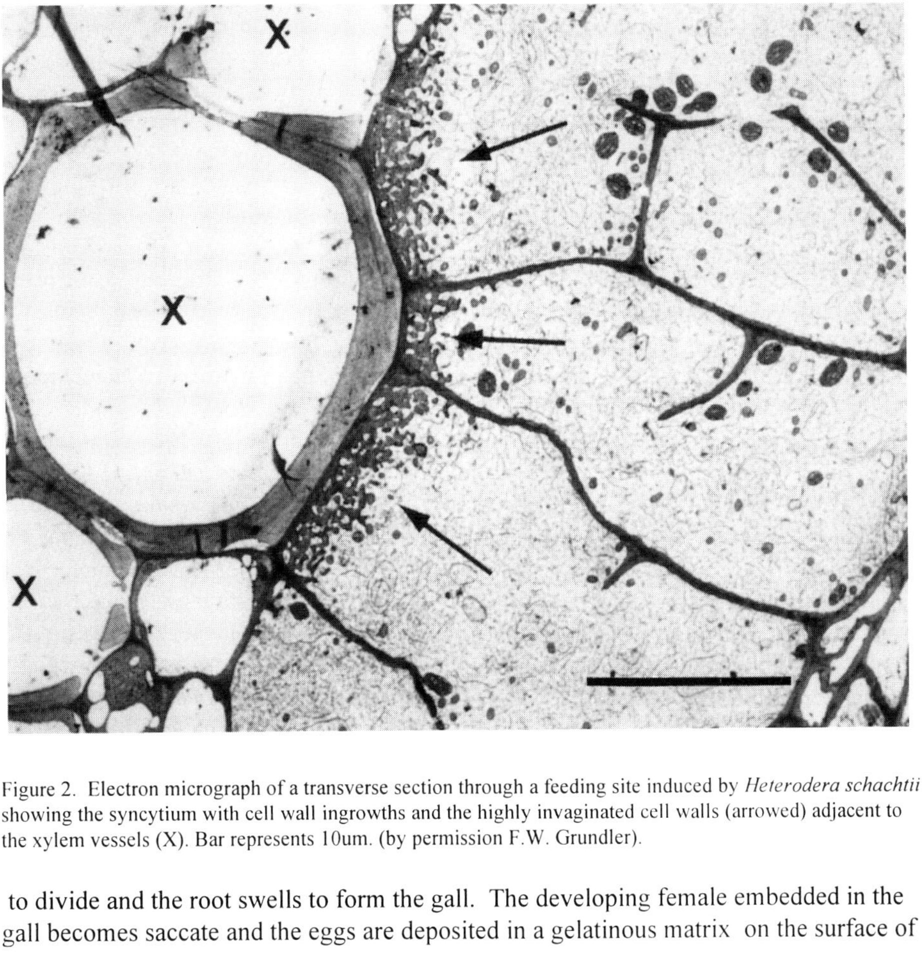

Figure 2. Electron micrograph of a transverse section through a feeding site induced by *Heterodera schachtii* showing the syncytium with cell wall ingrowths and the highly invaginated cell walls (arrowed) adjacent to the xylem vessels (X). Bar represents 10um. (by permission F.W. Grundler).

to divide and the root swells to form the gall. The developing female embedded in the gall becomes saccate and the eggs are deposited in a gelatinous matrix on the surface of the gall. Access to the exterior is maintained by means of enzymes which form a channel to the outer surface of the gall. Eggs hatch within the matrix thereby maintaining a continuous flow of juveniles to re-infect host roots (Sijmons *et al.*, 1994).

The cyst nematodes (*Heterodera* and *Globodera* spp.) invade intra-cellularly, using their stylets and gland cell secretions to penetrate the plant cell walls as they cut a path to their feeding site. These are usually initiated by modifying a single cortical cell close to the endodermis. The cell wall is penetrated carefully in a manner similar to that employed by *Meloidogyne*. *Heterodera-Globodera* also use oesophageal and/or amphidial secretions to seal the point of penetration. During feeding a feeding tube is also formed. However, the feeding site differs in certain important ways from that of *Meloidogyne* and the modified cells are called syncytia to distinguish them from the giant cells. The feeding cell expands mainly longitudinally, by the breakdown of adjacent cell walls. on the opposite side of the cell from the nematode and the cell. Several cortical, endodermal and cells within the stele can be involved with their contents fuse thereby creating a multinucleate cell. Cell walls adjacent to the xylem become thickened and are deeply invaginated (Fig. 2) possibly to maintain the transport of solutes into the syncytium (Jones, 1981; Sijmons *et al,* 1994). Nuclear division has not been observed but the highly active nuclei become polyploid. The female becomes sedentary in a similar manner to that described for Meloidogyne but mainly retains the eggs she produces within her swollen body. As the plant senesces the female dies and her cuticle tans to form the cyst that protects the eggs. In *Globodera* the cysts are very robust and have been known to remain in soil with viable eggs for up to 20 years.

2.2 ECTOPARASITIC NEMATODES - *Longidorus* and *Xiphinema*

Longidorus and *Xiphinema* spp., feed with their bodies outside the root (Fig 1) and use long, hollow, slender stylets to penetrate deep into the meristem or additionally in the case of *Xiphinema*, the vascular region of the differentiated tissue. Typically, *Longidorus* pushes its stylet deep into the meristem or the vascular tissue of the zone of elongation and initiates a feeding site by injecting material secreted from the oesophageal glands. This secretion occurs prior to ingestion and is seen as a period of inactivity lasting for up to one hour. Observations on *L. caespeticola* (Towle and Doncaster, 1978) and *L. elongatus* and *L. leptocephalus* (Robertson *et al.*, 1984) on ryegrass revealed that nematodes would move between root-tips, initiating feeding sites and then returning to exploit those modified earlier. This behaviour suggested that these species require a delay while cell changes occur. In ryegrass (*Lolium perenne*) gall formation can proceed in an organised manner. At the root-tip *Longidorus* induces the development of enlarged, uninucleate, amoeboid shaped nuclei with little or no vacuole. In addition to hyperplasia and hypertrophy effects which give rise to the gall, cell division is synchronised and the plane of cell division is turned through ninety degrees (Fig. 3). During feeding the stylet is inserted into a single cell deep within the gall but the cell contents are removed from a group of surrounding cells (Griffiths & Robertson, 1984; Andres *et al.*, 1989) This is achieved by inducing perforations to form (probably with enzymes) in adjacent cell walls with subsequent solubilisation of the cell cytoplasm to assist its removal (Fig. 4). Curiously, observations on ryegrass showed that the contents of all except the epidermal cells could be removed during this process.

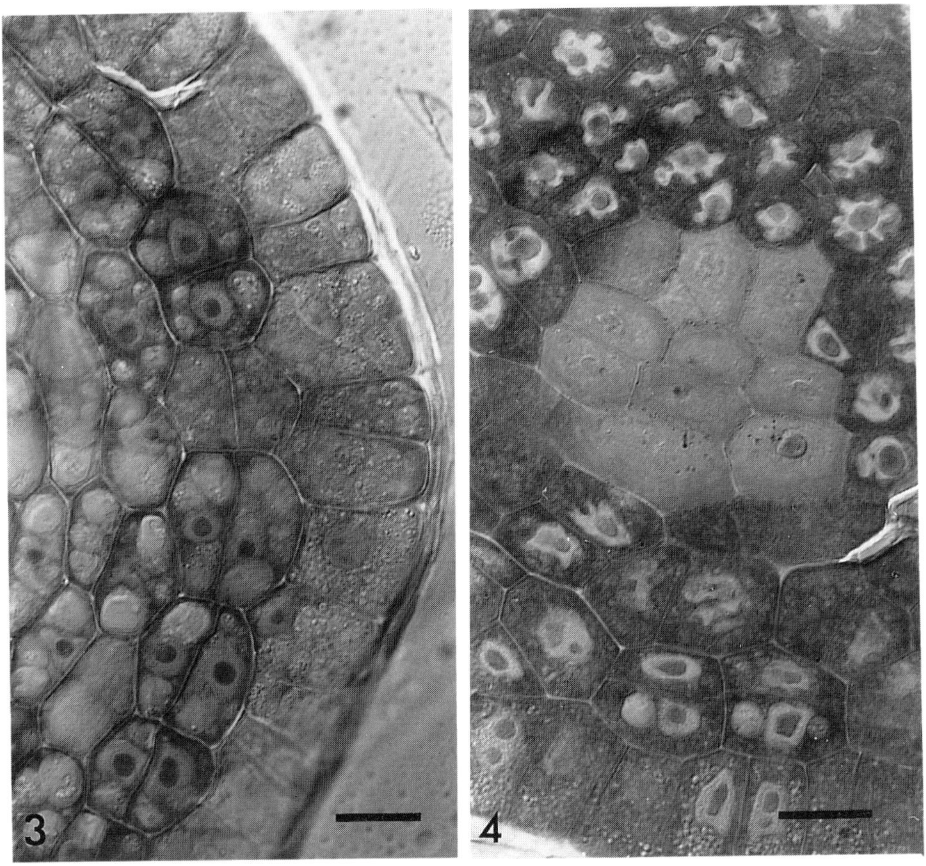

Figures 3 & 4. 3) Transverse section through an early stage of gall induced by *Longidorus elongatus* in *Lolium perenne* showing synchronised cell division in the outer layer of cortical cells. 4) Transverse setion of a mature gall showing a feeding site surrounded by enlarged, cytoplasm enriched cells each containing a single amoeboid nucleus. Bars represent 50 um. (from Griffiths and Robertson, 1984)

In contrast, *Xiphinema index* and *X. diversicaudatum* feed on a column of cells progressively penetrated by the stylet. During feeding they have been observed to rapidly inject the contents of their dorsal gland ducts into the root cells being attacked. This was followed by immediate withdrawal of the cell contents (Trudgill, 1976; Wyss, 1977). However, once a suitable site is found deep within the root tip *Xiphinema* adopts a quiescent behaviour similar to that observed during salivation by *Longidorus* spp. During this behaviour only small movements within the oesophageal bulb are seen and the plant responds by generating a gall containing enlarged, multinucleate, cytoplasm - enriched cells with no vacuole and an increased rate of division. These complexes form a feeding site from which nematodes can leave and return. The multinucleate condition occurs because regular mitosis without cytokinesis is induced and scanning microdensitometry of samples of nuclei stained by the Feulgen reaction has shown a significant increase in their nuclear DNA content (Fig. 5; Griffiths *et al.*, 1982). In many cases root elongation stops because the plane of cell division is disrupted. The

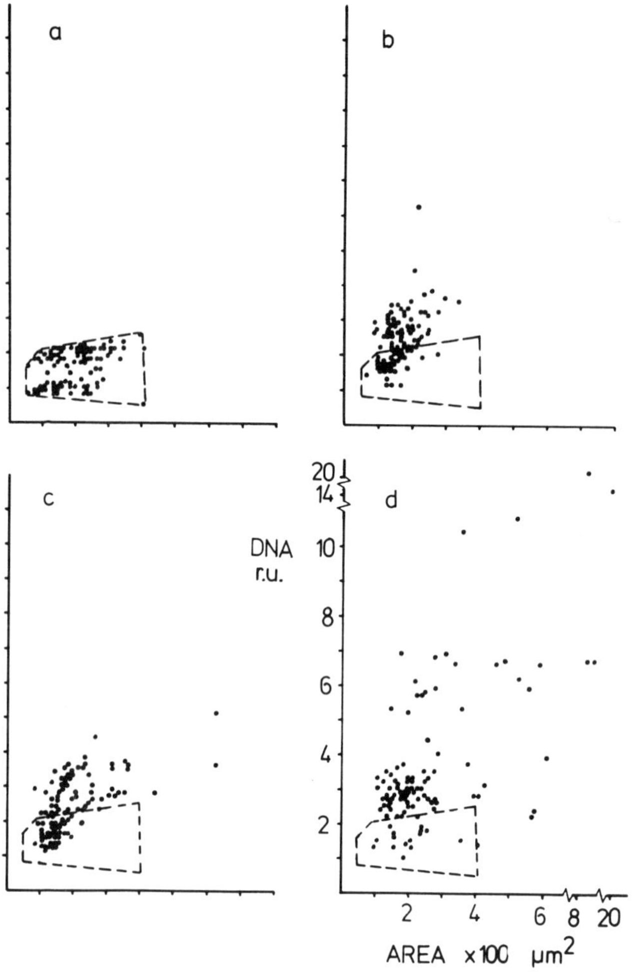

Figure 5. The relationship between area and DNA content of nuclei from galls induced in *Lolium perenne* by *Xiphinema diversicaudatum*, (a) control root-tip (b) after 1 day (c) 2 days (d) 4 days. The dashed line represents the 'control area'. (from Griffiths et al., 1982)

development of galls gives the nematodes a stationary food source, the formation of which is thought to be essential for the females to maximise their egg production (Wyss, 1978).

3. Discussion

There are several interesting parallels, but also some differences between these endo- and ectoparasitic nematodes. Both *Meloidogyne* and *Xiphinema* spp. induce multinucleate polyploid cells in their hosts and are able to do this in a wide range of plant species. However, whilst the inductive factors injected by *Xiphinema* spp. is diffusible and a group of cells around the feeding site respond by being multinucleate, only the cells penetrated directly by *Meloidogyne* respond. The inductive factors injected by *Longidorus* and *Globodera - Heterodera* also seem to be diffusible, and cause cell enlargement, the nuclei become polyploid without nuclear division. Both groups also appear to be able to induce breakdown of cell walls at a distance from the point of cell penetration.

The nematode secretions involved in inducing these changes were initially thought to come from the dorsal gland cell within the oesophagus. This cell actively produces secretions and these are injected into the plant. In *Xiphinema* they have a digestive role, and in *Globodera* and *Heterodera* they are probably involved in the formation of the feeding tube. *Xiphinema* spp., during a certain type of feeding behaviour involving prolonged ingestion also has a ball-like structure at its stylet tip which may also act as a molecular sieve preventing ingestion of the cytoplasm of the developing multinucleate cell. Whether *Longidorus* spp. produce a similar structure is uncertain because the stylet tip is always too deep within the gall for it to be seen. Hence whether the dorsal gland cell secretions are involved in the induction or exploitation of the feeding site, or both is unclear.

The alternative source of the secretions involved in the induction process are the two subventral gland cells Their ducts open within the food pump chamber and for a long while their products were thought to be passed backwards into the gut. However, there is growing evidence that these products are also capable of being passed forwards into the plant. The amphidial glands, two chemoreceptors on either side of the stylet opening, are the only other possible source of the inductive factor, and then only with the endoparasites.

The secretions involved in the induction of these complex feeding sites is only one part of the current interest world-wide. Choice of a specific feeding site, one where the cells still have the capacity to respond is probably a crucial aspect. *Longidorus* and *Xiphinema* spp. feed on cells at some distance from the sense organs on their heads and contain sense organs thought to be gustatory within their stylet; presumably to help them locate their feeding sites deep within the root gall (Robertson, 1976).

The nematode secretions must be involved, also, in any resistant reaction displayed by the plant in response to nematode attack. Resistance appears to be more widespread to cyst nematodes, and perhaps this is largely a reflection of their intracellular invasion which exposes them to a wide range of host receptors involved in defence. The intercellular invasion of *Meloidogyne* spp. and the ectoparasitic behaviour of *Longidorus* and *Xiphinema* appears to be associated with a wider host range, but the closely related *X. index* is largely confined in agricultural systems to grapevine and fig. Consequently there are many different gene/allele products and interactions/ recognition processes occurring between these nematodes and their hosts; an understanding of which we are only now starting to develop. It is likely that a better understanding of the fundamental mechanisms of nematode - host interactions is more likely to be achieved using *Arabidopsis thaliana* as a model host (Grundler *et al.*, 1994)

4. References

Andres, M., Bleve-Zacheo, T. and Arias, M. (1989) The ultrastructure of cereal and leguminous root tips parasitized by *Longidorus belloi*. *Revue de Nematologie* **17**, 365-374.

Griffiths, B.S. and Robertson, W.M. (1984) Morphological and histochemical changes occurring during the lifespan of root-tip galls on *Lolium perenne* induced by *Longidorus elongatus*. *J. of Nematology* **16**, 223-229.

Griffiths, B.S. and Robertson, W.M. (1983) Nuclear changes induced by the nematode *Longidorus elongatus* in root-tips of ryegrass, *Lolium perenne*. *Histochemical Journal* **15**, 927-934.

Griffiths, B.S., Robertson, W.M. and Trudgill, D.L. (1982) Nuclear changes induced by the nematodes *Xiphinema diversicaudatum* and *Longidorus elongatus* in root-tips of ryegrass, *Lolium perenne*. *Histochemical Journal* **14,** 719-730.

Grundler, F.M.W., Bockenoff, A, Schmidt, K.-P., Sobczak, M., Golinowski, W. and Wyss, U. (1994) *Arabidopsis thaliana* and *Heterodera schachtii* : A versitile model to characterize the interaction between host plants and cyst nematodes in F. Lamberti, C. De Giorgi, and D. McK. Bird, (eds.) *Advances in Molecular Plant Nematology,* NATO ASI Series, Plenum Press, New York, pp. 171-180.

Jones, M.G.K. (1981) Host cell responses to endoparasitic attack: structure and function of giant cells and syncytia. *Annals of Applied Biology* **97**, 353-372.

Robertson, W.M. (1976) A possible gustatory organ associated with the odontophore in *Longidorus leptocephalus* and *Xiphinema diversicaudatum*. *Nematologica* **21**, 443-448.

Robertson, W.M., Trudgill, D.L. and Griffiths, B.S. (1984) Feeding of *Longidorus elongatus* and *L. leptocephalus* on root-tip galls of ryegrass (*Lolium perenne*). *Nematologica* **30**, 222-229.

Sijmons, P.C., Atkinson, H.J. and Wyss, U. (1994) Parasitic strategies of root nematodes and associated host cell responses. *Annual Review of Phytopathology* **32**, 235-259.

Towle, A. and Doncaster, C.C. (1978) Feeding of *Longidorus caespeticola* on ryegrass, *Lolium perenne*. *Nematologica* **24**, 277-285.

Trudgill, D.L. (1976) Observations on the feeding behaviour of *Xiphinema diversicadatum*. *Nematologica* **22**, 417-423.

Wyss, U. (1971) Der Mechanismus der Nahrungsafnahme bei *Trichodorus similis*. *Nematologica* **17**, 508-518.

Wyss, U. (1977) Feeding phases of *Xiphinema index* and associated processes in the feeding apparatus *Nematogica* **23**, 463-470.

Wyss, U. (1978) Root and cell response to feeding by *Xiphinema index*. *Nematologica* **24**, 159-166

IN SITU DETECTION OF GRAPEVINE FLAVESCENCE DOREE PHYTOPLASMAS AND THEIR INFECTION CYCLE IN EXPERIMENTAL AND NATURAL HOST PLANTS

J. LHERMINIER [1] and E. BOUDON-PADIEU [2]
(1) Service commun de Microscopie Electronique, (2) Station de Recherche sur les Mycoplasmes, INRA, BV 1540, 21034 Dijon Cedex, France.

1. Phytoplasmas

Phytoplasmas are unculturable wall-less prokaryotes classified in the class Mollicutes. They are characterized and differentiated by symptoms on plants, host plant specificity and insect vectors [1, 2, 3]. Based on electron microscope observations, especially in thin sections, these microorganisms were described as being morphologically similar to animal mycoplasmas and, consequently, they were referred to for a long time as mycoplasma-like organisms. They contain DNA, ribosomes, but have no vacuole or internal membranes. They are heterogeneous in shape, size (80 to 1000 nm diameter) and electron density. However molecular analysis based on the 16S rRNA genes [4] and 16S rDNAs [5] indicate that they are a homogeneous phylogenetic group close to acholeplasmas and different from mycoplasmas. This genetic homogeneity has led to the recent classification of these organisms in a single group named "phytoplasmas" (International Committee on the Taxonomy of Mollicutes, International Organization for Mycoplasmology, Bordeaux France, 1994). Phytoplasmas are transmitted by grafting diseased plants onto healthy ones, by dodder (*Cuscuta*) bridges and by phloem feeding vector insects, especially leafhoppers and psylla. Phytoplasma cells in host tissues are commonly detected *in situ* by classical transmission electron microscopy and fluorescent microscopy using DNA-specific staining.

In the past two decades, immunological tools followed by nucleic acid-based techniques have led to specific identification and differentiation of different phytoplasma isolates [reviewed in references 5, 6]. Apart from one report using *in situ* hybridization [7], immunocytochemistry has been a powerful tool to detect [reviewed in reference 6, 8, 9, 10, 11] and to unequivocally identify these membrane-limited microorganisms within host cells, avoiding any confusion with host organelles or membrane artefacts caused by fixation and embedding procedures.

M. Nicole and V. Gianinazzi-Pearson (eds.), Histology, Ultrastructure and Molecular Cytology of Plant-Microorganism Interactions, 245-255.

2. Grapevine Flavescence dorée phytoplasma

A pathogenic phytoplasma is the aetiological agent of flavescence dorée (FD), an economically important grapevine disease of which the symptomatology and epidemiology have been fully characterized [12]. The FD phytoplasma is transmitted by the leafhopper *Scaphoideus titanus* Ball in the field. For laboratory studies, in greenhouse conditions, the FD phytoplasma is experimentally maintained on broadbean plants (*Vicia faba* L.) to which it is transmitted by the leafhopper *Euscelidius variegatus* [13]. Production of specific polyclonal [14] or monoclonal [15] antibodies and, more recently, DNA probes [16], has permitted *in vitro* visualization [14], characterization and differentiation of FD phytoplasmas in extracts of experimental or natural insect and plant hosts [17, 18, 19]. *In situ* detection and infection site determination has been performed on whole dissected salivary glands of the vector using immunofluorescence [9] but this method is limited because of the autofluorescence of insect tissues such as the alimentary tract. Histological *in situ* detection of FD phytoplasma has also been performed on infected broadbean tissues by cryosectioning followed by indirect immunoenzyme labelling (Figure 1). However, to gain understanding into the infection cycle of FD phytoplasmas within experimental insect and plant hosts, we developed a post-embedding immunogold labelling technique [10]. This has allowed sensitive and specific localization of FD phytoplasmas in tissue sections and, consequently, provided information about the morphology of the phytoplasma cells during the course of infection. The propagation of the FD phytoplasma in the experimental insect vector has been described [20] and two multiplication sites have been located, the alimentary tract and some specific acini of the salivary glands. The occurrence of attachment sites for FD phytoplasma on these main organs has been investigated and preliminary results have shown that a lectin/glycoconjugate system could be involved in the recognition between phytoplasma cells and the acini [21].

3. *In situ* detection of FD-phytoplasma by immunocytochemistry in host-plants.

Since there are very few studies reporting the pattern of colonization of phytoplasmas in plants [22], we have used ELISA to monitor the movement of FD phytoplasma during the process of *Vicia faba* infection after inoculation by infective leafhoppers [23]. Moreover, we associated this with post-embedding indirect immunolabelling to specifically detect and identify the FD phytoplasma cells in tissue sections during the main stages of infection [23]. Our aim was to provide information about 1) the characteristics of the antigenic sites recognized by the antibodies, 2) the morphology of the phytoplasma cells, and 3) their cellular and histological distribution in plant tissues before symptom expression. In the present paper, we also present preliminary results of phytoplasma immunolabelling in naturally infected grapevine.

It is well known that phytoplasma antigenic sites are very sensitive to fixatives commonly used for immunocytochemistry. Tissue processing for electron and light

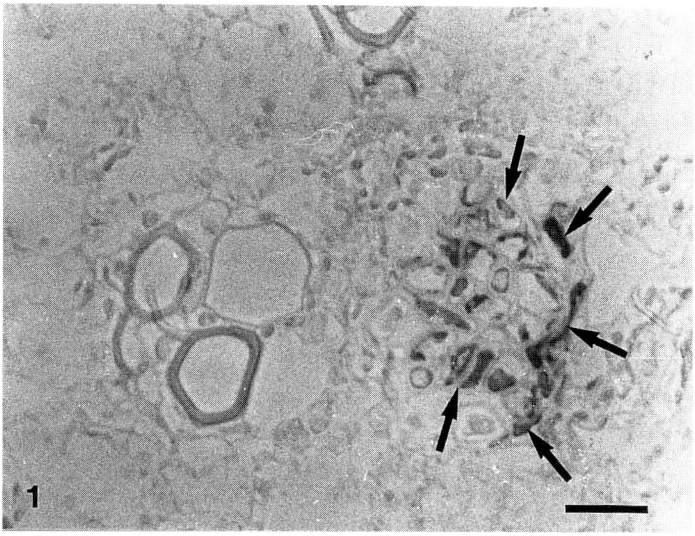

Figure 1. Indirect immunoenzyme localisation of FD phytoplasma in a cryostat section of a broadbean leaf petiole. Rabbit polyclonal antibodies were used followed by sequential incubations with biotinylated anti-rabbit IgG and alkaline phosphatase-labelled streptavidin. Staining with a subtrate-chromogen (naphtol phosphate-New Fuchsin) system (Dako LSAB+ kit, France). Bar = 20 μm.

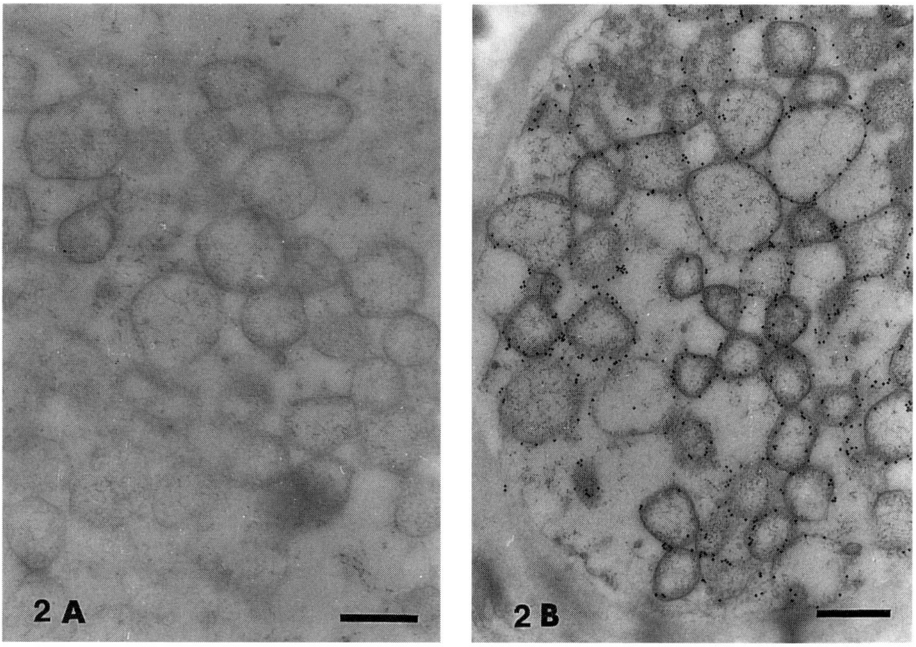

Figures 2A and 2B. Thin sections on grids of FD infected broadbean root tissues embedded in LR White and treated with either A) pronase or B) sodium metaperiodate, prior to immunolabelling. Bars = 0.5 μm.

microscopy studies, as well as the way of screening combinations of fixatives and polyclonal or monoclonal antibodies, have already been described [23]. Briefly, it was necessary to use a 2 % paraformaldehyde and 0.1 % glutaraldehyde fixative solution followed by low temperature embedding in LR White resin to conserve antigenic sites. Polyclonal rabbit antibodies prepared against phytoplasma-enriched FD-infected broadbean extracts and carefully cross-absorbed by repeated incubations with the extracts of healthy broadbeans were used as primary antibody. Resulting gold labelling was mainly associated with the phytoplasma membrane. Pronase treatment of thin sections on grids before immunolabelling eliminated the gold labelling (Figure 2A). On the other hand, the antigenic reaction was not affected by a periodate treatment (Figure 2B) suggesting that the majority of the epitopes recognized by the polyclonal antibodies were not associated with sugar residues in the membrane proteins.

In investigations of the movement of phytoplasmas through a host plant, FD-infective leafhoppers *Euscelidius variegatus* were caged on individual broadbean seedlings for a one week feeding access period and samples of different plant parts were taken the first day of feeding access and through the period required for symptom expression. The delay for the appearence of symptoms is, at least, 28 days under controlled greenhouse conditions.The first labeled phytoplasma cells were detected in root samples 17 days after the beginning of the inoculation (Figure 3A) indicating that the process of infection spread had begun. They were scattered either as conventional phytoplasma cells in mature sieve tubes or as filamentous or vesicular bodies, suggesting reproductive forms, in the cytoplasm of immature phloem elements (Figure 5A). 20 to 24 days after the first day of the feeding access period, light microscopy (silver enhancement of gold particles) showed that the lumen of some phloem cells of root (Figure 3B) and collar (Figure 3C) tissues was filled by FD phytoplasma cells. Budding forms were mainly observed within the root phloem cells, suggesting a continuous replication of phytoplasma cells (Figures 5B, 5C).

In the same period (20 to 24 days after the first inoculation), a basal axillary shoot had emerged. When histological immunolabelling was performed on the leaf petioles and on the apical part of the axillary shoot, prior to symptom expression, FD phytoplasmas were located in non functional phloem cells on the periphery of the vascular cylinder, in mature sieve tubes of the metaphloem and in phloem cells with early stages of necrosis (Figure 3D). FD phytoplasmas were never detected in phloem companion cells nor parenchyma cells and uninfected cells adjacent to those invaded by the pathogen were not affected. In the lower part of basal axillary shoot, some mature phloem cells were infected and collapsed (Figure 4A). Examination of the terminal bud of the axillary shoot revealed that the meristematic zone was free of phytoplasma (Figure 4B) but that recently differentiated sieve tubes in the future leaf midvein (Figure 4C) and petiole (Figure 4D) were largely infected and exhibited necrosis. Necrotic phloem cells were observed in the root as well as in the basal axillary shoot samples on symptom bearing plants, 28 days after inoculation. Electron dense and distorted, elongated bodies were gold labelled in the lumen of these collapsed cells (Figure 5D).

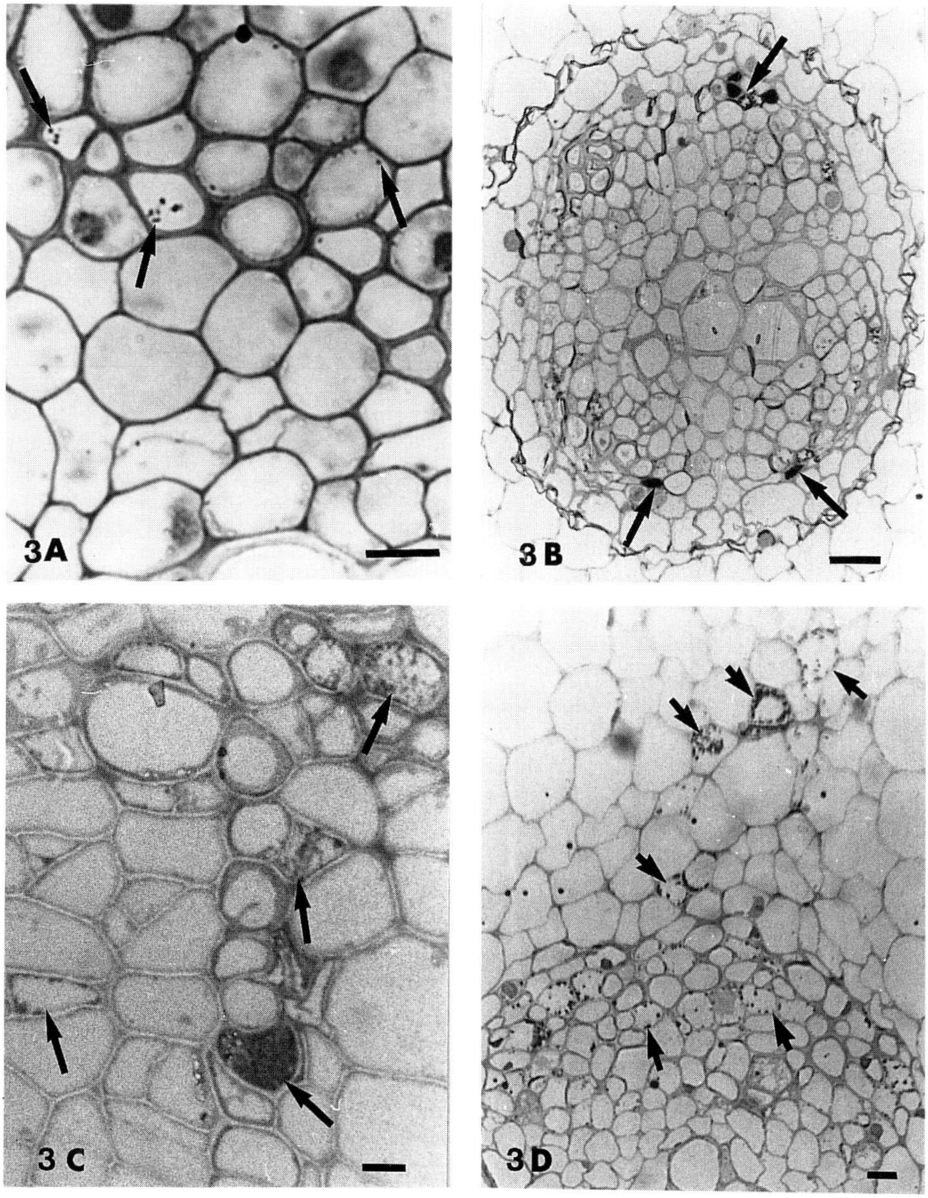

Figure 3. Light micrographs of root (A, B), collar (C) and apical axillary shoot (D) tissues from symptomless broadbean. Gold label was amplified by the silver enhancement technique and black points or areas allow to locate FD phytoplasma cells (arrows). A) Initial visualisation of FD phytoplasma scattered in mature phloem root cells at the beginning of the infection process, 17 days after inoculation. Bar = 10 μm. B) FD phytoplasma are packed in some phloem root cells, 20 days after inoculation. Bar = 20μm. C) Cross-section of the collar showing invasion of this organ, 24 days after inoculation. Bar = 10 μm. D) 24 days after inoculation, FD phytoplasma are detected in metaphloem and in non-functional phloem cells at the periphery of the vascular bundle of an apical axillary shoot. Bar = 10 μm.

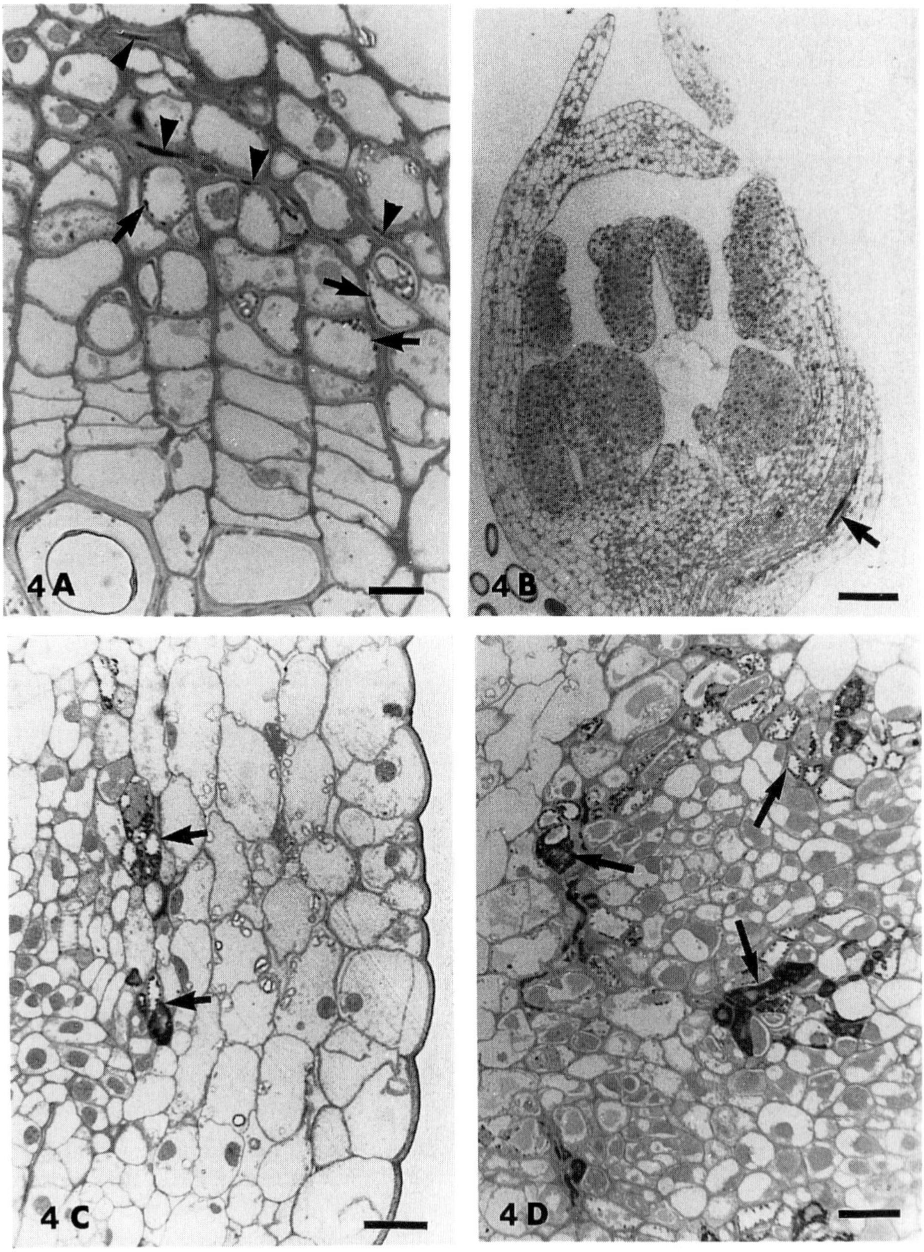

Figure 4. Light micrographs of symptomless broadbean, 24 days after inoculation. A) Cross-section of the lower part of the basal axillary shoot. Note the collapsed infected phloem cells (arrrowheads). Bar = 10μm. B) Longitudinal section of the axillary shoot apex. The meristematic zone is free of phytoplasma but they are immunolocated as black areas in the future leaf midvein. Bar =105 μm. C) Detail of a vein in a subtending leaf where immature and early differentiated sieve tubes are infected. Bar =25 μm. D) Detail of vascular bundles at the base of the terminal bud showing phloem cells exhibiting necrosis. Bar = 25 μm.

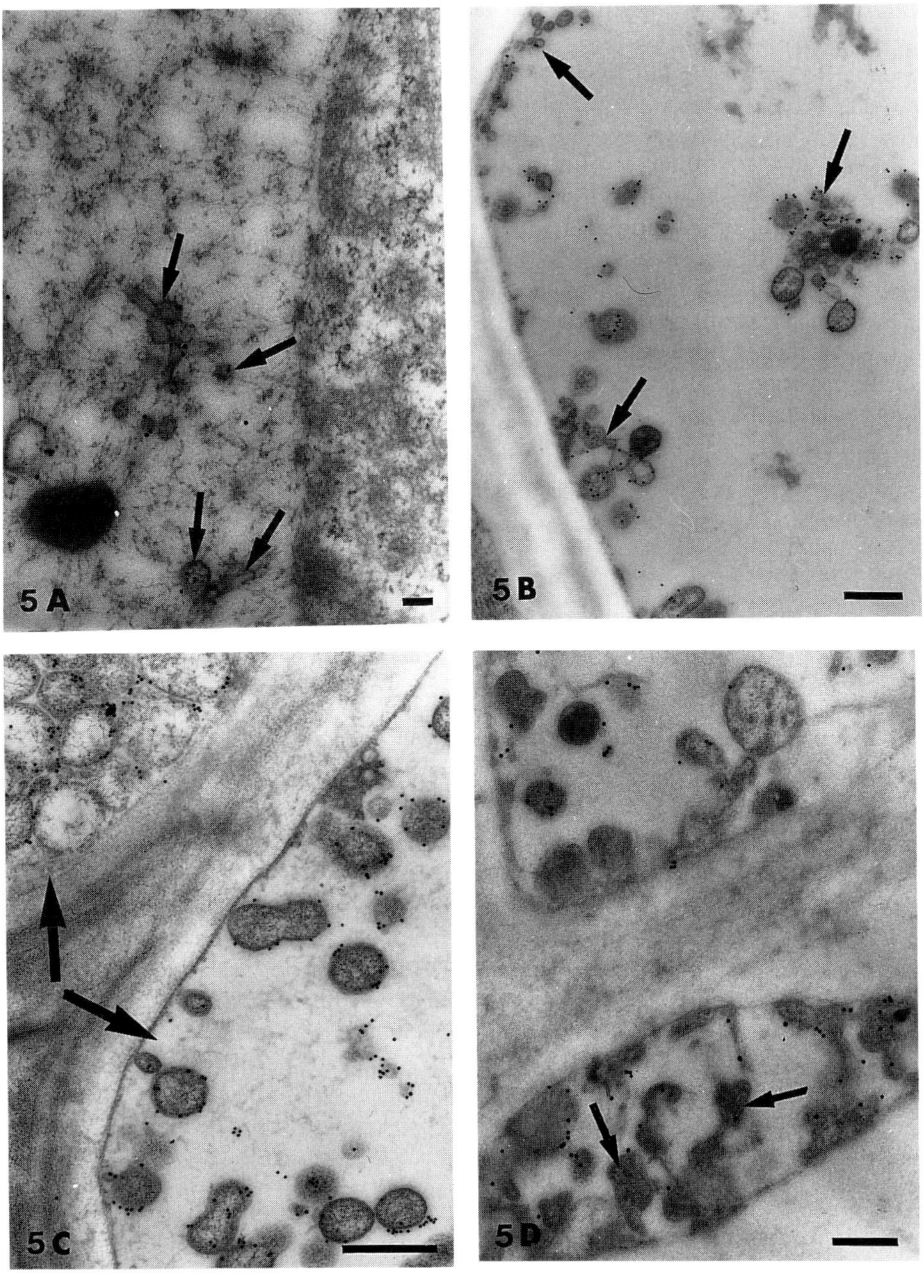

Figure 5. Transmission electron micrographs of immunolabelled FD phytoplasma in phloem cells of broadbean root, 17 (A), 20 (B), 24 (C) and 28 (D) days after inoculation. A) Vesicular and tubular bodies are gold labelled in the cytoplasm of an immature sieve element. Bar = 150 nm. B) Budding replicative forms are mainly observed. Bar = 0.5 μm. C) The lumen of some phloem cells is filled with FD phytoplasma. Bar = 0.5 μm. D) Dense and distorted FD phytoplasma cells can be observed. Bar = 0.3 μm.

Although the pattern of systemic colonization of FD phytoplasma in their host plant *Vicia faba* has already been described [23], this study indicated for the first time the essential role of the root system as a phytoplasma multiplication site. Moreover, observations of the morphology of phytoplasma cells during the process of infection of host roots has provided valuable information about phytoplasma replicating and senescent forms of the phytoplasma. Replicating forms (80-120 nm diameter) were either budding from mature spherical phytoplasma bodies or they were free, as thin tubes or round bodies, in the cytoplasm of immature phloem cells or within the lumen of mature sieve tubes. Senescent phytoplasma cells occurred in mature sieve tubes at the end of the infection process, before symptom expression.

In contrast to those on herbaceous and insect hosts, investigations on FD-infected grapevine tissue sections using immunocytochemistry have repeatedly failed whatever the combinations of fixative mixtures/polyclonal or monoclonal antibodies used. Since ELISA procedures for phytoplasma detection in FD-diseased grapevine have only succeeded with detergent-containing extraction medium [18], we can infer that phytoplasma antigens must not be easily accessible in infected grapevine phloem cells. Recently, affinity purified FD phytoplasma from infected leafhopper extracts were obtained [24] and used to raise polyclonal antibodies in rabbits (A. Seddas and R. Meignoz, unpublished results). Successful *in planta* immunolabelling of FD phytoplasma in grapevine has been obtained using these antibodies but required the fixation of plant tissues with 0.1M buffered sodium phosphate 4% paraformaldehyde, and fixative mixtures with glutaraldehyde completely inhibited the labelling. The *in situ* immunological reaction visualized with a 5 nm gold-conjugated goat anti-rabbit IgG followed by silver enhancement amplification of the signal is shown in Figures 6A and 6B.

These first results of *in situ* immunological detection of FD phytoplasmas in their natural host, grapevine, now open the possibility of screening for the most suitable, available antibodies and of improving conditions for fixation, labelling and visualization. Hopefully, this approach will be of great help in investigations of the cytological, histological and biochemical events which happen in grapevine varieties after FD infection. It is known that the reactions of different cultivars range from a very sensitive behaviour (death of the plant) to the ability to recover after infection. In addition, infected rootstock varieties are tolerant and symptomless, leading to infection of young grafted plants in nurseries. A comprehensive understanding of these phenomena relies on reliable analytical tools to locate multiplication and propagation of FD phytoplasma in vine cultivars with different responses to FD infection.

References

1. Kirkpatrick, B.C. (1989) Strategies for characterizing plant pathogenic mycoplasma-like organisms and their effects on plants, in T. Kosuge and E. Nester (eds), Plant-microbe interactions. Molecular and genetic perspectives, vol. **3**, McGraw-Hill Publishing Co., New-York, pp. 241-293.

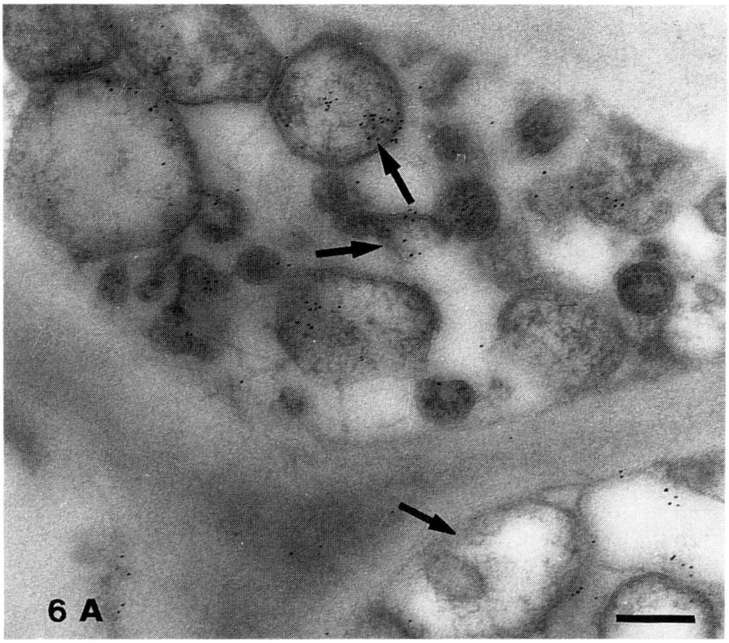

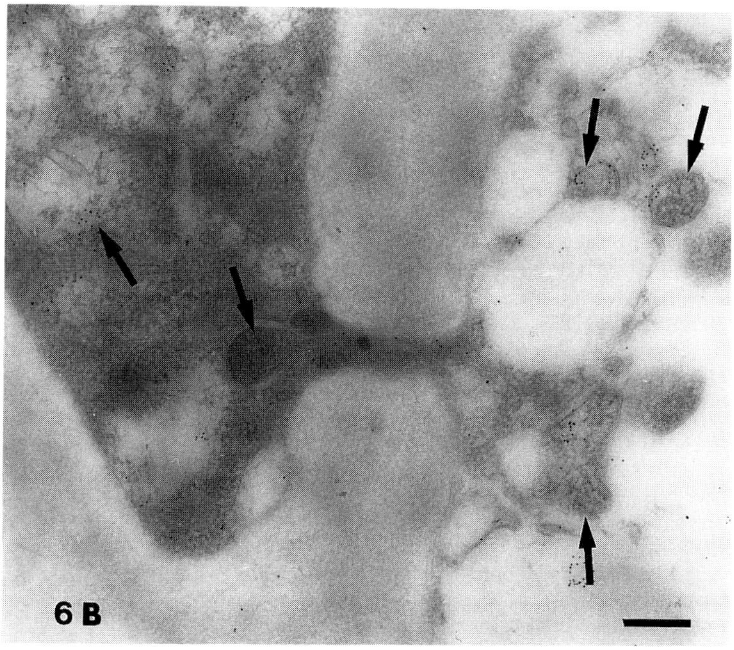

Figures 6A and 6B. Indirect immunogold labelling of FD phytoplasma in their natural host, grapevine, cultivar Alicante Bouschet, naturally infected in the field and grown in greenhouse conditions for symptom expression. Polyclonal rabbit antibodies were obtained after purification of FD phytoplasma by immunoaffinity and used as primary antibody. Gold label was amplified by the silver enhancement technique. Bars = 0.2 μm.

2. Chiykowski, L.N. and Sinha R.C. (1990) Differentiation of MLO diseases by means of symptomatology and vector transmission, Zentralblatt für Bakteriologie, Supplemente **20**, 280-287.
3. Marwitz, R. (1990) Diversity of yellows disease agents in plant infections, Zentralblatt für Bakteriologie, Supplemente **20**, 431-434.
4. Lim, P.O. and Sears B.B. (1989) 16S rRNA sequence indicates that plant-pathogenic mycoplasmalike organisms are evolutionary distinct from animal mollicutes, Journal of Bacteriology **171**, 5901-5906.
5. Seemüller, E., Schneider B., Mäurer R., Ahrens U., Daire X., Kison H., Lorentz K.H., Firrao G., Avinent L. Sears B.B. and Stackbrandt E. (1994) Phylogenetic classification of phytopathogenic mollicutes by sequence analysis of 16S ribosomal DNA, International Journal of Systematic Bacteriology **44**, n°3, 440-446.
6. Lee, I.M. and Davis, R.E. (1992) Mycoplasmas which infect plants and insects, in J. Maniloff (ed), Molecular Biology and Pathogenesis, Washington, pp 379-390.
7. Deng, S. and Hiruki, C. (1991) Localization of pathogenic mycoplasma-like organisms in plant tissue using *in situ* hybridization, Proceedings of the Japan Academy **67,** 197-202.
8. Cousin, M.T., Dafalla, G., Demazeau, E., Theveu, E. and Grosclaude J. (1989) *In situ* detection of MLOs for *Solanaceae* Stolbur and faba bean phyllody by indirect immunofluorescence, Journal of Phytopathology **124,** 71-79.
9. Lherminier, J., Terwissha van Sheltinga, T., Boudon-Padieu, E. and Caudwell A. (1989) Rapid immunofluorescent detection of the grapevine flavescence dorée MLO in salivary glands of the leafhopper *Euscelidius variegatus* Kbm, Journal of Phytopathology **125**, 353-360.
10. Lherminier, J., Prensier, G., Boudon-Padieu, E. and Caudwell A. (1990) Immunolabelling of grapevine flavescence dorée MLO in salivary glands of *Euscelidius variegatus*: a light and electron microscopy study, Journal of Histochemistry and Cytochemistry **37**, 79-85.
11. Milne, R.G., Masenga, V., Lenzi, R., Ramasso, E. and Sarindu N. (1991) Gold immunolabelling and electron microscopy of mycoplasma-like organisms in plant tissues using pre-embedding and post-embedding techniques, Phytoparasitica **19,** 263.
12. Caudwell, A. (1990) Epidemiology and characterization of flavescence dorée and other grapevine yellows, Agronomie **10**, 655-663.
13. Caudwell, A. and Larrue J. (1977) La production de cicadelles saines et infectieuses pour les épreuves d'infectivité chez les jaunisses à mollicutes des végétaux. L'élevage de *Euscelidius variegatus* Kbm et la ponte sur mousse de polyuréthane, Annales de Zoologie Ecologie **9**, 443-456.
14. Caudwell, A., Meignoz, R., Kuszala, C., Schneider, C., Larrue, J. Fleury, A. and Boudon E. (1981) Observation de l'agent pathogène de la Flavescence dorée de la vigne en milieu liquide par immunosorbant électromicroscopie (ISEM), Progrès Agricole et Viticole **24,** 835-838.
15. Schwartz, Y., Boudon-Padieu, E., Grange, J., Meignoz, R. and Caudwell A. (1989) Obtention d'anticorps monoclonaux spécifiques de l'agent pathogène de type mycoplasme (MLO) de la Flavescence dorée de la vigne, Research in Microbiology **140**, 311-324.
16. Daire, X., Boudon-Padieu, E., Berville, A., Schneider, B. and Caudwell, A. (1992) Cloned DNA probes for detection of grapevine flavescence dorée mycoplasma-like organism (MLO), Annals of Applied Biology **121**, 95-103.
17. Boudon-Padieu E., Larrue, J. and Caudwell A. (1989) Elisa and dot-blot detection of Flavescence dorée MLO in individual leafhopper vectors during latency and inoculative state, Current Microbiology **19**, 357-364.
18. Caudwell, A. and Kuszala, C. (1992) Mise au point d'un test Elisa sur les tissus de vignes atteintes de flavescence dorée, Research in Microbiology **143**, 791-806.
19. Daire, X., Clair, D., Larrue, J., Boudon-Padieu, E. and Caudwell A. (1993) Diversity among mycoplasma-like organisms inducing grapevine yellows in France, Vitis **32**, 159-163.
20. Lefol, C., Lherminier, J., Boudon-Padieu, E., Larrue, J., Louis, C. and Caudwell, A. (1993) Propagation of the Flavescence dorée MLO (Mycoplasma-like organism) in the leafhopper vector *Euscelidius variegatus* Kbm, Journal of Invertebrate Pathology **63**, 285-293
21. Lefol, C. , Lherminier, J., Boudon-Padieu, E., Meignoz, R., Larrue, J., Louis, C., Roche, A.C.and Caudwell A. (1994) Presence of attachment sites accounting for recognition between the Flavescence dorée MLO and its leafhopper vector, 10th International Congress of the IOM, Bordeaux, France, IOM letters **3**, 282-283.

22. Kuske, C.R. and Kirkpatrick, B.C. (1992) Distribution and multiplication of western aster yellows mycoplasma-like organisms in *Catharanthus roseus* as determined by DNA hybridization analysis, Phytopathology **82**, 457-462.
23. Lherminier, J., Courtois, M. and Caudwell, A. (1994) Determination of the distribution and multiplication sites of Flavescence dorée mycoplasma-like organisms in the host plant *Vicia faba* by ELISA and immunocytochemistry, Physiological and Molecular Plant Pathology **45**, 125-138.
24. Seddas, A., Meignoz, R., Daire, X., Boudon-Padieu, E. and Caudwell A. (1993) Purification of grapevine Flavescence dorée MLO (mycoplasma-like Organism) by immunoaffinity, Current Microbiology **27**, 229-236.

AUTHOR INDEX

SUBJECT INDEX

Developments in Plant Pathology

1. R. Johnson and G.J. Jellis (eds.): *Breeding for Disease Resistance.* 1993 ISBN 0-7923-1607-X
2. B. Fritig and M. Legrand (eds.): *Mechanisms of Plant Defense Responses.* 1993 ISBN 0-7923-2154-5
3. C.I. Kado and J.H. Crosa (eds.): *Molecular Mechanisms of Bacterial Virulence.* 1994 ISBN 0-7923-1901-X
4. R. Hammerschmidt and J. Kuć (eds.), *Induced Resistance to Disease in Plants.* 1995 ISBN 0-7923-3215-6
5. C. Oropeza, F.W. Howard, G. R. Ashburner (eds.): *Lethal Yellowing: Research and Practical Aspects.* 1995 ISBN 0-7923-3723-9
6. W. Decraemer: *The Family Trichodoridae: Stubby Root and Virus Vector Nematodes.* 1995 ISBN 0-7923-3773-5
7. M. Nicole and V. Gianinazzi-Pearson (eds.): *Histology, Ultrastructure and Molecular Cytology of Plant-Microorganism Interaction.* 1996 ISBN 0-7923-3886-3

KLUWER ACADEMIC PUBLISHERS – DORDRECHT / BOSTON / LONDON